Extreme Wildfire Events and Disasters

Extreme Wildfire Events and Disasters

Root Causes and New Management Strategies

Edited by

Fantina Tedim
Faculty of Arts, University of Porto, Porto, Portugal; Charles Darwin University, Darwin, Australia

Vittorio Leone
Faculty of Agriculture, University of Basilicata (retired), Potenza, Italy

Tara K. McGee
Department of Earth and Atmospheric Sciences, University of Alberta, Edmonton, AB, Canada

ELSEVIER

Elsevier
Radarweg 29, PO Box 211, 1000 AE Amsterdam, Netherlands
The Boulevard, Langford Lane, Kidlington, Oxford OX5 1GB, United Kingdom
50 Hampshire Street, 5th Floor, Cambridge, MA 02139, United States

Notices
Knowledge and best practice in this field are constantly changing. As new research and experience broaden our understanding, changes in research methods, professional practices, or medical treatment may become necessary.

Practitioners and researchers must always rely on their own experience and knowledge in evaluating and using any information, methods, compounds, or experiments described herein. In using such information or methods they should be mindful of their own safety and the safety of others, including parties for whom they have a professional responsibility.

To the fullest extent of the law, neither the Publisher nor the authors, contributors, or editors, assume any liability for any injury and/or damage to persons or property as a matter of products liability, negligence or otherwise, or from any use or operation of any methods, products, instructions, or ideas contained in the material herein.

Library of Congress Cataloging-in-Publication Data
A catalog record for this book is available from the Library of Congress

British Library Cataloguing-in-Publication Data
A catalogue record for this book is available from the British Library

ISBN: 978-0-12-815721-3

For information on all Elsevier publications visit our website at
https://www.elsevier.com/books-and-journals

Publisher: Candice Janco
Acquisition Editor: Marisa LaFleur
Editorial Project Manager: Redding Morse
Production Project Manager: Prem Kumar Kaliamoorthi
Cover Designer: Christian Bilbow

Cover Photo: João Mourinho, Vieira de Leiria wildfire, Portugal, 2017

Typeset by TNQ Technologies

Working together
to grow libraries in
developing countries

www.elsevier.com • www.bookaid.org

Contents

Contributors   ix
Acknowledgments   xi

**Part One   Extreme Wildfire Events and Disasters: Concept
and Global Trends   1**

1   **Extreme wildfire events: The definition**   **3**
*Fantina Tedim, Vittorio Leone, Michael Coughlan, Christophe Bouillon,
Gavriil Xanthopoulos, Dominic Royé, Fernando J.M. Correia
and Carmen Ferreira*
   1.1   Extreme wildfires: A true challenge for societies   3
   1.2   EWE definition and rationale   7
   1.3   A wildfire classification: Integrating fire intensity with potential
        consequences   16
   1.4   Conclusion   24
        References   24

2   **Extreme wildfires and disasters around the world: Lessons
to be learned**   **31**
*Luís M. Ribeiro, Domingos X. Viegas, Miguel Almeida, Tara K. McGee,
Mário G. Pereira, Joana Parente, Gavriil Xanthopoulos,
Vittorio Leone, Giuseppe Mariano Delogu and Holy Hardin*
   2.1   Introduction   31
   2.2   Extreme wildfire cases in Portugal   31
   2.3   Extreme wildfire cases in the world   41
   2.4   Conclusion   49
        References   49

**Part Two   Extreme Wildfire Events and Disasters:
The Root of the Problem   53**

3   **The role of weather and climate conditions on extreme wildfires**   **55**
*Mário G. Pereira, Joana Parente, Malik Amraoui, António Oliveira
and Paulo M. Fernandes*
   3.1   Introduction   55
   3.2   The influence of climate   55

3.3   The role of weather   57
3.4   The role of climatic and weather extreme events   59
3.5   Fire weather danger and risk rating   60
3.6   Climate change: The future of extreme wildfires   62
3.7   Concluding remarks   63
       References   64

4   The relation of landscape characteristics, human settlements,
      spatial planning, and fuel management with extreme wildfires   73
      Christophe Bouillon, Michael Coughlan, José Rio Fernandes,
      Malik Amraoui, Pedro Chamusca, Helena Madureira, Joana Parente
      and Mário G. Pereira
4.1   Introduction   73
4.2   France   74
4.3   Portugal   77
4.4   The United States of America   81
4.5   Conclusion   85
       References   85

5   Safety enhancement in extreme wildfire events   91
      Fantina Tedim, Vittorio Leone, Sarah McCaffrey, Tara K. McGee,
      Michael Coughlan, Fernando J.M. Correia and Catarina G. Magalhães
5.1   Wildfire disasters: Trends and patterns   91
5.2   Causes and circumstances leading to fatalities   95
5.3   The safety protocols   99
5.4   BESAFE: A safety framework for citizens   103
5.5   Conclusion   107
       References   107

6   Firefighting approaches and extreme wildfires   117
      Gavriil Xanthopoulos, Giuseppe Mariano Delogu, Vittorio Leone,
      Fernando J.M. Correia and Catarina G. Magalhães
6.1   Wildfire fighting approaches   117
6.2   Effectiveness and efficiency considerations   127
6.3   Firefighting approaches regarding extreme wildfires   128
6.4   Conclusions   130
       References   130

Part Three   Towards a New Approach to Cope with Extreme
                Wildfire Events and Disasters   133

7   The suppression model fragilities: The "firefighting trap"   135
      Gavriil Xanthopoulos, Vittorio Leone and Giuseppe Mariano Delogu
7.1   The dominant fire management approach today: The wildfire
       suppression model   135

7.2   Assessment of the fire suppression model   **136**
7.3   A proactive model as a possible alternative   **145**
7.4   Conclusions   **148**
      References   **148**

**8   Understanding wildfire mitigation and preparedness in the
      context of extreme wildfires and disasters   155**
      *Sarah McCaffrey, Tara K. McGee, Michael Coughlan
      and Fantina Tedim*
8.1   Introduction   **155**
8.2   Social science theoretical insights into preparedness and
      mitigation   **156**
8.3   Factors that influence individual protective action decisions,
      with reference to specific fire research findings   **163**
8.4   Diffusion of innovations   **168**
8.5   Risk and crisis communication   **169**
8.6   Conclusion   **171**
      References   **172**

9   Resident and community recovery after wildfires   175
      Tara K. McGee, Sarah McCaffrey and Fantina Tedim
9.1   Introduction   **175**
9.2   Disaster recovery frameworks   **175**
9.3   Wildfire recovery: Residents   **177**
9.4   Wildfire recovery: Community   **179**
9.5   Conclusion   **182**
      References   **182**

**Part Four   How to Cope with the Problem of Extreme
              Wildfires and Disasters   185**

10 Wildfire policies contribution to foster extreme wildfires   187
      *Paulo M. Fernandes, Giuseppe Mariano Delogu, Vittorio Leone
      and Davide Ascoli*
10.1 Introduction   **187**
10.2 Shaping wildfire disasters through misguided policy   **189**
10.3 The perpetuation of misguided fire policies   **193**
10.4 Transforming fire management policies   **195**
10.5 Conclusion   **195**
      Acknowledgments   **196**
      References   **196**

**11   Fire Smart Territory as an innovative approach to wildfire risk
      reduction**   **201**
      Vittorio Leone, Fantina Tedim and Gavriil Xanthopoulos
      11.1   The wildfire paradoxes   **201**
      11.2   Wildfires: An unsolved problem   **202**
      11.3   Communities and wildfires: How to reduce losses?   **204**
      11.4   Fire Smart Territory as a model to *"thrive with fire"*   **206**
      11.5   Conclusion   **211**
             References   **212**

12   How to create a change in wildfire policies   **217**
      Vittorio Leone and Fantina Tedim
      12.1   Introduction   **217**
      12.2   The origins of suppression policy   **217**
      12.3   Wildfire research and policies   **219**
      12.4   How to create a change in wildfire policies   **227**
      12.5   Conclusions   **228**
             References   **229**

**13   What can we do differently about the extreme wildfire problem:
      An overview**   **233**
      *Fantina Tedim, Sarah McCaffrey, Vittorio Leone,
      Giuseppe Mariano Delogu, Marc Castelnou, Tara K. McGee
      and José Aranha*
      13.1   Introduction   **233**
      13.2   Looking for a new paradigm of wildfire management:
             existing ideas and proposals   **235**
      13.3   The Shared Wildfire Governance paradigm and framework   **240**
      13.4   Next steps   **258**
             References   **258**

Index   **265**

Contributors

Miguel Almeida Forest Fire Research Centre of ADAI, University of Coimbra, Coimbra, Portugal

Malik Amraoui Centre for the Research and Technology of Agro-Environmental and Biological Sciences (CITAB), University of Trás-os-Montes and Alto Douro, Vila Real, Portugal

José Aranha Centre for the Research and Technology of Agro-Environmental and Biological Sciences (CITAB), University of Trás-os-Montes and Alto Douro, Vila Real, Portugal

Davide Ascoli DISAFA Department, University of Torino, Grugliasco, Italy

Christophe Bouillon National Research Institute of Science and Technology for Environment and Agriculture (IRSTEA), Risks Ecosystems Environment Vulnerability Resilience (RECOVER) research unit, Aix-en-Provence, France

Marc Castelnou Bombers Generalitat.DGPEiS. DI., Barcelona, Spain; University of Lleida, Lleida, Spain

Pedro Chamusca Centre of Studies of Geography and Spatial Planning (CEGOT), University of Porto, Portugal

Fernando J.M. Correia Faculty of Arts and Humanities, University of Porto, Porto, Portugal

Michael Coughlan Institute for a Sustainable Environment, University of Oregon, Eugene, OR, United States

Giuseppe Mariano Delogu Former Chief Corpo Forestale e di Vigilanza Ambientale (CFVA), Autonomous Region of Sardegna, Italy

Paulo M. Fernandes Centre for the Research and Technology of Agro-Environmental and Biological Sciences (CITAB), University of Trás-os-Montes and Alto Douro, Vila Real, Portugal

José Rio Fernandes Centre of Studies of Geography and Spatial Planning (CEGOT), University of Porto, Portugal

Carmen Ferreira Faculty of Arts and Humanities, University of Porto, Porto, Portugal

Holy Hardin   Public Affairs Science and Technology (PAST) Fusion Cell, Argonne National Laboratory, Lemont, IL, United States

Vittorio Leone   Faculty of Agriculture, University of Basilicata (retired), Potenza, Italy

Helena Madureira   Centre of Studies of Geography and Spatial Planning (CEGOT), University of Porto, Portugal

Catarina G. Magalhães   Faculty of Arts and Humanities, University of Porto, Porto, Portugal

Sarah McCaffrey   Rocky Mountain Research Station, USDA Forest Service, Fort Collins, CO, United States

Tara K. McGee   Department of Earth and Atmospheric Sciences, University of Alberta, Edmonton, AB, Canada

António Oliveira   Centre for the Research and Technology of Agro-Environmental and Biological Sciences (CITAB), University of Trás-os-Montes and Alto Douro, Vila Real, Portugal

Joana Parente   Centre for the Research and Technology of Agro-Environmental and Biological Sciences (CITAB), University of Trás-os-Montes and Alto Douro, Vila Real, Portugal

Mário G. Pereira   Centre for the Research and Technology of Agro-Environmental and Biological Sciences (CITAB), University of Trás-os-Montes and Alto Douro, Vila Real, Portugal; Instituto Dom Luiz, University of Lisbon, Lisbon, Portugal

Luís M. Ribeiro   Forest Fire Research Centre of ADAI, University of Coimbra, Coimbra, Portugal

Dominic Royé   University of Santiago de Compostela, Santiago de Compostela, Spain

Fantina Tedim   Faculty of Arts and Humanities, University of Porto, Porto, Portugal; Charles Darwin University, Darwin, NWT, Australia

Domingos X. Viegas   Forest Fire Research Centre of ADAI, University of Coimbra, Coimbra, Portugal

Gavriil Xanthopoulos   Hellenic Agricultural Organization "Demeter", Institute of Mediterranean Forest Ecosystems, Athens, Greece

Acknowledgments

This work was prepared in the frame of the project 'FIREXTR — Prevent and prepare society for extreme fire events: The challenge of seeing the "forest" and not just the "trees"' (FCT Ref: PTDC/ATPGEO/0462/2014), co-financed by the European Regional Development Fund (ERDF) through the COMPETE 2020 — Operational Program Competitiveness and Internationalization (POCI Ref: 16702) and national funds by the Foundation for Science and Technology (FCT), Portugal.

Part One

Extreme Wildfire Events and Disasters: Concept and Global Trends

Extreme wildfire events: The definition

1

Fantina Tedim [1,2], Vittorio Leone [3], Michael Coughlan [4], Christophe Bouillon [5], Gavriil Xanthopoulos [6], Dominic Royé [7], Fernando J.M. Correia [1], Carmen Ferreira [1]

[1]Faculty of Arts and Humanities, University of Porto, Porto, Portugal; [2]Charles Darwin University, Darwin, NWT, Australia; [3]Faculty of Agriculture, University of Basilicata (retired), Potenza, Italy; [4]Institute for a Sustainable Environment, University of Oregon, Eugene, OR, United States; [5]National Research Institute of Science and Technology for Environment and Agriculture (IRSTEA), Risks Ecosystems Environment Vulnerability Resilience (RECOVER) research unit, Aix-en-Provence, France; [6]Hellenic Agricultural Organization "Demeter", Institute of Mediterranean Forest Ecosystems, Athens, Greece; [7]University of Santiago de Compostela, Santiago de Compostela, Spain

1.1 Extreme wildfires: A true challenge for societies

1.1.1 An escalating worldwide problem

In the absence of human activity, wildfires are a natural phenomenon in many types of vegetation cover and forest ecosystems, but their current manifestation around the world is far from "natural." Humans make them worse at every step; their activities are becoming the predominant cause of fires and are increasing the available forest fuels (e.g., planting inappropriate species for high-risk areas); by building next to or inside forests, they increase the risk to people and property and contribute to the risk of fire spread [1]. Notwithstanding escalating management costs, increased knowledge, development of technological tools and devices, improvement of training, and reinforcement of resources, wildfires continue to surprise us, largely because the aforementioned social activities increase the likelihood of extreme fire behavior and impacts. Climate change processes will further escalate the associated risk and costs.

Almost every year, wildfires of unprecedented size and intensity occur around the globe. Many of them provoke massive evacuation, fatalities and casualties, and a higher toll of damage, exceeding all previous records. These powerful wildfires represent a minority among all wildfires, but they create a disproportionately large threat to firefighter crews, assets, natural values, societies, and their members [2]. Some countries such as Australia, United States, and Canada have a long history of these powerful and often destructive phenomena [3—10].

With its abundant forests and extremely hot and dry climate, since European settlement Australia has suffered from extremely deadly fire events; a long series starting in 1851 with *Black Thursday* [11], when fires covered a quarter of what is now Victoria.

The series of ferocious bushfires continued with 1st February 1898 *Red Tuesday* that burned out 260,000 ha, caused the death of 12 people, and destroyed more than 2000 buildings in South Gippsland; then in 1926 in Gippsland, Eastern Victoria *Black Sunday*, with 60 fatalities and widespread damage to farms, homes, and forests. Finally, the series peaked with the 1939 *Black Friday* blaze in Victoria, which killed 71 people, destroyed more than 650 structures, and burned 1.5 to two million hectares. Many years later, in 1983, in the *Ash Wednesday* bushfires, in Victoria and South Australia, more than 22 fires burned about 393,000 ha and killed 75 people [12]. Then in 2009, *Black Saturday* fires become the worst in Australia's history with 500 injured and 173 fatalities, far exceeding the loss of life from any previous bushfires. For southeastern Australia (one of the three most fire-prone landscapes on Earth [13]), bushfires exhibit an abrupt increase in the frequency of pyrocumuloninbus (pyroCb; according to the World Meteorological Organization, pyrocumuloninbus is the unofficial name for *cumulonimbus flammagenitus*) events over the last decade [10] and a bigger value of fire intensity and total power in GW [14].

The history of wildfire in the United States is peppered with stories of disaster and destruction leading back into the 19th century [15]. From 1871 to 1918, wildfires in the Midwestern states sparked by steam-powered machinery and fueled by the waste of early industrial logging frequently engulfed whole settlements. Even if lack of evidence prevents us from defining the biophysical severity of these historic fire disasters, the contemporary context of severe wildfire disasters is nevertheless tied to this history as a consequence of the fire suppression policy that emerged in that period. Toward the end of this period (e.g., ∼ 1910), wildfire suppression policy became doctrine in the United States as the federal government set out to protect its new National Forests from fire. This proved especially challenging in the Western US where forests are prone to large, stand replacing fires. Perhaps the first well-documented extreme fire in the United States is the 1933 Tillamook Burn. The fire started in hot and dry weather in locations characterized by carelessly left logging slash and burned 16,000 ha in the first 10 days. When firefighters appeared to have it under control, the onset of gale-force winds changed fire behavior abruptly. Within 20 hours, the fire burned an additional 97,000 ha. This fire produced a pyroCb cloud 12.9 km high [15]. Firefighters were overwhelmed and helpless in their efforts to stop the blaze. It was only extinguished two weeks later by heavy rain. The burnt-over landscape of standing dead trees provided fuel for additional catastrophic fires over the following 20 years. The burn also left its legacy in the fire suppression landscape serving as the impetus for the *10 a.m. policy* whereby it became policy on National Forest to extinguish fires by 10 a.m. the morning after they were reported.

Currently, in many parts of the United States, wildfires are fueled by a legacy of fire suppression practices. These have contributed to the build-up of dense fuels in many forests after the disturbance of old growth forests by logging [4,7]. Recent powerful and disastrous wildfires in states as far apart as Tennessee (2016) and California (2017—18) have also been attributed mainly to extreme weather events, specifically co-occurrence of drought and high winds [3,9]. The 2016 Chimney Tops 2 Fire in Tennessee killed 14 people and destroyed 1684 structures. The blaze was fueled by drought conditions, 70 years of fire suppression, and gusting high winds. In the October of 2017, the Nuns Fire in Northern California killed 42 people, destroyed nearly 1355 buildings, and burned over 225,000 ha [16]. The Nuns Fire was started by wind-damaged electrical and gas utilities and spread very fast fanned by seasonal winds called the "Diablo winds," with gusts of up to 110 km h^{-1}. In July and August of 2018, the Carr Fire burned over 92,000 ha. It destroyed 1604 structures and has been blamed for at least eight deaths, including three firefighters [16]. Finally, the Camp Fire ignited November 8, 2018, became California's most destructive and deadliest wildfire. Fueled by 20 m per second winds, the fire burned 40,000 ha in the first two days, and the fire had burned over 62,862 ha; it destroyed 13,696 residences and 4821 other structures, and killed at least 85 people in and around the town of Paradise, California [17].

Canada, similar to the United States, has its own long engagement with infrequent, large, high-intensity, crown fires [5]. Their occurrence is an increasing concern [8]. The most destructive wildfires in terms of loss of lives and structures occurred between 1825 and 1938 [8]. In 1911, 1916, and 1922 fires destroyed multiple towns in Ontario, killing more than 500 people [18]. The most significant loss of life occurred during the 1916 Matheson fire with probably 223 fatalities [19]. A reduced number of structures have been destroyed since the 1938 Dance Township Fire [19]. In 2003, in British Columbia, more than 338 structures and businesses were destroyed or affected, and three operational staff lost their lives [20]. In 2011, the Flat Top Complex Wildfires destroyed about 340 homes, six buildings with several apartments, three churches, and 10 businesses, as well as affected the government center [8]. Sometimes these extreme fires are characterized by long duration and substantial impacts. The devastating Fort McMurray Horse River Wildfire in Alberta, which started on May 1st, 2016, and was declared out after 15 months, forced some 90,000 to flee the city of Fort McMurray and nearby communities in the Regional Municipality of Wood Buffalo and destroyed 3244 residential and other buildings [21]. It burned and destroyed some 589,552 ha of forest [22].

In the last few decades, this reality emerged in several countries, most notably in Southern Europe and Southern America, including Greece (2007 and 2018), Portugal (2003, 2005, and 2017), and Chile (2017). Chile and Portugal experienced, in 2017, the worst fire season ever recorded, with unprecedented events of extreme fire behavior. The 2017 fires in Chile were of unusual size and severity for the austral Mediterranean regions, affecting a total of 529,974 ha; four individual fires burned over 40,000 ha of land. They affected large extensions of exotic forest plantations of *Pinus radiata* and *Eucalyptus* [23].

In Portugal, after the disastrous fire seasons of 2003 and 2005, 2017 brought the most catastrophic season ever with 112 fatalities and wildfires that reached fireline intensities (FLIs) of 80,000 kWm^{-1}, rate of spread (ROS) of 15.2 km h^{-1}, and several episodes of downdraft that explain most of the loss of lives [24,25]. In Pedrógão Grande Fire (June 2017) most of the people (45 out of 66) died on the roads, overtaken by the sudden, scaring, and extraordinary fire manifestations spreading with amazing speed in a continuous artificial forest cover of *Pinus* and *Eucalyptus*.

In another Mediterranean country, Greece, the fire problem has been worsening steadily, in spite of increased investments in firefighting personnel and resources. After 17 fires caused fatalities in 1993, and 16 fatalities in 2000, in 2007 Greece faced its worst fire season in terms of burned area accompanied by numerous fatalities. In that dry year, and following three heat waves, the conditions became explosive between August 23 and 27 when a series of almost simultaneous very aggressive fires in Peloponnese, Attica, and Evia escaped initial attack, overwhelming the firefighting forces. They brought the burned area to approximately 270,000 ha of forest, olive groves and farmland, more than 5 times the average. The death toll reached 78 people. More than 100 villages and settlements were affected, and more than 3000 homes and other structures were destroyed. The financial damage by some estimates reached five billion US$ [26,27].

A second wildfire disaster hit Greece in 2018, in a seemingly "easy" fire season, with unusually high precipitation until mid-July. On July 23, the first day with predicted very high fire danger due to expected extremely strong westerly winds, an intense fire started at mid-day in western Attica, approximately 50 km west of Athens. While firefighting efforts were concentrated there, a second fire started at 16:41 in Eastern Attica, 20 km northeast of Athens. Fanned by a west-northwest wind of $45-70$ km h^{-1} with gusts that exceeded even 90 km h^{-1}, the fire first hit the settlement of Neos Voutzas and then, moving with approximately 4.0 km h^{-1}, spread through the settlement of Mati, burning most homes in its path until it reached the sea. Neither the ground forces that were slow to respond nor the aerial resources that had to face the extreme wind could do much to limit the disaster. Many people who tried to escape by running toward the sea were trapped by the fire on the steep cliff above the water and lost their life. There were also fatalities among the hundreds of people who managed to reach the water, either due to the effects of heat and smoke or due to drowning. Although the burned area was only 1431 ha, 100 people lost their lives, making this Eastern Attica Fire the second-deadliest wildfire in the 21st century, after the Kilmore East Fire in 2009 in Victoria (Australia) that killed 120 people [28].

In North Europe, extraordinary wildfires can assume large size in areas normally characterized by the relative absence of fires. For instance, the wildfire of Västmanland in central Sweden that started on July 31st, 2014, burned 13,800 ha of forest mainly covered by wind-fallen trees. The wildfire caused one fatality and required French and Italian water bombers to come and help fight the fire. More than 1000 people and 1700 animals (cattle and sheep) were evacuated, and thousands of people were prepared for evacuation when the fire approached towns. Approximately, 1.4 million cubic meters of wood and 71 buildings were damaged or destroyed by the fire [29].

The growing incidence of such large-scale and disastrous fire events around the globe makes it important to develop a method to classify and define them. Doing so is an essential precursor to the development of a common international approach to their study and to the development of the risk reduction and response capabilities required to manage risk that will only increase in the coming decades, namely due to climate change.

1.1.2 The need of a standardized definition

The aforementioned wildfires have captured the attention of the scientific community and have been studied using different analytical approaches from a series of disciplines (e.g., fire ecology, forestry, engineering, geography, anthropology, psychology, or social sciences), benefiting from the technological advances in computer sciences, remote sensing data, the development of software tools such as Geographical Information Systems (GIS), and fire behavior and spread modeling; however, almost every discipline has its own definition of these wildfires which seem to possess, by themselves, no intrinsic identity [30].

In the scientific literature, we found 25 terms to label these wildfires. This plurality of terminology is accompanied by a diversity of descriptors covering fire behavior, postfire metrics, impacts, and fire environment (Table 1.1). Furthermore, even where the same term is used, no agreement on the descriptors used was found. Some of the descriptors do not present quantitative thresholds, and, for the same descriptor, thresholds are greatly variable and influenced, among other things, by the distribution of fire sizes within each region or ecoregion, geographical conditions, and landscape vegetation composition [31].

Among the terms described in Table 1.1, extreme wildfire event (EWE) best captures the nature of wildfires that exhibit characteristics of extreme behavior manifestations with extremely high power and, frequently, unusual size, thus exceeding the capacity of control even in the most prepared regions of the world. They cause major negative socioeconomic impacts and undesirable environmental effects, if they occur in areas of social concentration or environmental importance.

The term "extreme" is usually used as the top value of a range of categories or "extreme values" of a statistical data set, exhibiting a typical heavy tailed distribution of data. In this case what can be called extreme event can vary from region to region in an absolute sense. This method of analysis is correct but does not contribute to understanding the identity of EWE. In contrast, the definition of EWE proposed by Tedim et al.[31] is an attempt to create an overarching categorization, adopting selected attributes of wildfire hazard. This approach was possible by using an interdisciplinary approach and evidence from analyses of wildfire events in different regions of the world.

1.2 EWE definition and rationale

1.2.1 EWE definition

The definition of EWEs by Tedim et al. ([31], p.10) is: "*a pyro-convective phenomenon overwhelming capacity of control (fireline intensity currently assumed 10,000 kWm⁻¹; rate of spread >50 m/min), exhibiting spotting distance > 1 km, and erratic and unpredictable fire behavior and spread. It represents a heightened threat to crews, population, assets, and natural values, and likely causes relevant negative socio-economic and environmental impacts.*" Fig. 1.1 presents a visual representation of the aforementioned definition, where all the components are considered. It is evident that some EWEs can generate huge impacts and thus turn into a disaster, requiring prolonged socioenvironmental recovery and reinstatement of socioenvironmental resources, amenity values, and cultural heritage.

Table 1.1 Descriptors used in scientific literature to label and characterize powerful and extraordinary wildfires.

Category	Terms	Fire behavior								Postfire metrics			Consequences		Fuel	Fire environment			
		Capacity of control	Extreme phenomena[b]	FLI & FL	FRP	ROS	Spotting	Simultaneous ignition	Rapid evolution/ sudden changes	Size of burned area	Duration	Fire severity	Impacts	Relief efforts	Fuel load and structure	Wind speed	Wind direction change	Atmospheric instability	Distance to WUI
Size	Extensive fire	X								X									
	Extremely large fires									X									
	Very large fires						X			X									
	Large fires			X						X									
	Large infrequent fires									X									
	Megafires	X		X						X			X						
	Megablaze									X				X					
	Megaburning									X	X								
Size and duration	Fires of concern	X																	
Fire behavior	Extreme wildfire event[a]	X	X	X	X	X	X		X	X	X	X	X		X	X	X	X	X
	Firestorm		X																
	Area fires		X																

Sudden changes of behavior									
Blow-up	X	X	X					X	X
Conflagration	X	X	X					X	
Eruptive fire	X	X	X						
Generalized blaze flash	X	X	X	X					
Mass fires	X								
Mass blazes	X								
Relevant impacts									
Catastrophic fires						X	X		
Disaster-fire						X	X		
Disasters						X	X		
Disastrous fires						X	X		
Social disaster						X	X		
Socially disastrous fires						X	X		
Natural disaster						X	X		

[a] And also extreme wildfire, extreme bushfire event, extreme fire event, extreme fire.

[b] It includes high level of energy, chaos, nonlinearity, mass spotting, eruptive fire behavior, vorticity-driven lateral spread, violent pyroconvective activity.

An EWE is a very complex process, and the thresholds used in its definition result from a deep analysis of the state of fire science. EWEs are characterized not just by their scale but also by their erratic and unpredictable behavior. The latter reflects how, as fires become more extreme, their complex patterns of interaction with the ecological, forest, agricultural, and built environments influence their behavior [32]; each of its physical attributes is related and influence each other in a concrete way and creates feedbacks. However, if the fire is above $10,000 \, \mathrm{kWm^{-1}}$ but the ROS is $40 \, \mathrm{m \, min^{-1}}$ instead of over $50 \, \mathrm{m \, min^{-1}}$, as proposed in the definition, can we still classify the fire as EWE? The answer is affirmative because the most socially relevant EWE attribute is the FLI that precludes any effort to control the fire, and in the example presented, the value is above the threshold. Future measurements of EWEs will help to validate and to adapt the selected thresholds.

EWEs are increasingly frequent events that exhibit non-instantaneous extreme behavior for at least several hours, but *normal fires* [31] too can have punctual manifestations of extreme fire behavior (e.g., fire whirls) because of the combination of several conditions, for instance, the influence of topography; they can create local situations that explain the entrapment of firefighter crews with tragic end.

From the natural hazards field, it has been shown that risk perception is influenced by the physical characteristics of the hazard [33]; thus, a good understanding of EWEs is fundamental to decrease their impacts.

A complete understanding of the EWE definition is easier if we discuss its rationale, valorizing: The physical properties of EWE; the duration of an EWE; the size; the consequences; and the fact that an EWE does not necessarily create a disaster. Fig. 1.2 helps to make clear this discussion; it depicts how physical properties of fires interact with

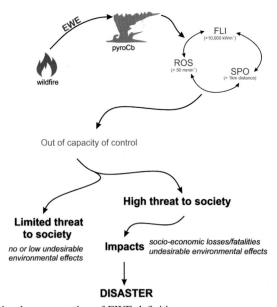

Figure 1.1 The visual representation of EWE definition.

human dimensions, i.e., human actions, residential development patterns (RDPs), and Wildland–Urban Interface (WUI) characteristics, in the context of socioecological systems [34], where wildfires occur, thus influencing fire behavior and its consequences.

1.2.2 EWE definition rationale

1.2.2.1 The physical properties of EWEs

The pillar of EWE definition are physical attributes concerning the pyroconvective nature of the phenomenon and fire behavior which result from the interaction of human and natural systems. It is evident from Fig. 1.2 that, after the fire outbreak, fire progression and spread are influenced by the interaction and feedback between the fire environment (characterized by several factors, such as fuel load and characteristics, existence of drought, and weather conditions), topography, human activities (e.g., land use options, land management, the use of prescribed burning, fire suppression actions), and WUI characteristics and RDPs.

The scientific community identifies as the main physical fire properties FLI, flame length (FL), and ROS, which can all be expressed in quantitative terms. Another important fire property which is expressed qualitatively is the presence or not of

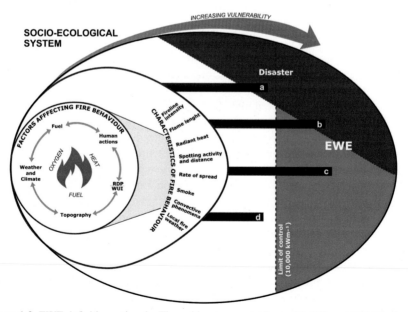

Figure 1.2 EWE definition rationale. The red bars represent fires with different FLI. The longer the bar, the higher the intensity. The impacts of fires result from the interaction and feedbacks between the physical characteristics of fires and the vulnerability of exposed elements. A wildfire becomes an EWE when its features exceed the capacity of control (10,000 kWm^{-1}). The letters a, b, c, and d, respectively, indicate (a) fire under control capacity (i.e. a *normal fire*) but affecting a vulnerable area provoking a disaster; (b) an EWE turning into a disaster; (c) an EWE; and (d) a wildfire below the capacity of control (i.e. a *normal fire*).

spotting; however, when present, it is further described by frequency, distance, and acceleration of spot fires. Less frequently used fire properties are smoke and radiant heat that are the main killers [35] and convective phenomena that increase the extremeness of the event. Whereas convective activity is included in the definition, local fire weather, smoke, and radiant heat are not considered because of the impossibility to establish thresholds that could have worldwide representativeness.

FLI is the pivotal fire behavior parameter and can be determined from measurements or observations of ROS and fuel consumption [36] or alternatively from FL, using simple equations [37]. FLI influences the capacity of control, and the value of 10,000 kWm^{-1} is currently accepted as the threshold of impossibility of control [38−40]. Beyond 10,000 kWm^{-1}, it is well accepted that even heavy water bombers are ineffective [41], and fire control is not possible with current day technology and technical resources [42]. With an increasing FLI, the quantity of water required as an extinguishing agent to contain the flames grows. In addition, fire intensity influences the pattern of fire severity throughout the affected area [43,44] and the resistance of the structures, consequently acting on the losses and the number of fatalities.

ROS is important to firefighting strategy and fire size [45] affecting the capacity of suppression and the ability to move away from a fire safely [2]. ROS is dependent on the type, load and continuity of fuel, topography, and weather conditions (mainly wind velocity). The higher the ROS, the wider the spread of wildfire, thus increasing the perimeter of flames [46]. ROS can reach values of about 20 km h^{-1} [47]; it is considered extreme when it is ≥ 3 km h^{-1} [47], and is often augmented by massive spotting. ROS affects deployment of crews and resources, the efficacy of the suppression operations with cascade effects on the impacts induced by the arrival of a fire front: Decisions to stay and defend assets or leave, and evacuation of people and animals. This shows the importance of societal and community responses. The more the people have acted to prepare land/properties (e.g., community wide defensible spaces can create fire breaks to slow fire spread, increase efficacy of suppression), the more the time crews will have to respond and reduce localized risk; this can make it easier for them to plan where they can deploy limited resources.

FL influences the ability of a fire to cross barriers and so spread in discontinuous fuels [48], reach canopies [49], and affect radiation load on buildings [2,50].

Spotting activity (i.e., the ignition of new fire starts outside the fire perimeter by firebrands of bark, needles, twigs, pine cones, acorns, moss, or larger embers launched from a primary fire, conveyed by the convective column and landing on receptive fuels [51−53]) acts to carry fire across gaps in vegetation (firebreaks, agricultural fields, etc.) and to ignite buildings [50,54]. In EWEs the powerful convection column, the strong prevailing wind and/or the hot downdrafts from the column and the preheated dead fuels result in spotting distances sometimes exceeding 10 or more kilometers, rendering all preventive efforts to create horizontal fuel discontinuity useless [2,52,53,55,56]. Rapid acceleration of spot fires, due to high wind and very dry fuels, can sharply increase the risk of firefighter entrapment [56]. In the presence of species with loose fibrous bark (e.g., *E. obliqua* and *E. macrorhyncha*) a concentrated short distance spotting is the main fire propagation mechanism, while the firebrands responsible for long-range spotting are long ribbons of decorticating bark of some smooth-barked eucalypt species, such as *E. viminalis, E. globulus, and E. delegatensis* [56]. Long-range spotting requires a strong upward motion by the buoyant plume to transport relatively large firebrands several kilometers above the ground and high winds aloft to transport fuel particles for extended distances downwind [56].

EWEs are violent pyroconvective phenomena, for which a large integral of instantaneous energy release (IIER) or magnitude is required [57]. Pyroconvective activity and the creation of a pyroCb occur when a very intense fire generates heat, moisture, gases, and particles, in the form of a plume with an upward movement, powered by the amount of energy released by the combustion, integrated by the latent heat released when vapor condensation occurs [6,57−59]. When the vertical upward motion ceases, convective updrafts can reverse and become downdrafts, which hit the ground and spread out, in all directions, so inducing changeable wind direction and becoming responsible for the erratic fire spread [60], with implications for firefighters' and people safety. Rarely, a pyroCb produces downbursts, which are gusty, erratic, and intense winds causing unpredictable changes in fire intensity, FL, ROS, direction of fire spread, and ember spotting [61].

The level of threat a fire represents to society is thus strongly influenced by the characteristics of fire behavior. For instance, a fire with an FL of 0.5 m in any context cannot produce intensities that overwhelm the capacity of suppression, whereas the radiant heat of flames that leap 100 m above the tree line, as in Australia in 2009, can kill from 300 m [62].

An EWE definition that focuses on the physical attributes of a fire can be used in any part of the world because it is not place-dependent as the fire process is the same everywhere. At same time, presenting the capacity of control as a threshold to define an EWE, it is possible to understand the limitations of suppression capacity resources and systems (much less than 10% of the fire intensity spectrum observed in wildfires [55]), considering that the highest FLI record so far can be 150,000 kWm^{-1} for 2009 Kilmore and Murrundindi Victorian fires [63]). This does not mean that this value represents the maximum FLI that a fire can reach.

In addition, it is possible to have an absolute scale of fire power with the result that, independent of geographical and cultural context, an EWE is an EWE anywhere in the world, regardless of the variability of consequences. This supports the importance of including complementary social response capability (e.g., household and community actions on vegetation management, creation of defensible spaces, increase house resistance, physical and mental preparedness) that can slow fire spread, reduce fire intensity and sources of spotting to some extent, and reduce the risk to dwellings and mainly to people.

1.2.2.2 The duration of an EWE

A wildfire classified as an EWE requires that in some moment, it presents the characteristics and the critical thresholds contained in the definition. To reach them, it is necessary for fire to spread for some time to gain energy potential which depends on several conditions (e.g., amount and state of the fuel, weather conditions, atmospheric instability), but this can occur shortly after the fire outbreak. In Pedrógão

Grande (2017, Portugal), the fire took just 2 hours to assume the characteristics of an EWE, although the maximum intensity occurred about 6 hours after the fire outbreak; most of the losses and fatalities occurred in less than an hour [25].

The period when the fire exhibits violent behavior and when most damage occurs is a relatively short interval, although these periods can recur in close succession [64]. In the Tasmania fire of 7 February 1967, in just 5 hours, 226,500 ha, or 85% of the total area of 264,000 ha, burned causing extensive damage. The remaining 36,000 ha burned over a period of several weeks afterward [65].

EWEs can persist over prolonged periods of time, but this cannot be a defining characteristic, as it strongly depends on the continuity and amount of fuels and persistence of favorable weather conditions. *Black Friday* fire in Australia ($\sim 36,000$ kWm^{-1}), 1939, is considered one of the worst events in the country but did not burn for a long time because of sudden weather changes, namely rainfall [66]. Kilmore East fire (2009, Victoria, Australia) burned nearly 100,000 ha and destroyed more than 2200 buildings in the first 12 h. The fire merged with the Murrindindi fire, burning approximately 400,000 ha over a period of 3 weeks [56].

As the duration of a wildfire is variable even in the same country because it is dependent on the fire environment, it is not possible to determine a threshold that could be used worldwide.

1.2.2.3   The constraints of wildfire size

EWEs as a pyroconvective phenomenon would most certainly be large, but size is not a suitable descriptor of this type of events as it is a very weak and misleading attribute. As a rule of thumb, the diameter of the flaming zone can be a rough measure of the height to which the convection column resists, mixing with the air it is passing through as it rises [58]. At the point of pyroCb onset, the flaming zone is unusually large and flame-front intensity is simultaneously great [67]. In the EWE of Pedrógão Grande (2017, Portugal), the highest intensity (up to 60,000 kWm^{-1}) with a downburst phenomenon occurred about 6 hours after the fire outbreak, when the fire was only about 3799 ha out of 29,354 ha; in 1 hour (from 8 to 9 p.m.), the fire burned 4459 ha [25].

There is a strong inverse correlation between the probability of containing wildfires and fire size and intensity. An attempt to quantify this relationship for initial attack crews in Canada was made by Hirsch and Martell [40] and Hirsch et al. [68], and although not formally documented, the same type of relationship exists for suppression activities by air tankers (K. Hirsch pers. com.).

Size and severity do not have a direct correlation, as the latter reflects the interaction between landscape characteristics and fire behavior. Even very-high-intensity fires create an heterogenous pattern of burning severity inside the affected area [44,69]. Size tells us little about losses which depend on the affected area characteristics, and, mainly on the vulnerability of the exposed population and assets. For instance, in the northern California wildfire event in October 2017, the Tubbs Fire burned 14,895 ha, destroyed 5636 structures, and damaged 317 ones, while Nuns Fire was a larger event (burned 22,877 ha) with lower destruction (it destroyed 1355 structures and damaged 172 ones) [9]. The Eastern Attica Fire (Greece, 2018) is another excellent example of a wildfire that provokes huge destruction and loss of lives, while it only burned 1431 ha because it stopped when reached the sea.

Finally, size can also be the result of wildland fire use, which is an accepted fire management practice in areas where naturally caused wildfires are not a threat to assets, homes, or people as it happens in the United States.

As fire sizes vary tremendously around the world, selection of an absolute fire size for EWE at world level is difficult to establish or even impossible, as it is place-dependent.

1.2.2.4   EWE consequences are place-dependent

Although the quantitative metrics used in our definition are solely based on the physical attributes of the wildfires, we do not ignore that context matters. Thus, in Fig. 1.2 we depict the influence of factors affecting fire behavior and their threat to society. The role of the context is indirectly present in its influence on the physical characteristics of fire (e.g., land/forest use practices at commercial, community and regional levels of analysis) and, directly, in the human actions that influence fire behavior (e.g., defensible space), as well as the impacts on crews, population, assets, natural values, socio-economic activities, and environment.

Sometimes fire consequences have been used to define extreme events. However, the integration of consequences in the definition of extreme is not consensual and in our opinion is inadequate. Recent and specialized literature finds the introduction of impacts in the definition of extreme events very concerning because if the purpose is reducing the losses and impacts and these are included in the definition, how is it possible to assess the efficiency of management or resilience to extreme events? [70].

The impacts of extreme events are governed not only by the physical properties of the hazard but also by the characteristics of places (e.g., the characteristics of the WUI) and people's preparedness. The physical attributes of EWEs (Fig. 1.2) characterize the level of threat and act as triggers on vulnerable societies and ecosystems [71,72]; between these two components, there is a loop [73], but it is not adequate to integrate impacts in the definition of EWE. If we do not establish a differentiation between EWE and its impacts, we become unable to address how prevention, and mitigation has decreased vulnerability and increased the efficacy of social and economic recovery post-event.

1.2.2.5   EWE and disaster

Tedim et al. [31] clearly differentiated an EWE from a disaster recognizing that (1) not all EWEs produce huge consequences because not all EWEs will affect people and societal resources; it is important to consider whether an event will impact vulnerable people and assets. Losses of social assets may be exacerbated by factors such as location, building materials, design, and communities urban planning [45,74−76]; and (2) *normal fires* can also produce huge consequences because of inadequate management or control actions (e.g., lack of resources, lack of coordination, wrong instructions and evaluation of situations), lack of preparedness by individuals and concerned communities, or poor land management.

The circumstances around fatalities are related to (1) people's attitudes and behavior, for instance, last-minute decision to evacuate or escape, lack of awareness and preparedness [77−79]; (2) spatial and environmental conditions [6,14,28]; and (3) entrapments and burnover [80−82].

For instance, in Canada, fires that exhibit extreme fire behavior occur each year. Throughout the boreal, many of these fires start naturally (and have been doing so for centuries) but have little direct negative impact on people or ecosystems although their behavior is extreme. If the Fort McMurray Horse River Fire of 2016 had started 50 km east of where it did, it would have had little social and/or economic impact and not have been on the national and international news for months. It most likely would have been comparable to the 2011 Richardson Fire that was 100−150 km north of Fort McMurray and burned roughly the same amount of area (∼ 500,000 ha) but was in no way seen as a disaster.

On the other hand, sometimes small but intense fires can cause damage to homes and people. An example could be the Tubbs Fire, in 2017, in California; it was not the largest wildfire but was the most destructive (at the time) as it moved downslope into the city of Santa Rosa and the surrounding communities [9]. In 1994, of the fires around Sydney (Australia), the most destructive was one of the smallest: With just 476 ha, and with a maximum intensity about 6650 kWm^{-1}, the Como-Janelli Fire provoked four fatalities and the loss of 99 houses [14,45]. The Mount Carmel Fire in December 2010, in Israel, burned about 4000 ha but claimed 44 lives.

Another excellent example is the 2018 wildfire disaster in Eastern Attica in Greece already described.

1.3   A wildfire classification: Integrating fire intensity with potential consequences

The wildfire classification proposed by Tedim et al. [31] is not the first one developed. Byram [83] was the first to come up with the idea of measuring fires by their rate of heat release [84]. Rather popular wildfires classifications have been proposed [38,79−90], all of them mainly based on measurable parameters of fire behavior, such as FL and FLI. Some of these classifications are integrated with narrative, interpretation of fire behavior, type of fire, and suppression interpretation [40]. These classifications only refer to fires below the threshold of 10,000 kWm^{-1} [86,87,89,91], so they cannot be applied to EWEs. In other cases, the upper limit is > 10,000 kWm^{-1} [38,92], a value which only represents the lower 10% of the fire intensity spectrum observed in wildfires [56] and cannot be applied to EWE, which cover the remaining 90%, precautionarily considering the upper limit of FLI in 100,000 kWm^{-1} [84,93,94]. The purpose of the classifications described earlier was to provide information on firefighting productivity and effectiveness of suppression to support planning and tactics.

Gill (1998) [84] proposed a classification based on seven categories of fire intensity to facilitate the communication of the nature of fire variation, the highest one being between 35,000 and 100,000 kWm^{-1}, as it was supposed the fires could not go over level 7. The recent values of intensity of about 150,000 kWm^{-1} provided by Tolhurst [63] for Victorian fires in Australia, 2009, confirm that EWEs are unpredictable phenomena that can evolve to unknown limits, especially considering the influence of climate change.

In France, a prototype of wildfires classification proposed by Lampin-Cabaret et al. [95] is based on intrinsic physical measurement of the phenomenon and its consequences. It has a posteriori purpose and no predictive character; it classifies fires depending on the effects and damage found on standard elements of the fire environment (vegetation, people, buildings, and infrastructures) but does not characterize the level of risk of forest fire hazard, nor the economic consequences of the event.

More recently, exploratory studies were conducted to support the development of a bushfire severity scale to provide a warning for potential losses of properties and human lives [6]. These studies recognize the importance of fire power to explain its destructive potential, as a stronger relationship exists between community loss and the power of a fire [6,14,96]. In addition, these studies demonstrate that the current fire danger scale used to assess the probability for fire occurrence and the spread dynamics as well as the difficulty of control do not adequately reflect the destructive potential of a fire [6].

The wildfire classification proposed by Tedim et al. [31] recognizes the importance of fire power to explain the losses, but it does not only focus on FLI, adopting a multicriteria approach to make the scale precise, objective, and operational. The parameters used to create this classification are real-time measurable behavior parameters (FLI, ROS, and FL) and real-time observable manifestations of extreme fire behavior (presence of pyroCb and downdrafts, spotting activity, and distance) [31]. It presents seven categories of fires, four labeled as *normal fires* (categories one to four) and three at the highest end (categories five to seven) that cover different categories of EWEs. Whereas the current classifications of some natural hazards are descriptive (e.g., Richter and Mercalli scales for earthquakes) or prospective (e.g., Saffir Simpson scale for hurricanes), the classification by Tedim et al. [31] is both descriptive and prospective.

One of the possible criticisms is whether the classification proposed by Tedim et al. [31] develops a holistic view of EWEs as a socioecological phenomenon; it seems to be based solely on measurable fire behavior parameters and suppression difficulty. The second sentence of the EWE definition includes statements related to the potential impact of wildfires which reveal it as a key aspect of the definition. People can easily understand that a fire of category 7 can provoke more losses and injuries than one of category 2, applying the same logic of other natural hazards classifications. Considering that many times people need to face a fire by themselves, it is crucial that citizens are aware about the threats and the potential consequences to protect themselves in fires with different intensities.

To improve the classification by Tedim et al. [31], in this chapter we have enhanced it by integrating the multiple fire attack mechanisms (i.e., smoke, radiant heat, flames, spotting, and wind) and their potential consequences to people, crews, and assets, in each of the seven categories of fires (Table 1.2). The proposed integration is not based

Table 1.2 Wildfire classification integrating potential physical and psychological consequences.

Category		Physical		Psychological		Assets/Structures	Ecosystems
		People	Crew	People	Crew		
Normal fires	1	Null to minimum.	1) Null to minimum. 2) Minor problems from lack of safety rules compliance.	Possible but rare episodes of stress.	Low because all small fires include the potential to escalate.	Null to minimum.	Null to minimum.
	2	Minimum to medium because there is generally adequate time to react.	1) Possible, but rare, minor problems because there is generally adequate time to react. 2) Problems in the case of inappropriate or missing use of personal protective equipment (PPE).	Possible but uncommon episodes of stress because there is time for rational thinking, although higher if people are unprepared or events occur during the day when parents are at work, children are at school. Family separation and uncertainty regarding the location and safety of others increase stress.	Low/moderate because of the potential of fire to escalate.	Possible but rare damage.	Low to medium burning severity.
	3	1) Possible accidents by inappropriate attitude and behavior concerning evacuation or sheltering in place. 2) Possible injuries from direct flame contact if people are not adequately trained and prepared to face the fire by themselves. 3) Possible but rare fatalities. 4) Autonomous defense using hoses and water is possible but risky, especially in the case of an electrical power failure (or shut-off by the power company) while there is no back-up of petrol driven generators. 5) Likely inadequate response for evacuation of disabled or elderly people.	1) Possible entrapment and burnover, especially from not observing what the fire is doing at all times and thinking of what it will be doing next. 2) Problems from lack of safety rules compliance. 3) Serious problems in the case of inappropriate or missing use of PPE.	1) Sporadic episodes of stress because of the reduced time for rational thinking, lack of information, and the absence of firefighters' support. 2) Stress can be higher if people are unprepared or events occur during the day when parents are at work and children are at school. Family separation and uncertainty regarding the location and safety of others increase stress.	1) Possible occurrence of stress due to lack of information, weaknesses in the control and command chain, resource availability, and deployment to tackle fires. 2) Fatigue and exhaustion can increase risk.	Possible but not frequent.	Low to high burning severity.

4	1) Major accidents may be caused by inappropriate attitude and behavior concerning timely evacuation or passive sheltering. 2) Difficulty to find adequate response for evacuation of nonmobile people: Children, disabled, or elderly people. 3) High difficulty of autonomous defense without adequate awareness and preparedness to face the aggressive assault of the flames. 4) Impossibility of autonomous defense using hoses and water in the case of lack of electricity and the inexistence of petrol-driven generators. 5) Injuries and fatalities from direct flame contact. 6) Accidents on the road from smoke, flames, radiant heat, and falling trees. 7) Difficulty of circulation by car from lack of visibility. 8) The survival conditions are very difficult so requiring additional precautions.	1) The likelihood of entrapment and burnover increases. 2) Fatal problems from lack of compliance with safety rules, especially from not knowing what the fire is doing at all times and inability to predict what it will be doing next. 3) Fatal problems in case of inappropriate or missing use of PPE. 4) Difficulty of circulation of vehicles from lack of visibility. 5) Difficulty to impossibility of aerial operations from smoke, wind, and convective activity. 6) Relevant problems of radiant heat with consequent exhaustion of firefighters.	1) Feeling of fear, loss, or lack of control, anxiety, affecting the decision of passive or active sheltering and timely evacuation. 2) Diffuse episodes of stress because there is no time for rational thinking and the urgency to take decision facing more extreme fire manifestations. 3) Stress can be higher if people are unprepared or events occur during the day when parents are at work and children are at school. Family separation and uncertainty regarding the location and safety of others increase stress.	1) Possible occurrence of stress due to lack of information, weaknesses in the control and command chain, and the incapacity to cope with all the simultaneous fires started by spotting, and issues with resource availability and deployment to tackle fires. 2) Fatigue and exhaustion can become risk factors.	1) Frequent and relevant. Strong wind making buildings more vulnerable to mechanisms of fire attack. 2) Potential damages depending on building design and type of RDPs and WUI.	Heterogeneity in burning severity pattern, with wide areas of medium to high.
5 Extreme wildfire events	1) Difficulty to find adequate response for evacuation of nonmobile people: Children, disabled, or elderly people. 2) Possible fatalities in case of escape or outside houses. 3) Extreme difficulty of autonomous defense. 4) Impossibility of autonomous defense using hoses and water in the case of lack of electricity and the inexistence of a petrol-driven generators. 5) Relevant problems of health from radiant heat and smoke. 6) High difficulty of	1) Likely entrapment and burnover danger aggravated by spotting. 2) Sometimes reaching a designated safety zone in time may not be possible. 3) Relevant problems of radiant heat with consequent exhaustion of firefighters. 4) Problems from smoke. 5) Fatal problems from lack of compliance with safety rules. 6) Fatal problems in case of inappropriate or	1) Strong feeling of fear, loss or lack of control, anxiety, namely in the case of roads cut off by fast-moving fire fronts or closed by authorities, incapacity to communicate with family, friends, and authorities. 2) Diffuse and persistent psychological stress due to the fire experience and consequences. 3) High stress if people are unprepared or events occur during the day	1) Possible diffuse and persistent psychological stress due to the incapacity to contain the flames spread under the urgency of immediate multiple decisions; issues with resource availability and deployment to tackle fires, and coordination and devolution of responsibility, situational awareness if events	1) Massive spotting exacerbates losses occurrence. 2) Potential damages likely exacerbated by building design and type of RDPs/WUI. 3) Strong wind making buildings more vulnerable to mechanisms of fire attack.	1) Heterogeneous severity patterns inside fire perimeter. 2) Evidence of areas with high burning severity.

Continued

Table 1.2 Wildfire classification integrating potential physical and psychological consequences.—cont'd

Category	Physical		Psychological		Assets/Structures	Ecosystems
	People	Crew	People	Crew		
	circulation from lack of visibility. 7) Road accidents resulting from falling trees provoked by winds or by fast moving fire fronts. 8) Difficulty or impossibility of communication by failure of power lines and lifelines.	missing use of PPE. 7) Road accidents resulting from falling trees provoked by winds. 8) Extreme difficulty of communication. 9) Difficulty to impossibility of circulation of terrestrial vehicles from lack of visibility. 10) Difficulty to impossibility of aerial operations from smoke, wind, and convective activity.	when parents are at work and children are at school. Family separation and uncertainty regarding the location and the safety of others increase stress.	escalate and evolve. 2) Fatigue and exhaustion can become risk factors. 3) Issues arising from working with unfamiliar firefighter teams or interaction between professional and volunteer firefighters.		
6	1) The survival conditions are very extreme, but it is possible to survive if adequately prepared. 2) Difficulty to find adequate response for evacuation of nonmobile people. 3) Likely fatalities in case of escape, or outside houses or in case of passive sheltering. 4) The erratic and unpredictable fire behavior can preclude evacuation and make dangerous evacuation or escape. 5) Time to make decisions may be limited and inadequate for carrying out an order. 6) Impossibility of autonomous defense using hoses and water in the case of lack of electricity and the inexistence of petrol-driven generators. 7) Last-minute evacuation can conduct to dramatic end 8) Roads are likely to be cut off for those who were late in deciding and by fast-moving fire fronts. 9) Extreme difficulty of	1) Likely entrapment and burnover danger aggravated by spotting. 2) Injuries from imprudent but generous attempts to contain fire spread also due to profuse spotting. 3) Relevant problems of radiant heat with consequent exhaustion of firefighters. 4) Problems provoked by smoke. 5) Fatal problems in case of inappropriate or missing use of PPE. 6) Fatal problems with noncompliance with safety rules. 7) Road accidents resulting from falling trees provoked by winds or by fast-moving fire fronts. 8) Extreme difficulty of communication and compliance with orders of the control and command chain. 9) Difficulty to impossibility of circulation of	1) Decisions are made under physical stress/suffering due to heat and smoke. 2) Strong feeling of fear, total loss or lack of control, anxiety, affect evacuation or escape. 3) Stress increased because of the collapse of phone and mobile communication network with subsequent incapacity of communicate with family, friends, and authorities. 4) Lack of psychological preparedness conducting to unsafe decision-making (e.g., last-minute evacuation) can precipitate the occurrence of fatalities. 5) People focus on the threat and not on the choice of the adequate behavior. 6) Diffuse and persistent psychological stress even for well-aware and well-	1) Diffuse and strong persistence of stress also because of involvement in rescue activities of victims. 2) Increasing issues with resource availability and deployment to tackle fires. 3) Fatigue and exhaustion can become risk factors. 4) Issues arising from working with unfamiliar firefighter teams or interaction between professional and volunteer firefighters. 5) Coordination and devolution of responsibility, situational awareness if events escalate and evolve.	1) Long-distance massive spotting. 2) Potential damages likely exacerbated by building design and type of RDPs and WUI. 3) Wind making buildings more vulnerable to fire.	1) Heterogeneous severity patterns inside fire perimeter. 2) Medium- to large-scale extreme burning severity.

#						
	autonomous defense. 10) Very relevant problems of health from radiant heat and smoke. 11) High difficulty of circulation from lack of visibility. 12) Lack of information to support emergency decisions. 13) Impossibility of communication caused by failure of power lines and lifelines. 14) High probability of car accidents in case of escape in very difficult environmental conditions. 15) Strong winds can make trees fall and provoke car accidents or traffic jams with loss of lives. 16) Survival outside a shelter made very difficult by wind and ember transport.	vehicles from lack of visibility. 10) Difficulty to impossibility of aerial operations from smoke, wind, and convective activity.	prepared people, can precipitate the occurrence of fatalities. 7) Very high stress if people are unprepared or events occur during the day when parents are at work, children are at school. Family separation and uncertainty regarding the location and safety of others increase stress.			
7	1) The survival conditions are very extreme, but it is possible to survive if adequately prepared. 2) Difficulty to find adequate response for evacuation of nonmobile people. 3) Maximum probability of fatalities and casualties directly caused by radiant heat, smoke, and flames and indirectly by escape in very difficult environmental conditions. 4) Last-minute evacuation can conduct to dramatic end. 5) Roads are likely to be cut off by fast-moving fire fronts or by authorities frustrating last-minute evacuation. 6) The erratic and unpredictable fire behavior can preclude evacuation, make fatal escape and dangerous passive sheltering in place. 7) Extreme difficulty of autonomous defense using hoses and water in the case of lack of electricity and the inexistence of petrol-driven generators. 8) Entrapment and	1) Likely entrapment and burnover danger aggravated by spotting. 2) Fast and wise decision-making is needed but is difficult because of poor information due to smoke, other adverse conditions. 3) Relevant problems of radiant heat with consequent exhaustion of firefighters. 4) Last-minute evacuation can conduct to dramatic end. 5) Problems from smoke. 6) Fatal problems in case of inappropriate or missing use of PPE. 6) Fatal casualties from noncompliance with safety rules. 7) Extreme difficulty of communication. 8) High probability of failure in communication systems and lack of information to take the adequate and timely decisions. 9) Difficulty of circulation of terrestrial vehicles from lack of	1) Strong feeling of fear, total loss or lack of control, and anxiety affect decisions mainly for evacuation or escape. 2) Lack of psychological preparedness conducts to unsafe decision-making (e.g., last-minute evacuation) facing the extreme manifestations of fire and can precipitate the occurrence of fatalities. 3) Fear makes people focus on the threat and not on the choice of the adequate behavior; high levels of stress can precipitate the occurrence of fatalities. 4) Extreme psychological stress even for well-aware and well-prepared people. 5) Stress increased because of the collapse of phone and mobile communication network with subsequent incapacity of	1) Diffuse and strong persistence of psychological stress because of involvement in rescue activities of victims. 2) Increasing issues with resource availability and deployment to tackle fires. 3) Fatigue and exhaustion can become risk factors. 4) Issues arising from working with unfamiliar firefighter teams or interaction between professional and volunteer firefighters. 5) Coordination and devolution of responsibility, situational awareness if events escalate and evolve.	1) Long-distance massive spotting. 2) Potential damages exacerbated by building design and type of RDPs and WUI. 3) Ember attack, and radiant heat can induce loss of tenability in structures. 4) Wind damage can render building more vulnerable to fire attack.	1) Heterogeneous severity patterns inside fire perimeter. 2) Large-scale extreme burning severity. 3) Unburned patches can be found.

Continued

Table 1.2 Wildfire classification integrating potential physical and psychological consequences.—cont'd

Category	Physical		Psychological		Assets/Structures	Ecosystems
	People	Crew	People	Crew		
	burnover danger aggravated by long-distance massive spotting and by the loss of tenability of the houses. 9) Lack of monitoring of both the internal and external conditions of buildings can make difficult decision to move toward an exit as houses lose tenability. 10) Strong winds make difficult the survival outside of the protection of a shelter. 11) Strong winds can make trees and power lines fall and provoke car accidents or traffic jams with loss of lives. 12) Impossibility of communication caused by failure of power lines and lifelines.	visibility. 10) Difficulty to impossibility of aerial operations from smoke, wind, and convective activity.	communicate with family, friends and authorities. 6) Very high stress if people are unprepared or events occur during the day when parents are at work, children are at school. Family separation and uncertainty regarding the location and safety of others increase stress. 7) Issues arising from evacuation and possibly dealing with short- and long-term relocation, and livelihood issues.			

Adapted from Tedim et al. [31]

on a scale of adjectives (for instance none, slight or minimal, considerable, substantial), which are considerably arbitrary and subjective dependent on expert judgment [73,97], and therefore would not have any conceptual and practical utility and efficiency in our point of view.

Our proposed scale does not consider the amount of consequences but the type of consequences, as the amount is influenced by the characteristics of the area (e.g., WUI, rural area, prevention and mitigation measures) and the level of preparedness. Thus, we identify potential physical and psychological consequences for people and crews and the consequences on assets and ecosystems to inform people and crews about the scenario they can face and to assist with preparedness and response.

The integration of the physical description of wildfire categories with the narration of the potential consequences enhances the importance of the definition of each category. This is one more element of analysis to be considered when establishing priorities of intervention and to familiarize people with different fire scenarios to increase risk awareness.

This classification is not just addressed to the scientific community but also to citizens, operational staff, and policy decision-makers. Furthermore, it is important that stakeholders play complementary roles in fire management and that stakeholder representatives at each level of analysis (e.g., staff, citizens) understand their role and take responsibility for developing their capability to contribute to the overall strategy and action. The overall goal is to enhance people's' preparedness and to assist fire management activities. It is important to include in planning the fact that the development of capability at each level of analysis requires its own dedicated strategy. Irrespective of the quality of the scientific analyses that contributes to developing risk profiles, this will not in itself motivate policy makers (e.g., their actions involve reconciling scientific information with economic and political criteria) to act or community members to prepare (e.g., strategies are needed to encourage people to understand their risk and motivate them to take responsibility to manage their local risk). Complementary strategies are needed to give operational staff a tool to communicate with citizens and inform in a way that can be easily understood following what happens with other natural hazards' classifications and to provide them with the tools to collaborate with scientists and to communicate with policy makers.

With the current state of scientific research and its dissemination, science communicators and emergency planners should cooperate to provide people with a known, reliable metric at their disposal to gauge the intensity of a wildfire and the potential consequences, similar to what happens with storms and earthquakes. For instance, people living in coastal areas prone to hurricanes are very familiar with the Saffir-Simpson Scale and easily understand that a category 5 storm causes more wind and surge damage than a hurricane of category 1. This information is used to help people to prepare and respond appropriately. In the case of fires and mainly EWEs, similar information could be crucial to help people to understand the physical properties of fire, the different levels of threat, the potential consequences, and how they can protect themselves. This could represent a common reference marker for people across political boundaries to make shared decisions [98], also in terms of regional and transboundary emergency cooperation level.

The importance of the proposed wildfire classification is high, knowing that communication of information on wildfire risk can be challenging. Key factors influencing community preparedness for wildfires include constructs such as outcome expectancy and fatalism [99−101]. The former means that if people do not believe that they can influence the causes and the consequences of fires, then there is nothing they can do to prepare. These beliefs are reinforced by media coverage and the dissemination of information from civic and scientific sources that discuss the magnitude of potential events; the bigger the event, the greater the people's sense of helplessness. Fatalism can have similar implications. Recognition of these impediments to action have fostered the development of community-based disaster risk reduction strategies that need to be included in planning [99−101]. This is especially important for EWEs as they represent incidents that fall at the extreme end of events that could put people at the limits of a possibility to survive.

1.4 Conclusion

All wildfires are not the same, and they have intrinsic physical characteristics that are influenced by the environmental, socioeconomic, and political context where fires occur.

EWEs are the most challenging wildfires because of the threats they represent to society and environment. They are not an ecological inevitability, and even though they have power to create huge amount of losses, it is possible to prevent their occurrence and mitigate their impacts.

The science of EWEs is in its early stages but is developing very fast, as this type of event is increasing in frequency because of more hazardous fire regimes, as a consequence of climate, landscape, and societal changes.

The proposed standardized definition of EWEs and wildfire classification integrating the physical attributes of fires with the potential physical and psychological consequences to people, crews assets, and ecosystems, are crucial for informing citizens about different fire scenarios and the distinctive challenges they present for human safety. Both the concept of EWE and the wildfires classification are excellent instruments to enhance wildfire risk and crisis communication programs, as well as in the definition of appropriate prevention, mitigation, response, and recovery actions adapted to the characteristics of the areas affected by wildfires.

References

[1] U. Irfan, California's wildfires are hardly "natural" — humans made them worse at every step, Vox (2018). https://www.vox.com/2018/8/7/17661096/california-wildfires-2018-delta-mendocino-climate.
[2] M. Gill, S.L. Stephens, G.J. Cary, The worldwide "wildfire" problem, Ecol. Appl. 23 (2013) 438−454.

[3] L.M. Andersen, A Vulnerability Assessment of Extreme Drought and Unprecedented Wildfire in the Southern Appalachian Mountains, Appalachian State University, 2017.

[4] D.E. Calkin, J.D. Cohen, M.A. Finney, M.P. Thompson, How risk management can prevent future wildfire disasters in the wildland-urban interface, Proc. Natl. Acad. Sci. 111 (2014) 746−751.

[5] W.J. Groot, M.D. Flannigan, A.S. Cantin, Climate change impacts on future boreal fire regimes, For. Ecol. Manage. 294 (2013) 35−44.

[6] S. Harris, W. Anderson, M. Kilinc, L. Fogarty, The relationship between fire behaviour measures and community loss: an exploratory analysis for developing a bushfire severity scale, Nat. Hazards 63 (2012) 391−415.

[7] J.B. Kauffman, Death rides the forest: perceptions of fire, land use, and ecological restoration of western forests, Conserv. Biol. 18 (2004) 878−882.

[8] T. McGee, B. McFarlane, C. Tymstra, Wildfire: a Canadian perspective, in: Wildfire Hazards, Risks and Disasters, Elsevier, 2015, pp. 35−58.

[9] N. Nauslar, J. Abatzoglou, P. Marsh, The 2017 north bay and southern California fires: a case study, Fire 1 (2018) 18, https://doi.org/10.3390/fire1010018.

[10] J.J. Sharples, G.J. Cary, P. Fox-Hughes, S. Mooney, J.P. Evans, M.-S. Fletcher, M. Fromm, P.F. Grierson, R. McRae, P. Baker, Natural hazards in Australia: extreme bushfire, Clim. Change 139 (2016) 85−99.

[11] VicGov, Bushfires Em Victoria, 2018. https://guides.slv.vic.gov.au/bushfires/1851.

[12] T.H. Dorgelo, The ash wednesday fires of 1983 in South Eastern Australia, prog. Fight. Fires catastrophes from air, in: Lect. Discuss. from Second Int. Sci. Symp. Minist. Res. Technol., 1984, pp. 38−54.

[13] M. Adams, P. Attiwill, Burning Issues: Sustainability and Management of Australia's Southern Forests, CSIRO Publishing, 2011.

[14] S. Harris, A. Anderson, M. Kilinc, L. Fogarty, Establishing a Link between the Power of Fire and Community Loss: The First Step towards Developing a Bushfire Severity Scale, Victorian Government Department of Sustainability and Environment, 2011.

[15] S.J. Pyne, J. Fire in America: A Cultural History of Wildland and Rural Fire, Princeton University Press, Princeton, New Jersey, 1982.

[16] CalFire, Incident Information, 2018. http://www.fire.ca.gov/current_incidents.

[17] Cal_Fire, No Title, (n.d.). http://cdfdata.fire.ca.gov/pub/cdf/images/incidentfile2277_4326.pdf (Febuary, 18, 2019).

[18] M. Barnes, Killer in the Bush: The Great Fires of Northeastern Ontario, Cobalt, Ont, Highway Book Shop, 2004.

[19] M. Alexander, 'Lest we forget': Canada's major wildland fire disasters of the past, 1825−1938, in: 3rd Fire Behav. Fuels Conf, Spokane, Washington, Oct, 2010, pp. 25−29.

[20] G. Filmon, Firestorm 2003 Provincial Review: A Report to the Province of British Columbia, 2004. http://bcwildfire.ca/history/reportsandreviews/2003/firestormreport.pdf.

[21] T.K. McGee, Residents' experiences of the 2016 Fort McMurray wildfire, Alberta, in: Adv. for. Fire Res. 2018, Imprensa da Universidade de Coimbra, Coimbra, 2018, pp. 1155−1159, https://doi.org/10.14195/978-989-26-16-506_129.

[22] L. Krugel, Fort McMurray wildfire named Canadian Press news story of 2016, in: Not Even a Hollywood Script Could Match the Terror, Uncertainty, and Heroism, Can. Press · CBC News, 2016.

[23] F. de la Barrera, F. Barraza, P. Favier, V. Ruiz, J. Quense, Megafires in Chile 2017: monitoring multiscale environmental impacts of burned ecosystems, Sci. Total Environ. 637 (2018) 1526−1536.

[24] Adai/Laeta, O complexo de incêndios de Pedrogão grande e concelhos limítrofes, iniciado a 17 de junho de 2017, Faculdade de Ciências e Tecnologia Universidade de Coimbra, 2017.

[25] CTI (Comissão Técnica Independente), Análise e apuramento dos factos relativos aos incêndios que ocorreram em Pedrógão Grande, Castanheira de Pêra, Ansião, Alvaiázere, Figueiró dos Vinhos, Arganil, Góis, Penela, Pampilhosa da Serra, Oleiros e Sertã entre 17 e 24 de junho de 2017 (2017).

[26] M. Diakakis, G. Xanthopoulos, L. Gregos, Analysis of forest fire fatalities in Greece: 1977–2013, Int. J. Wildland Fire 25 (2016) 797–809.

[27] G. Xanthopoulos, Olympic flames, Wildfire 16 (2007) 10–18.

[28] R. Blanchi, J. Leonard, K. Haynes, K. Opie, M. James, F.D. de Oliveira, Environmental circumstances surrounding bushfire fatalities in Australia 1901–2011, Environ. Sci. Policy 37 (2014) 192–203.

[29] R. Lidskog, D. Sjödin, Extreme events and climate change: the post-disaster dynamics of forest fires and forest storms in Sweden, Scand. J. For. Res. 31 (2016) 148–155.

[30] S.J. Pyne, Problems, paradoxes, paradigms: triangulating fire research, Int. J. Wildland Fire 16 (2007) 271–276.

[31] F. Tedim, V. Leone, M. Amraoui, C. Bouillon, M. Coughlan, G. Delogu, P. Fernandes, C. Ferreira, S. McCaffrey, T. McGee, J. Parente, D. Paton, M. Pereira, L. Ribeiro, D. Viegas, G. Xanthopoulos, Defining extreme wildfire events: difficulties, challenges, and impacts, Fire (2018), https://doi.org/10.3390/fire1010009.

[32] P.A. Werth, B.E. Potter, C.B. Clements, M.A. Finney, J.A. Forthofer, S.S. McAllister, S.L. Goodrick, M.E. Alexander, M.G. Cruz, Synthesis of Knowledge of Extreme Fire Behavior: Volume I for Fire Managers, 2011.

[33] R. Tripathi, S.K. Sengupta, A. Patra, H. Chang, I.W. Jung, Climate change, urban development, and community perception of an extreme flood: a case study of Vernonia, Oregon, USA, Appl. Geogr. 46 (2014) 137–146.

[34] J. Liu, T. Dietz, S.R. Carpenter, M. Alberti, C. Folke, E. Moran, A.N. Pell, P. Deadman, T. Kratz, J. Lubchenco, Complexity of coupled human and natural systems, Science 317 (2007) 1513–1516, https://doi.org/10.1126/science.1144004.

[35] R. Blanchi, J. Whittaker, K. Haynes, J. Leonard, K. Opie, Surviving bushfire: the role of shelters and sheltering practices during the Black Saturday bushfires, Environ. Sci. Policy 81 (2018) 86–94.

[36] G.M. Byram, Combustion of Forest Fuels, for. Fire Control Use, 1959, pp. 61–89.

[37] M.E. Alexander, M.G. Cruz, Interdependencies between flame length and fireline intensity in predicting crown fire initiation and crown scorch height, Int. J. Wildland Fire 21 (2012) 95, https://doi.org/10.1071/WF11001.

[38] M.E. Alexander, R.A. Lanoville, Predicting Fire Behavior in the Black Spruce-Lichen Woodland Fuel Type of Western and Northern Canada, Northern Forestry Centre, 1989.

[39] P.M. Fernandes, H.S. Botelho, A review of prescribed burning effectiveness in fire hazard reduction, Int. J. Wildland Fire 12 (2003) 117–128.

[40] K.G. Hirsch, D.L. Martell, A review of initial attack fire crew productivity and effectiveness, Int. J. Wildland Fire 6 (1996) 199–215.

[41] B.M. Wotton, M.D. Flannigan, G.A. Marshall, Potential climate change impacts on fire intensity and key wildfire suppression thresholds in Canada, Environ. Res. Lett. 12 (2017) 95003.

[42] C.R.C. Bushfire, Victorian 2009 Bushfire Research Response, Final Report, 2009.

[43] J.E. Keeley, Fire intensity, fire severity and burn severity: a brief review and suggested usage, Int. J. Wildland Fire 18 (2009) 116–126.

[44] S.J. Leonard, A.F. Bennett, M.F. Clarke, Determinants of the occurrence of unburnt forest patches: potential biotic refuges within a large, intense wildfire in South-eastern Australia, For. Ecol. Manage. 314 (2014) 85−93. https://doi.org/10.1016/j.foreco.2013.11.036.

[45] A.M. Gill, P.H.R. Moore, Big versus Small Fires: The Bushfires of Greater Sydney, January 1994, in: J.M. Moreno (Ed.), Large Forest Fires: Studies from Four Continents, Backbuys Publishers, The Netherlands, 1998, pp. 49−68.

[46] D. Alexander, Death and injury in earthquakes, Disasters 9 (1985) 57−60.

[47] M.G. Cruz, J.S. Gould, M.E. Alexander, A.L. Sullivan, W.L. McCaw, S. Matthews, A Guide to Rate of Fire Spread Models for Australian Vegetation, Australasian Fire and Emergency Service Authorities Council Limited and Commonwealth Scientific and Industrial Research Organisation, 2015.

[48] J.H. Scott, R.E. Burgan, Gen. Tech. Rep. RMRS-GTR-153, Standard Fire Behavior Fuel Models: A Comprehensive Set for Use with Rothermel's Surface Fire Spread Model, vol. 72, Fort Collins, CO US Dep. Agric. For. Serv. Rocky Mt. Res. Station., 2005, p. 153.

[49] M.A. Finney, J.D. Cohen, I.C. Grenfell, K.M. Yedinak, An examination of fire spread thresholds in discontinuous fuel beds, Int. J. Wildland Fire 19 (2010) 163−170.

[50] J.D. Cohen, What Is the Wildland Fire Threat to Homes?, Thompson Meml. Lect. Sch. for. North, vol. 10, Arizona Univ, Flagstaff, AZ, 2000. April 2000.

[51] E. Koo, P.J. Pagni, D.R. Weise, J.P. Woycheese, Firebrands and spotting ignition in large-scale fires, Int. J. Wildland Fire 19 (2010) 818−843.

[52] J. Martin, T. Hillen, The spotting distribution of wildfires, Appl. Sci. 6 (2016) 177.

[53] K.G. Tolhurst, D.M. Chong, Incorporating the Effect of Spotting into Fire Behaviour Spread Prediction Using PHOENIX-Rapidfire, Bushfire CRC Ltd., East Melbourne, VIC, Australia, 2009.

[54] G.C. Ramsay, N.A. McArthur, V.P. Dowling, Building in a Fire-Prone Environment: Research on Building Survival in Two Major Bushfires, 1996.

[55] F.A. Albini, Potential Spotting Distance from Wind-Driven Surface Fires, US Department of Agriculture, Forest Service, Intermountain Forest and Range Experiment Station, 1983.

[56] M. Cruz, A.L. Sullivan, J.S. Gould, N.C. Sims, A.J. Bannister, J.J. Hollis, R.J. Hurley, Anatomy of a catastrophic wildfire: the Black Saturday Kilmore East fire in Victoria, Australia, For. Ecol. Manage. 284 (2012) 269−285.

[57] J.H. Scott, Off the Richter: magnitude and intensity scales for wildland fire, in: Proc. 3rd Int. Fire Ecol. Manag. Congr. Chang. Fire Regimes Context Consequences, San Diego, CA, USA, 2006, pp. 13−17.

[58] R. McRae, Extreme Fire − A Handbook, ACT Government and Bushfire Cooperative Research Centre, 2010.

[59] K.G. Tolhurst, N.P. Cheney, Synopsis of the Knowledge Used in Prescribed Burning in Victoria, Department of Natural Resources and Environment, East Melbourne, Victoria, 1999.

[60] N.P. Lareau, C.B. Clements, Cold Smoke: smoke-induced density currents cause unexpected smoke transport near large wildfires, Atmos. Chem. Phys. 15 (2015) 11513−11520.

[61] W. Tory, K.J. Thurston, Pyrocumulonimbus: A Literature Review, Bushfire Nat, Hazards CRC East Melbourne, VIC, 2015.

[62] C. Stewart, Australia bushfires of 2009, Encycl. Br. (2009).

[63] K. Tolhurst, Report on the Physical Nature of the Victorian Fires Occurring on 7th February 2009, 2009.

[64] N.P. Cheney, Bushfire disasters in Australia, 1945−1975, Aust. For. 39 (1976) 245−268.

[65] H.G. Bond, K. Mackinnon, P.F. Noar, Report on the Meteorological Aspects of the Catastrophic Bushfires in South-Eastern Tasmania, Bureau of Meteorology, Melbourne, 1967.

[66] VicGov, History and incidents - Black Friday 1939, (n.d.). https://www.ffm.vic.gov.au/history-and-incidents/black-friday-1939 (accessed October 16, 2018).

[67] J.J. Sharples, G.A. Mills, R.H.D. McRae, R.O. Weber, Foehn-like winds and elevated fire danger conditions in Southeastern Australia, J. Appl. Meteorol. Climatol. 49 (2010) 1067−1095.

[68] K.G. Hirsch, P.N. Corey, D.L. Martell, Using expert judgment to model initial attack fire crew effectiveness, For. Sci. 44 (1998) 539−549.

[69] F. Tedim, D. Royé, C. Bouillon, F. Correia, V. Leone, Understanding unburned patches patterns in extreme wildfire events: evidences from Portugal, in: D.X. Viegas (Ed.), Adv. for. Fire Res. 2018, ADAI/CEIF, Coimbra, 2018, pp. 700−715. https://doi.org/10.14195/978-989-26-16-506.

[70] L.E. McPhillips, H. Chang, M. V Chester, Y. Depietri, E. Friedman, N.B. Grimm, J.S. Kominoski, T. McPhearson, P. Méndez-Lázaro, E.J. Rosi, Defining extreme events: a cross-disciplinary review, Earth's Futur. 6 (2018) 441−455.

[71] S.L. Cutter, The landscape of disaster resilience indicators in the USA, Nat. Hazards 80 (2016) 741−758.

[72] F. Tedim, Enhance Wildfire Risk Management in Portugal: The Relevance of Vulnerability Assessment, Wildfire Community Facil. Prep. Resilience, Charles C. Thomas Publ, Springfield, Ill, 2012, pp. 66−84.

[73] D. Alexander, A magnitude scale for cascading disasters, Int. J. Disaster Risk Reduct. (2018).

[74] M. Holland, A. March, J. Yu, A. Jenkins, Land use planning and bushfire risk: CFA referrals and the February 2009 Victorian fire area, Urban Policy Res. 31 (2013) 41−54.

[75] R. Blanchi, J. Leonard, Investigation of bushfire attack mechanisms resulting in house loss in the ACT bushfire 2003, Bushfire Coop. Res. Centre. (2005), 61pp.

[76] A. March, Y. Rijal, Reducing bushfire risk by planning and design: a professional focus, Plan. Pract. Res. 30 (2015) 33−53.

[77] J. Handmer, K. Haynes, Community Bushfire Safety, CSIRO publishing, 2008.

[78] K. Haynes, J. Handmer, J. McAneney, A. Tibbits, L. Coates, Australian bushfire fatalities 1900−2008: exploring trends in relation to the 'Prepare, stay and defend or leave early' policy, Environ. Sci. Policy 13 (2010) 185−194.

[79] S.J. O'Neill, J. Handmer, Responding to bushfire risk: the need for transformative adaptation, Environ. Res. Lett. 7 (2012) 14018.

[80] B.W. Butler, J.D. Cohen, Firefighter safety zones: a theoretical model based on radiative heating, Int. J. Wildland Fire 8 (1998) 73−77.

[81] N.P. Cheney, J. Gould, L. McCaw, The dead-man zone—a neglected area of firefighter safety, Aust. For. 64 (2001) 45−50.

[82] D.X. Viegas, D. Stipanicev, L. Ribeiro, L.P. Pita, C. Rossa, The Kornati fire accident—eruptive fire in relatively low fuel load herbaceous fuel conditions, WIT Trans. Ecol. Environ. 119 (2008) 365−375.

[83] G.M. Byram, Combustion of forest fuels, in: K.P. Davis (Ed.), Forest Fire: Control and Use, 1959, pp. 61−89.

[84] A.M. Gill, A Richter-type scale for fires? Research Letter #2 (1998).

[85] M.E. Alexander, F. V Cole, Predicting and interpreting fire intensities in Alaskan black spruce forests using the Canadian system of fire danger rating, in: Soc. Am. for. Conv., 1995.

[86] M.E. Alexander, W.J. De Groot, Fire Behavior in Jack Pine Stands: As Related to the Canadian Forest Fire Weather Index (FWI) System, Northern Forestry Centre, 1988.

[87] P.L. Andrews, R.C. Rothermel, Charts for interpreting wildland fire behavior characteristics, Gen. Tech. Rep. INT-131. Ogden, UT US Dep. Agric. For. Serv. Intermt. For. Range Exp. Station. 21 (1982) 131.

[88] J.E. Deeming, R.E. Burgan, J.D. Cohen, The national fire danger rating system-1978. USDA for. Serv., intermountain forest and range experiment station, Ogden, Utah, Gen. Tech. Report. INT. 38 (1977).

[89] R.C. Rothermel, How to Predict the Spread and Intensity of Forest and Range Fires, 1983.

[90] P.J. Roussopoulos, V.J. Johnson, Help in making fuel management decisions, Res. Pap. NC-112, in: St. Paul, MN US Dept. Agric. for. Serv. North Cent. for. Exp. Stn., vol. 112, 1975.

[91] J.E. Deeming, The National Fire-Danger Rating System, 1978, 1977.

[92] M.E. Alexander, R.A. Lanoville, Wildfires as a Source of Fire Behavior Data, 1987.

[93] A.M. Gill, P.H.R. Moore, Fire Intensities in Eucalyptus Forests of Southeastern Australia, 1990.

[94] R.H. Luke, A.G. McArthur, Bushfires in Australia, Australian Government Publishing Service for CSIRO, 1978.

[95] C. Lampin-Cabaret, M. Jappiot, N. Alibert, R. Manlay, R. Guillande, D.X. Viegas, Prototype of an intensity scale for the natural hazard: forest fire, in: For. Fire Res. Wildl. Fire Safety. Proc. IV Int. Conf. for. Fire Res., 2002, pp. 18—23.

[96] M. Kilinc, W. Anderson, D. Anderson, B. Price, On the Need for a Bushfire Scale that Represents the Bushfire Hazard, Monash Univ. Bushfire CRC Rep. to DSE, Melb., 2013.

[97] B.E. Potter, P.A. Werth, Introduction, in: Synth. Knowl. Extrem. Fire Behav, Fire Behav. Spec. Res. Meteorol., vol. 2, United States Department of Agriculture, Forest servive, 2016.

[98] B. Brettschneider, In Defense of the Saffir-Simpson Scale, Forbes, 2018.

[99] D. Paton, P. Burgelt, T.D. Prior, Living with bushfire risk: social and environmental influences on preparedness, Aust. J. Emerg. Manag. (2008).

[100] D. Paton, J. McClure, Preparing for Disaster: Building Household and Community Capacity, Charles C Thomas Publisher, 2013.

[101] D. Paton, F. Tedim, Enhancing forest fires preparedness in Portugal: integrating community engagement and risk management, Planet@ Risk 1 (2013).

Extreme wildfires and disasters around the world: lessons to be learned

2

Luís M. Ribeiro[1], Domingos X. Viegas[1], Miguel Almeida[1], Tara K. McGee[2],
Mário G. Pereira[3,8], Joana Parente[3], Gavriil Xanthopoulos[4], Vittorio Leone[5],
Giuseppe Mariano Delogu[6], Holy Hardin[7]

[1]Forest Fire Research Centre of ADAI, University of Coimbra, Coimbra, Portugal;
[2]Department of Earth and Atmospheric Sciences, University of Alberta, Edmonton, AB,
Canada; [3]Centre for the Research and Technology of Agro-Environmental and Biological
Sciences (CITAB), University of Trás-os-Montes and Alto Douro, Vila Real, Portugal;
[4]Hellenic Agricultural Organization "Demeter", Institute of Mediterranean Forest
Ecosystems, Athens, Greece; [5]Faculty of Agriculture, University of Basilicata (retired),
Potenza, Italy; [6]Former Chief Corpo Forestale e di Vigilanza Ambientale (CFVA),
Autonomous Region of Sardegna, Italy; [7]Public Affairs Science and Technology (PAST)
Fusion Cell, Argonne National Laboratory, Lemont, IL, United States; [8]Instituto Dom Luiz,
University of Lisbon, Lisbon, Portugal

2.1 Introduction

Following the definition proposed in the study by Tedim et al. [1], we describe in this
chapter some cases of extreme wildfire events (EWEs) with their associated personal
accidents and the lessons that should be learned from them. As can be seen in Table 2.1,
this study covers 16 cases of EWEs from six different countries with a wide range of
fire sizes and number of fatalities. The analysis of these cases provides insight into the
weather and fire management conditions that were associated to their ignition and
spread and into the operation of the fire suppression and civil protection agencies.
Owing to space limitations, the description of each case is necessarily short, but the
readers are referred to bibliography to obtain more details on the cases of interest.

2.2 Extreme wildfire cases in Portugal

2.2.1 The fires of Picões and Caramulo (2013)

In Portugal the fires of 2013 were marked by a relatively large burned area (around
160,000 ha) and by the death of 8 firefighters and a Mayor while fighting the fires
[1]. Two particular events marked the fire season negatively, the *Picões* fire, the largest
one of that year, and the *Caramulo* complex of fire events (CFEs), in which four firefighters lost their lives.

Extreme Wildfire Events and Disasters. https://doi.org/10.1016/B978-0-12-815721-3.00002-3

Table 2.1 Cases of EWE that are described.

Country	Place	Date	Burned area (Ha)	Fatalities
Portugal	Picões	8/12-July.2013	14,000	-
	Caramulo	20/30-August.2013	9,146	4
	P.Grande (+Góis)	17/24-June.2017	45,329	66
	Seia	15/22-October.2017	17,000	2
	Lousã	15/22-October.2017	50,000	14
	O.Hospital	15/22-October.2017	50,000	23
	Sertã	15/22-October.2017	31,000	2
	Vouzela	15/22-October.2017	16,000	10
Greece	South Greece	24-27-August.2007	270,000	84
Italy	Peschici	24-July.2007	535	3
	Laconi	7/8August.2013	2,450	-
Australia	Victoria and South Australia	February.1983	200,000	75
	Southeast Australia	8-January/7-March.2003	1,120,000	4
	Victoria	7-February.2009	400,000	160
USA	Gatlbinburg	28-Nov./9-Dec.2016	7,200	14
Canada	Horse river	1-May/4-July.2016	589,550	-

2.2.1.1   Picões (July 8th to 12th)

The region where this fire took place is characterized by continuous mountains with steep slopes and several waterlines running in between. It is an aged and low populated region, where most of its inhabitants dedicate their few resources to subsistence agriculture and olive and almond tree plantations. Some *Eucalyptus globulus* L. and *Pinus pinaster* Ait. are also present.

On July 8th, 2013, at 14h44, a fire ignited on the right bank of river *Sabor*, in the northeast region of Portugal, near a small village called *Cilhade* (municipality of *Torre de Moncorvo*, *Bragança* District). This fire was controlled on that evening, at 20h53 with an estimated burned area of 180 ha. During the night and the next morning, a couple of firefighter teams stayed on watch, monitoring the extinguished fire line. Despite the vigilance, at 14h00 of July 9th, the fire rekindled and spread rapidly, with several episodes of extreme behavior, until it was considered extinct early in the morning of July 12th, leaving an area of approximately 14,000 ha burned.

Topography and wind are always determinant in fire spread, and their effect was noticeable in *Picões*, where the open valleys and wind direction were parallel to the fire propagation during the most important parts of the fire development. Fuels near the origin were mostly herbaceous and dense shrub lands (Northern Forest Fire Laboratory (NFFL) models 2 and 4 [2]), which favor a very quick spread of the fire. As the fire progressed to east, herbaceous fuels dominated the landscape on the margins of the water courses. Apart from that, the majority of the central region of the fire was covered by shrubs of different heights and loads (NFFL models 4, 5, and 6 [2]). In these fuel models, fire propagates rapidly and intensely, even with high moisture contents. During the days of the fire, fuel moisture for fine fuels was estimated to be around a very low value of 7% between 14h00 and 16h00, doubling during the night.

In terms of meteorology, 2013 began with precipitation values above average, until the end of May, but were below the climatic mean for the period 1971–2000 in June and July. On July 3rd a heat wave affected Portugal, staying in the northeast until July 13th. Between July 9th and July 11th, temperatures reached up to 40°C with minimum values above 20°C. Relative humidity was also extremely low (around 10%) during the afternoons. When the fire rekindled, on July 9th at around 14h00, a nearby weather station (from the Portuguese Meteorological Institute) recorded 38°C of temperature and 13% of relative air humidity. Along with the preheat and desiccation of fuels, induced by the fire of the previous afternoon, the conditions were set for a rapid and intense fire propagation toward east/northeast, the same direction as the 20 km hour^{-1} average wind (with gusts up to 40 km hour^{-1}). In Fig. 2.1, which depicts the global estimated propagation, as reported by Viegas et al. [1], the brown area to the west refers to the initial *Cilhade* fire. The origin and rekindle are marked with a star and a triangle, respectively. The green areas near the triangle represent the fire propagation in the first 3 hours after the rekindle. At this stage, the fire rate of spread averaged between 1 and 2 km hour^{-1}.

After 17h00, fire behavior changed, with an increase in velocity and intensity. In approximately 3 hours, it spread for about 20 km, not only by surface but also with dozens of flying embers originating new spot fires. This represents an average of almost 7 km hour^{-1}, well above the commonly observed rate of spread for similar fuels. For instance, using BehavePlus [3], a well-known fire behavior simulation tool, the predicted rate of spread is 6–7 times lower (Table 2.2).

The rapid propagation during the afternoon affected a large construction site, which had to be evacuated, and a small village, called *Quinta das Quebradas*. This village, of less than 100 inhabitants, was impacted by the main fire front, and an important number of means were allocated for its protection: 30 vehicles and 99 men. As a consequence of both situations, during certain periods, there were insufficient firefighting resources dealing with fire suppression.

During the night of July 9th and morning of July 10th, the fire was slowly burning in favor of the now mild west wind and was nearly considered as contained. Around 8000 ha had burned so far. At around 14h00 of July 10th, the wind shifted to north, increasing in speed, and a large section of the right flank (south of the fire) became an active front. The fire then burned for one more day and consumed another 6000 ha, ending with a total of 14,000 ha of burned area.

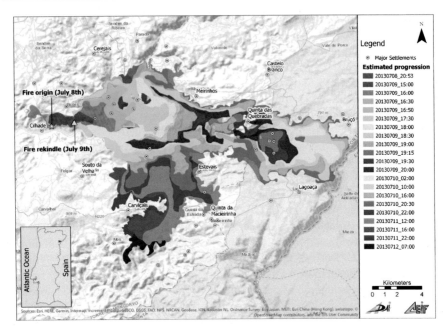

Figure 2.1 Picões fire estimated evolution.
Source: ADAI/CEIF.

On the aftermath of this fire, we could observe that, by that time, there was a marked lack of self-defense culture among most urban areas, despite their size, as well as in isolated houses. This was the case not only for managers, or politicians, but also for the inhabitants. The major consequence was that in most cases, fighting the fire became secondary for the crews, as they had to protect people and houses first. Although nowadays there is a shift in paradigm, related to self-protection, we still observe this, maybe in a more reduced scale. In the case of *Picões* fire, the population of the affected area had a high aging index, on average 2 times higher than the national average but in some regions close to 10 times higher. This aging contributed to the abandonment of agricultural practices and a lower availability to implement self-protection measures near the small urban centers. This was particularly noticeable in the small villages, such as *Quinta das Quebradas*. During the first day of the fire, especially when that

Table 2.2 Predicted rate of spread (km hour^{-1}) according to BehavePlus.

	Slope		
	20%	**30%**	**40%**
Herbaceous fuels	1.25	1.33	1.45
Shrubs	1.05	1.10	1.19

village was impacted, fire propagation often exceeded the firefighters control capacity, possibly reaching the level 5 in the classification system used by Tedim et al. [4].

This was the largest wildfire in Portugal during 2013, but, fortunately, there were no major personal incidents, apart from a burned firefighting vehicle.

2.2.1.2 Caramulo (August 20th to 30th)

The *Caramulo* CFE that developed in the *Caramulo* Mountain area in the central region of Portugal consisted of a group of three apparently independent fire events [1] that began in two distinct periods: the alert for the fire events of *Alcofra* and *Silvares* occurred at 23h54 of August 20th and at 00h32 of August 21st, both caused intentionally by the same group of two people; the *Guardão* Fire event started on August 28th by 11h05 and was accidently caused in the course of a fuel management operation with a bush cutter. The final burned areas were about 1,346 ha (*Silvares*), 1,522 ha (*Alcofra*), and 6,548 ha (*Guardão*).

The fires developed in a mountainous area with steep slopes, often greater than 45%, mostly composed by poorly managed areas of bushland and forests with predominance of *Pinus pinaster*, *Eucalyptus globulus*, and some oak trees, in a context of average values of temperature and relative humidity of 29°C and 21% at 12h00 and 19°C and 54% during the night. The wind blew always with great intensity, mainly from the north quadrant, with gusts between 27 and 72 km hour^{-1}.

The fire evolution in the three fire events was approximately that presented in Fig. 2.2. In spite of the significant role of wind in the development of the fire, the steep orography oriented the local winds, so these fire events presented quite different fire spread directions, in spite of being close to each other. The fire event of *Alcofra* spread mainly to the north according to the dominant wind. The event of *Silvares* rotated on itself driven by local winds, which led to several fire fronts reaching areas previously burned by this fire event. In the fire event of *Guardão*, in which about 90% of the final area was burned in about 9 hours during the night of August 29th, the dry east wind with an average speed of about 45 km hour^{-1} was the major cause of the great acceleration of the fire during this period. By this time, the fire events of *Silvares* and *Alcofra* were dominated with some reactivations of already suppressed lines.

Besides the extensive area burnt and large economic impacts, this CFE was mainly noticeable by four fatalities among the operational community, which resulted from two accidents: (1) the "*Olival Novo* Accident," killing two firefighters in the fire area of *Silvares*, on August 22nd by 15h30; and (2) the "*São Marcos* Accident," which also occurred in the area of the *Silvares* fire event, by 11h00 of August 29th, taking the lives of two firefighters.

The *Olival Novo* Accident occurred in the final phase of a tactical operation that took place on a hillside with a steep slope. At a certain moment there was a blow-up episode in which an area with a thick duff layer (~ 40 cm) of dead pine needles, previously burned superficially, reactivated with intense flames that extended to the crown of the pine trees. The firemen had defined the previously superficially burned area as the area for escape and refuge, not realizing that it could rekindle. Thus,

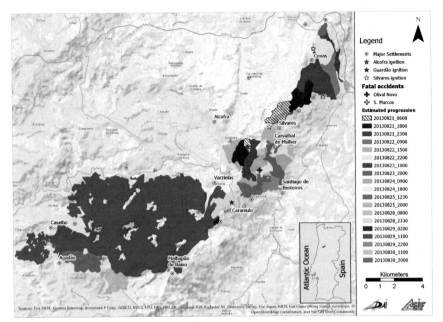

Figure 2.2 − Caramulo fire estimated evolution.
Source: ADAI/CEIF.

when this area reactivated, the firefighting team was caught by the flames. In addition to the two firefighters who died, several others were injured.

The *São Marcos* accident occurred during the first intervention to a reactivation of the *Silvares* fire event, in a local very close to the accident described above. During the operation in a canyon-shaped area, the fire assumed an eruptive extreme behavior going toward some firefighters that were improperly located on a mid-slope road at the top of the canyon. This dangerous situation had been detected a few minutes earlier, but those elements although warned of the danger did not retreat in time and were caught by the rapid and intense fire front eruption in the canyon. Besides the two fatalities, two other firefighters were seriously injured.

2.2.2 *The firestorms of 2017*

In 2017, Portugal was affected by several large fires that, in addition to a vast destruction, caused an unprecedented number of victims in this country. Among the several days with large fires in this year, we highlight the days of June 17th and October 15th as they have been associated with dramatic episodes of extreme fire behavior that caused 117 fatalities in total. Although there have been other large fires during 2017, the following description only addresses those that caused deaths.

2.2.2.1   Pedrógão Grande complex of fire events
(June 17th to 22nd)

The alert of this fire event that started near the village of *Escalos Fundeiros*, under an electric line that may have caused the ignition, was given around 14h43 when the burning area was already around 0.6 ha. In this period of intense heat (the air temperature was above 40°C), the wind was very strong with bursts that reached ≈ 40 km hour^{-1}, causing several spot fires that led to the loss of control of the situation in the first intervention. Moreover, there was a thunderstorm approaching from Southeast-East (SEE) producing erratic low-level winds.

By 14h53, at about 12 km from *Escalos Fundeiros*, another fire event started—the event of *Góis* (*Fonte Limpa*)—which, due to its great potential to become a large fire, diverged firefighting resources from the previous occurrence. This event ended up having a burnt area of more than 16,000 ha. Other smaller occurrences (e.g., *Moninhos* at 15h41) in the area also scattered firefighting resources.

At around 16h00, again under an electrical line, a new occurrence designated by "Fire of *Regadas*" started at about 3 km from *Escalos Fundeiros*. Both events developed independently at an initial phase. Owing to the lack of combat, means that were distributed in other fire events, and because due importance was not given to the Fire of *Regadas*, this event did not have a strong initial intervention and only one heavy vehicle and few human resources were allocated to it.

Around 18h00 there was a period of approximately 20 minutes with a very strong wind from north with speeds up to 70 km hour^{-1}. Shortly thereafter, by the 19h00, there was a new episode of southwest wind that approached the two fire fronts—*Escalos Fundeiros* and *Regadas*—which converged (Fig. 2.3) to create a unique very intense fire front, spreading very rapidly and producing a very intense smoke plume that rose up to about 18 km of altitude, surpassing the tropopause level. The reduction of this smoke plume, motivated by the fire and weather conditions, caused a downburst phenomenon that increased the fire intensity and accelerated the merged fire front producing uncountable spot fires, causing panic among the population. During this period, roughly between 19h30 and 20h30, the area in flames passed from ≈ 2,700 ha to ≈ 6,500 ha. Facing this scenario, many people tried to escape being caught by the flames in many cases. We highlight a situation in which 33 people died in or close to their vehicles in a section of road of only ≈ 400 m.

In total, 66 people died (1 firefighter) in this fire event that has a burnt area of ≈ 36,000 ha (excluding the burn area of the Fire of Góis). From these victims, only four died inside their homes. As June 17th was a Saturday, many of these people had come to spend the day or the weekend in this region, so many of the victims perished outside their area of residence. We believe that many of these people could have been saved if they had not fled the fire but had sought refuge in local safe areas such as houses, which in Portugal are typically built of stone, cement, or bricks. More details on this fire can be found in the study by Viegas et al. [5].

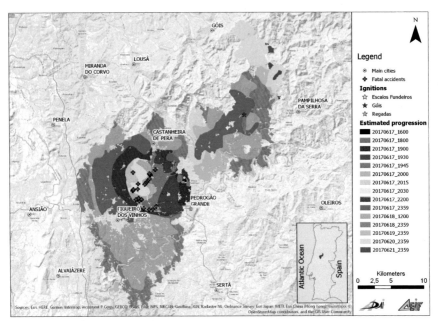

Figure 2.3 Pedrógão CFE estimated evolution. CFE, complex of fire event.
Source: ADAI/CEIF.

2.2.2.2   October 15th fires, in central Portugal (October 15th to 22nd)

The fires of October 15th are associated to severe meteorological conditions with unusually high temperatures (T > 30°C) for the Autumn season, very low fuel moisture content after a long dry period, and strong winds with gusts larger than 50 km hour^{-1}, mainly in the early afternoon (from south) and after 21h30 (from southwest) due to the passage of the *Ophelia* Hurricane off the Portuguese Coast. During this day, 507 fire events were registered in Portugal, which largely exceed the number of about 200 fire events that are considered the limit for which the Portuguese firefighting system is prepared. In the study by Viegas et al. [6], the main fires that occurred in this period are analyzed.

Among the set of fire events of October 15th, five EWEs (Fig. 2.4) are highlighted since they caused the death to 51 people among the population. These EWEs are described briefly in the following.

2.2.2.2.1   Extreme wildfire event of Seia

It started shortly before 06h00 near the village of *Sabugueiro*. The initial firefighting strategy was planned to protect the settlement and sequentially fight the three fire fronts that were slowly developing in a valley with short shrubs. Around 08h30, a spot fire emerged at about 1 km in the north direction, spreading very rapidly, starting to interact with the original occurrence. This caused the loss of operations control. Later,

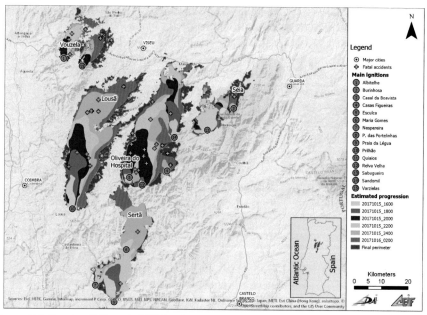

Figure 2.4 October 15th main EWE estimated evolution. EWE, extreme wildfire event. Source: ADAI/CEIF.

by 20h30, when the original fire was heading for a recently burned area, a new fire event—*Casal da Boavista*—started at about 12 km from *Sabugueiro*. This new occurrence not only invalidated the strategy that was being followed but also caused one death. This EWE consumed an area of 17,000 ha. The cause of the two ignitions is unknown.

2.2.2.2.2 Extreme wildfire event of Lousã

By 08h41, the alert for a fire event near the village of *Prilhão* was given. Apparently, this ignition was caused by an electric line, after the falling of a tree on the line producing its oscillation and causing an ignition about 200 m from the point of tree impact.

Owing to the strong winds, the firefighting was being very problematic when, by the 10h15, two spot fires appeared at about 300 and 750 m distance, creating a strong interaction among the fire fronts and threatening several populations that had to be protected by firefighters. Consequently, the fire developed almost freely to the north, traveling about 60 km in less than 20 hours, leaving a track of more than 50,000 ha of burned area, mostly shrubland and forest (pine and eucalyptus), and causing 14 fatalities.

2.2.2.2.3 Extreme wildfire event of Oliveira do Hospital

This EWE is the most complex of the five as it resulted from four main ignitions: (1) The fire of *Sandomil* started by 10h26 probably caused by a private fuel management activity close to a house; (2) The fire of *Esculca* by 12h28 possibly resulting from spotting with origin in an area burned during the previous week that was having several reactivations; (3) The fire of *Relva Velha* apparently resulted from several spot fires ($\sim$6 km) by 14h00, originated by the Fire of *Esculca*; and (4) The fire of *Casas Figueiras*, by 23h00, possibly resulted by spotting of 21 km from the EWE of *Sertã*.

The fires of *Esculca* and *Relva Velha* were the first to interact, around 19h00. The convergence of these two fire fronts intensified the flames that spread very rapidly toward north to converge with the fire front of the *Sandomil* event after 21h00. This new convergence which coincided with a sudden increase of the wind velocity caused a great acceleration of the fire, surprising many people—the largest number of deaths occurred during this period. The *Casas Figueiras* fire event did not have a clear interaction with any of the other occurrences of this EWE, but the important fire spread observed was due not only to the strong wind but also to the interaction with the hot area burned in the other occurrences.

The extreme behavior observed in this EWE, especially in the *Casas Figueiras* fire event, was still strongly influenced by the effects of the three other EWEs that surrounded it, especially the EWE of *Sertã*, as the wind was mainly from south. This EWE caused 23 fatalities in a confined area of around 50,000 ha.

2.2.2.2.4   Extreme wildfire event of Sertã

The EWE of *Sertã* had three main ignitions. The fire event of *Ponte da Portelinha* started shortly before 12h00. Like in the previously described events, a succession of spot fires disrupted the defined strategies, occupying many of the firefighting resources in the protection of threatened settlements.

Later, at 18h41 the alert was given for the ignition of *Nespereira*. This event not only stretched the limited firefighting resources but also drove to the interaction with the fire front of *Ponte da Portelinha* event, producing an episode of convergent fire fronts. This episode may be associated with several spot fires in *Maria Gomes*, by 19h30, originating a very intense fire front that developed to north reaching the area previously burned by the fire event of October 6th that initiated the *Esculca* Fire Event (cf. EWE *Oliveira do Hospital*). All the left flank of the EWE of *Sertã* was limited by the burned area of the *Góis* Fire Event (cf. EWE of *Pedrógão Grande*). This EWE resulted in two fatalities within a burned area of 31,000 ha.

2.2.2.2.5   Extreme wildfire event of Vouzela

By 17h21 two very close ignitions were detected near the village *Albitelhe*. The fire was already very well developed at the arrival of the firefighters to the scene. The intense initial spread and the several spot fires created allowed to immediately perceive the potential of that situation to become a large wildfire. Later, by 18h50, a new event—*Varzielas* Fire—started at 6 km from the *Albitelhe* Fire. This ignition was probably originated by a 21 km spotting from the EWE of Lousã.

These two fire fronts evolved in an apparently independent way as the strong wind diluted the potential of interaction. By 03h00 on October 16th, the two fronts merged without an abrupt increase in the intensity of the unique fire front. The final area bounded by this fire that caused 10 fatalities was 16,000 ha.

In these five EWEs, 19 people died inside the houses, 10 of them while sleeping as they were surprised by the fire during the night. Once again, most people died when they were escaping, many times in panic, when probably the best option would have been to take refuge in a safe place as most houses are.

The spotting, driven by strong winds, was the major mechanism of fire propagation producing several fire fronts that frequently interacted to originate episodes of extreme fire behavior. The cause of the ignitions was very diverse—e.g., long-distance spotting, arson fire, and electric lines, among other causes.

2.3   Extreme wildfire cases in the world

Over the years, multiple EWE cases occurred all around the world, with different outcomes and lessons to be learned. A small selection of cases is presented here, covering five fire-prone countries.

2.3.1   Greece

In 2007 Greece experienced its worst fire season on record. Approximately 270,000 ha of vegetation burned, and more than 110 villages were affected directly by the fire fronts. More than 3000 homes were totally destroyed or partially damaged. Eighty-four people, mostly civilians, lost their lives in a series of fire-related accidents. While the whole fire season was characterized by intense, hard to fight fires, due to drought, high air temperature (3 heat waves during the summer) and low relative humidity, the situation became explosive in the 4-day period of August 24th-27th.

A series of simultaneous, very intense wildfires in the south of Greece spread very aggressively, exceeding by far the suppression capacity of the firefighting mechanism. Among them the fatal fire of *Palaiohori-Artemida* in the prefecture of Ilia, Peloponnese, stands out [7].

The fire started at about 14:40, August 24th, 2007, in the yard of a house in the small settlement of *Paleohori*. It was started by an old woman cooking on an open stove and was reported immediately to the Fire Service as a house fire. However, as the fire caught on the vegetation close to the house, it started accelerating and moving in a southwest direction fanned by a strong wind. Field weather measurements in the general area showed that temperature was at 40°C, relative humidity varied between 10% and 17%, and the wind was blowing from a northeast direction at 12 km hour^{-1}, with gusts reaching 29 km hour^{-1}. Under such conditions, burning in a valley with olive groves, mostly with mowed but left in place dead grass under the olive trees, interspersed with clubs of pine trees, the fire exhibited profuse spotting and built extreme intensity. As the fire front moved, it exhibited an average rate of spread of roughly 5 km hour^{-1} with bursts reaching even 10 km hour^{-1}.

The first village reached by the fire, at about 14h55, was *Makistos* at a distance of 1.7 km from *Paleohori*. The people there were caught by surprise. All but a handful left the village in panic toward the nearby village of *Artemida* (older name *Koumouthekras*) with the fire front running after them. The fire was spreading at an average speed of 5 km hour^{-1}, enough to catch a walking man, and it did overtake an older couple who were fleeing with their donkey toward the village of *Artemida*. In the abandoned village, most houses were destroyed or seriously damaged. Only 14 of the 60 homes did not suffer damages. Some of them were saved by the efforts of those who stayed behind and fought to save their property.

The village of *Artemida* was reached next by the fire at about 15h15. The people arriving from *Makistos*, and most of the people of *Artemida*, in panic, fled the village in two groups. The first group formed a convoy of cars and moved toward the town of *Zaharo*. A few minutes later the second group fleeing the village chose to drive to higher ground, toward the village of *Smerna*. They probably realized that the quick spread of the fire in the valley would cut off the route to *Zaharo*. This group as well as the people who stayed in the village survived without a problem. Only 17 of the 70 homes of the village were destroyed or heavily damaged.

The first group, accompanied by a Fire Service firetruck with a crew of three seasonal firefighters, were trapped by the fast-moving fire, at about 15h30, between the front and the spot fires that were numerous up to a few hundred meters ahead. The death toll reached 24 people. It was the worst accident of this type that ever happened in Greece until that time, only surpassed by the July 23rd, 2018 fire in eastern *Attica* with 100 fatalities.

The lessons that can be learned from this accident can be summarized as follows:

- Evacuation of settlements in case of forest fires is not necessarily a good option. In particular, when an evacuation is done spontaneously, without planning, the results can be catastrophic especially if realized at the last moment.
- The best option seems to remain in the houses when they and their occupants are prepared for such an event. This is especially true for the majority of homes in Greece and the rest of the Mediterranean region, which are generally built with nonflammable materials. There are no reports of lives lost for citizens that remained in their homes.
- The people in fire-prone countries need to be educated about forest fires, the need and measures to prevent them, the risks they pose, and the ways to protect themselves and their property.
- Firefighter safety training should be a priority.

2.3.2 Italy

In Italy individual wildfire reports do not contain values of flame length, intensity, rate of spread, or spotting activity. Given the impossibility to propose the most relevant case studies, we present two examples that are representative of EWEs: the wildfire of *Peschici* (*Parco Nazionale del Gargano*) and the one of *Isili-Nurallao-Laconi* (Sardinia).

2.3.2.1 Peschici Fire

In *Peschici*, a celebrated summer holiday site in the Gargano National Park, very severe fire weather conditions, synthesized by FWI>50 during all July 2007, except in 2 days, gradually increasing to >60 on July 19th, to >70 on July 22nd, and to >80 on July 23rd, reached the top of 98.14 on July 24th. In the same day, all FWI components exhibited values well above the threshold of extreme, namely FFMC-98.70, DMC-381.21, DC-960.45, ISI-43.86, and BUI-382.69, thus preparing the perfect disaster that occurred at about 09h00.

Fanned by westerly *Föhn* winds, with an average speed of 37 km hour^{-1} and gusts up to 53.5 km hour^{-1}, with temperature of 37.8°C and relative humidity of 22%, flames started from negligently burning agricultural residues and tore toward the promontory where *Peschici* is settled, surrounded by hills with sparse natural stands of *Pinus halepensis* with thick undergrowth of *Rosmarinus* and *Pistacia lentiscus*.

Fire gained momentum and rapidly reached the cemetery and peripheral buildings, spreading down slope toward a beach where about 3000 tourists, at the peak of the summer season, were surprised by flames that destroyed about 700 cars and two camping cars.

About 1000 people were rescued by boat by Coastal Guard, *Carabinieri*, *Vigili del Fuoco*, volunteers, and fishermen. Three fatalities occurred, 300 were injured or intoxicated by smoke, and 3538 persons were evacuated in close by municipalities. The burned area was only 535 ha and included buildings, camping sites, and hotels, randomly scattered in a disordered Wildland Urban Interface (WUI). The blaze lasted more than 5 days and was extinguished with the help of 148 firefighters, four Canadair, and 10 helicopters [8].

Although *Peschici* is one of the most severely affected municipalities in *Apulia*, with 821 fires registered (1974−2007, excluding 1995 and 2002), and a rate of voluntary fires of 59% (1998−2010), no prevention measures were activated.

As we do not have ground data for fire behavior, we simulated it with BehavePlus [3] and fuel model 4 [2] and interspersed with a sparse canopy of *Pinus halepensis*, with results shown in Table 2.3.

These values fit well with the behavior observed in media and permit to define an EWE in the classification by Tedim et al. [4], class 5 when cover is 55%−60% and class 6 for cover 30%−50%. It can certainly be defined as a disaster. Although not relevant for its dimension, the *Peschici* Fire was upsetting for its awesome behavior; the number of dead, injured, and evacuated people; complexity of rescue; and amount of damage.

Table 2.3 Fire predictions according to BehavePlus.

Fuel model	Flame length (m)	Intensity (kW.m^{-1})	ROS (m.minute^{-1})	Spotting (km)
4 (shrubs) with a canopy covering 30%−60%	6.50 to 11.40	15,217 to 52,011	31.60 to 107.90	1.60

ROS, rate of spread

2.3.2.2 Laconi (Bidda Beccia) fire

Owing to synoptic conditions at 850 hP, on seventh and eighth of August 2013, the center and south of *Sardinia* were characterized by high temperatures (38°C) and very low values of relative humidity, <20% for 80% of the island and in some cases <15%. Following the RI.SI.Co. system of the National Civil Protection of Italy, the day was declared at high risk.

From 11h25 to 18h30 of August 7th, seven great wildfires occurred in sequence from south to center of *Sardinia*, generating a "fifth generation" [9] situation of simultaneous great wildfires, which caused the collapse of the regional extinction system.

A deliberate ignition started the *Bidda Beccia* wildfire (Municipalities of *Isili-Nurallao-Laconi*) at 12h11, rapidly spreading in a shrubland and *Quercus ilex* vegetation exposed to hot, dry south winds; after 1 hour, flames extended to *Pinus radiata* and *Pinus brutia* plantations in a plateau morphology. When this wildfire was declared, 15 terrestrial means, four light regional helicopters, and two Canadair were dispatched.

The Incident Command adopted the strategy of tightening the hips to proactively block the fire head with an indirect attack using backfire, applied by the specialized "*Mastros de Fogu*" team of regional forest corps.

Wildfire exhibited an aggressive behavior, spreading in the early afternoon at the rate of 380 ha hour^{-1} or an rate of spread (ROS) of 3 km hour^{-1}, with massive spotting. An estimated flame length of 45−50m made the aerial drops very difficult.

During the night, owing to reduced teams, fire aggressiveness increased and behavior changed, entering into logging slash dried at soil.

The eighth of August, at 02h00, wildfire spread in a *maquis* and old, dense pine yards, with a rate of 100 ha hour^{-1}.

The nearby village was evacuated. Extinction operations were completed by 20h00. The total burned area was 2,450 ha (Figure 2.5).

Figure 2.5 The right flank of Laconi fire during the first phase of spread in plantations of *Pinus radiata*, *P. brutia*, and *P. halepensis*; average height of canopy 15−16 m, height of dominant trees 20 m.
Photo by G. Delogu.

2.3.3 Australia

Australia has a long history of extreme wildfire events. The conditions, causes, behavior, and consequences of three of the most disastrous cases will be briefly described in the following paragraphs.

The first episode occurred in February of 1983 when several wildfires were registered in Victoria and South Australia and the McArthur Forest Fire Danger Index (FFDI) reached a record value of 145 [10]. During these Ash Wednesday fires, about 200,000 ha were burnt, more than 2000 buildings were destroyed [10,11], 27,000 heads of stock were lost [12], and 75 persons had lost their lives [11]. Researchers showed [13] that these fatalities had occurred because of insufficient warning and the victims implemented an ineffective survival strategy or were incapable of implementing an effective survival strategy without support.

The second case took place between January 8th and March 7th, 2003 throughout Southeastern Australia and was preceded by six consecutive years of drought characterized by lack of rain, high air temperatures, and consequent dryness of vegetation [14]. A high number (80) of wildfires started with lightning strikes from dry thunderstorms activity in the predominantly forested and alpine areas [12,14,15] and burned more than 1.12 million ha of public and private forests and agricultural lands in the northeastern and Alpine districts of Victoria [12]. In the beginning of that period, drought levels remained elevated and associated to mild, anticyclonic weather conditions, but on 18th January hot (maximum air surface temperature of 37.4°C), dry (minimum relative humidity of 8%), and strong north−westerly winds favored the wildfires with wind speed of 13 m s^{-1} and gusts of 22 m s^{-1} [14−16]. On this date, during a 10-minutes flaming peak, 3.5×10^{12} kJ were released [5] and the firestorms culminated with pyrocumulonimbus "eruptions" that produced a stratospheric smoke injection that perturbed the hemispheric background analogous to the theorized "nuclear winter" in Canberra skies [15]. On the seventh of March 2003, the wildfires were officially declared contained and controlled after a period of calmer conditions and raining during February and March [14]. At the end, more than 500 houses were lost [14,16], four people died [14], 11,000 head of stock were killed, tens of thousands of kilometers of fencing, sheds, machinery, and other agricultural equipment were also destroyed [12], and the wildfire had a cost of AUD$300 million [11]. This event had also destroyed the historic 1942 Mount Stromlo Observatory with an estimated rebuilding price at AUD$50 million [11].

The third case of the bushfires of February 2009 in Victoria was also preceded by a drought season, air temperatures above the average (e.g., peak above 45°C in Melbourne), and two heat waves observed over Southeastern Australia from late January to early February [17]. These weather conditions were combined with favorable synoptic patterns characterized by a high-pressure system located in Tasman Sea [18] and low-pressure system located over northern Australia, which had promoted the development of hot air in central Australia and in conjugation with a strong cold front created very strong hot north−westerly winds up to 65 km hour^{-1} and gusting to 90 km hour^{-1} in western and central Victoria [17]. These fires affected a wide range of forest types dominated by *Eucalyptus regnans* with high fuel loads leading to exceptional fire intensities (e.g., 100,000 kW m^{-1}) [14]. During these EWEs, extremely high FFDI (ranging between 120 and 190) had been recorded [10,19], more than 400,000 ha burned [17,18], 173 people had lost their lives, more than 2000 buildings were lost [10,20], and the event had a total cost of more than AUD$3.5 billion [18,19].

Until the Victorian bushfires of 2009, where a large number of people had perished within their homes, Australia had established the policy that residents should prepare, stay, and defend their home and property against bushfire or leave well ahead of the fire front [10] and discourage the unjustified dependence on emergency services [20]. Thereafter, the Bushfires Royal Commission recommended the adoption of the national "Prepare. Act. Survive." strategy, which emphasizes that leaving early is the safer option and propose improvements to risk communication, wildfire education, and warning system [10].

2.3.4   USA—Gatlinburg wildfires risk/crisis communications lessons learned

On November 28th, 2016, unprecedented wildfires ravaged Sevier County, Tennessee, home to a popular tourist attraction in the Gatlinburg area, killing 14, injuring 191, and damaging or destroying more than 2400 homes and businesses. "The Sevier County wildfire (was) the most catastrophic wildland-urban interface fire event in the history of Tennessee," Tennessee Emergency Management Agency (TEMA) Director Patrick Sheehan said at that time. Experts reviewing details of this historical event now say it has provided many serious but valuable lessons in the areas of emergency notifications, protective action communications, and risk and crisis communications [21].

Gatlinburg, Tennessee, is known as the gateway to the roughly 7,200 ha Great Smoky Mountains National Park (GSMNP), which is the most visited national park in the United States with more than 11-million visitors a year and home to 4126 permanent residents. However, visitors to the area can boost the daily population to more than 50,000 [22].

On November 10, 2016, a level 3 state of emergency was declared because of extreme drought conditions. On November 23, 2016, a fire ignited within the confines of the GSMNP and continued to burn within the confines of the GSMNP for several days. Initially, mountainous terrain and poor communication led fire personnel to decide to only monitor the fire, despite indications that severe drought conditions and weather might affect fire growth [23]. On the evening of November 28th, 2016, windy conditions spread the fires beyond the boundaries of the GSMNP and spread toward, eventually entering the City of Gatlinburg and causing the evacuation of 14,000 people.

After the fires, emergency management communications analysts examined emergency public information during the wildfires via interviews of Gatlinburg business owners, the clergy, tourists, residents, survivors, emergency management experts, and public information officers. This work suggested that there were significant issues with communication: as one expert stated, "Communication was the first causality" [21]. Many individuals received late or no official warning and instead learned of the fire and the need to evacuate from other community members. One part of the delay in official notification was due to electrical and cell-phone outages at the command center. In addition, those who did receive messages indicated that official agencies offered no actionable messaging throughout the day the wildfires directly impacted the City of Gatlinburg. Some tourists felt officials were too focused on the wildfire response itself with little attention to providing emergency public information. One tourist felt responders and official agencies were in a constant reactive stance rather than thinking proactively and summarized by stating "People live and die on what comes out of a good public information system" [21].

More than 33,000 people talked about the fire on Facebook. It became a trending topic on the night of November 28th, 2016, and later when the wildfires spread causing mandatory evacuations for areas populated by almost 14,000 people. In fact, the wildfires even triggered Facebook's safety check-in feature for those located near the wildfires. However, news reports indicate that the GSMNP did not use social media for emergency alerts and notifications [24]. Social media also were underutilized for notification and communication: Official agencies did not disseminate any notifications, evacuation messages, or emergency public information via social media channels [24].

Findings around timely notification and communication during the Gatlinburg wildfires highlight why a robust crisis communication plan inclusive of plans, policies, protocols, and procedures is a vital part of emergency preparedness and response [21]. Communities should have a plan to manage public information before, during, and after an emergency. Integrated and comprehensive crisis communication plans are critical to the emergency response process and should address both internal and external communication needs. An integrated notification plan, including the use of social media, can help centralize and accelerate notification, reduce redundancies, and connect with a wider range of stakeholders during an emergency event, which is especially critical in a high tourist environment.

The work also highlights the importance of public information personnel being trained and prepared for the inherent communication complexities encountered during extreme wildfires. It is important to build capabilities and capacities focused on protecting stakeholders and the environment by developing response-appropriate notification and risk/crisis communication plans and systems so that when and if the day comes, response entities will be well trained and well versed in deploying the appropriate system and how to adapt when communication systems fail. Personnel with the responsibility to craft messages during complex incidents should be trained and exercised to identify and address communication complexities.

In the end, a comprehensive crisis communication plan—which has been integrated with the crisis management or operations plan and is well tested and understood and practiced by agency employees—can save lives and lead to improved outcomes overall.

2.3.5   Canada—2016 Horse River (Fort McMurray) fire

Wildfire is a natural part of Canada's forests, and on average about 8400 wildfires burn over two million hectares of forested land every year [25]. The Regional Municipality of Wood Buffalo (RMWB), which is located in the boreal forest of northeastern Alberta, includes the urban center of Fort McMurray and rural communities including First Nation and Métis settlements. Fort McMurray is located where the Athabasca and Clearwater rivers meet, with the smaller Horse, Christina, and Hangingstone Rivers also flowing through the city [26]. In 2016, RMWB had a population of 71,589, with 66,573 living in Fort McMurray [27]. Fort McMurray is accessible year-round by Highway 63 which runs north and south. Some neighborhoods in Fort McMurray have one road in and out joining Highway 63.

The winter of 2015−16 was very dry and mild in northern Alberta, followed by an unusually hot and dry spring, which provided conditions for extreme fire [28]. In Fort McMurray, daily temperatures at the end of April were above 30°C, low humidity, and winds approaching 25 km hour^{-1} [29]. The Horse River fire started on May 1st in a forested area 7 km outside Fort McMurray [29] and quickly grew in size because of extreme conditions. At 20h00 on May 1st, the RMWB's emergency department warned people in Centennial trailer park in the south of Fort McMurray to leave their homes and put the neighborhoods of Beacon Hill and Gregoire on evacuation alert. At 22h00, a state of local emergency was declared, and a mandatory evacuation was ordered for residents of Centennial RV Park Campground, Prairie Creek and Gregoire. On Monday May 2nd, the fire grew and was almost 1 km from Highway 63, the road in and out of Fort McMurray. Overnight, the fire expanded west, away from the city. The next morning, the smoke and flames were not visible from the city, but at 11h00 the fire chief warned that winds could shift. At noon, the fire jumped the Athabasca River, and later that day, the Horse River wildfire entered Fort McMurray.

The neighborhoods of Beacon Hill, Abasand, and the MacKenzie Industrial Park were impacted first. Although some residents had already evacuated, a mandatory evacuation order was issued by the RMWB for all of Fort McMurray at 16h20. Later in the evening, the wildfire entered the subdivisions of Thickwood and Waterways [29]. Most residents left in their own vehicles and traveled south on Highway 63 toward the City of Edmonton and other communities in Alberta. However, owing to traffic impediments on Highway 63 as people were evacuating south, approximately 15,000 to 20,000 residents followed advice from the RMWB and drove north and stayed in work camps and Fort McKay First Nation. On May 4th, the province of Alberta declared a state of emergency and called for a mandatory evacuation of Saprae Creek at 16h00. At 22h00, a mandatory evacuation was ordered by the provincial government for Anzac, Gregoire Lake Estates, and Fort McMurray First Nation. A few days after May 3rd, the fire started heading north and supplies started to run out in the work camps, so evacuees were either flown out to Calgary or Edmonton, or drove south on Highway 63 back through Fort McMurray to stay elsewhere.

More than 88,000 residents of urban Fort McMurray and rural communities in RMWB evacuated their homes and communities, mainly by personal vehicle, with many having difficulties leaving because of traffic congestion [30]. Residents stayed in a variety of towns, cities, and communities for 1 month or longer, with many evacuees having to stay in more than one location [30]. The government of Alberta completed a phased re-entry process which allowed the first residents to return to the RMWB on June 1st. The fire was under control on July 4th, after requiring the evacuation of more than 88,000 residents, burning 589,550 ha, and destroying 2579 structures in Urban Fort McMurray. The fire was officially "out" on August 2nd, 2017. Research indicates that mental health problems increased after the wildfire [31], and the wildfire had a significant impact on air quality in Fort McMurray [28]. Help provided to evacuees helped them to leave Fort McMurray and arrive safely in another place and during their stay elsewhere [30].

2.4 Conclusion

An overview of EWEs around the world was presented based on 16 cases in six different countries, spanning a period of 34 years and showing different realities and approaches to the problem of fire management. In all of them, unusual or extreme forms of fire behavior were observed, demanding novel advances in fire science. Almost all cases involved the need of protecting the population, and many of them had casualties among operational and civilian personnel. It was shown that in some countries, it may be preferable to stay at home rather than run away in the case of fire. There are nevertheless conditions under which it is better to leave early, even if one is prepared to resist the fire. Last-minute evacuation with fire nearby is not recommended as it can lead to disaster.

It is important to work with the communities and prepare them to face fires, if necessary, without the support of the authorities. Communication plans and tools must be prepared and tested beforehand. This is required not only for permanent residents but also for tourists or visitors.

It is noticed that this type of EWEs tends to become more frequent and bring new challenges to fire science and management.

References

[1] D.X. Viegas, L.M. Ribeiro, M.A. Almeida, R. Oliveira, M.T.P. Viegas, J.R. Raposo, V. Reva, A.R. Figueiredo, S. Lopes, The Large Forest Fires and the Fatal Accidents Occurred in 2013, Forest Fire Research Centre (CEI/ADAI/LAETA), 2013 (In Portuguese).

[2] H.E. Anderson, Aids to Determining Fuel Models for Estimating Fire Behavior, Lntermt. for. And Range Exp. Stn., Ogden, Utah 84401, 1982.

[3] P.L. Andrews, BehavePlus Fire Modeling System, Version 5.0: Variables, General Technical Report RMRS-GTR-213WWW Revised, USDA Forest Service, Rocky Mountain Research Station, Fort Collins, CO, 2009.

[4] F. Tedim, V. Leone, M. Amraoui, C. Bouillon, M. Coughlan, G. Delogu, P. Fernandes, C. Ferreira, S. McCaffrey, T. McGee, J. Parente, D. Paton, M. Pereira, L. Ribeiro,

D. Viegas, G. Xanthopoulos, Defining extreme wildfire events: difficulties, challenges, and impacts, Fire 1 (2018) 9, https://doi.org/10.3390/fire1010009.

[5] D.X. Viegas, M.F. Almeida, L.M. Ribeiro, J. Raposo, M.T. Viegas, R. Oliveira, D. Alves, C. Pinto, H. Jorge, A. Rodrigues, D. Lucas, S. Lopes, L.F. Silva, The Fire Complex of Pedrógão Grande and Neighbouring Municipalities that Started in June 17, 2017, Forest Fire Research Centre (CEIF/ADAI/LAETA), 2017 (In Portuguese).

[6] D.X. Viegas, M.A. Almeida, L.M. Ribeiro, J. Raposo, M.T. Viegas, R. Oliveira, D. Alves, C. Pinto, A. Rodrigues, C. Ribeiro, S. Lopes, H. Jorge, C.X. Viegas, Analysis of the Forest Fires Ocurred in October 15, 2017, Forest Fire Research Centre (CEI/ADAI/LAETA), 2019 (In Portuguese).

[7] K. Kalabokidis, C. Karavitis, C. Vasilakos, Automated fire and flood danger assessment system, in: G. Xanthopoulos (Ed.), Forest Fires Wildland-Urban Interface Rural Areas Eur. An Integr. Plan. Manag. Challenge, 2003.

[8] European Commission - Joint Research Centre, EUR 23492 EN. Forest Fires in Europe 2007, Office for Official Publications of the European Communities, Luxembourg, 2008.

[9] P. Costa, M. Castellnou, L. Asier, M. Miralles, D. Kraus, La Prevención de los Grandes Incendios Forestales adaptada al Incendio Tipo, Unitat Tècnica del GRAF, Divisió de Grups Operatius Especials, Direcció General de Prevenció, Extinció d'Incendis i Salvaments, Departament d'Interior, Generalitat de Catalunya, Barcelona, 2011.

[10] R. Blanchi, J. Leonard, K. Haynes, K. Opie, M. James, F.D. de Oliveira, Environmental circumstances surrounding bushfire fatalities in Australia 1901−2011, Environ. Sci. Policy 37 (2014) 192−203, https://doi.org/10.1016/j.envsci.2013.09.013.

[11] B. Ashe, K.J. McAneney, A.J. Pitman, Total cost of fire in Australia, J. Risk Res. 12 (2009) 121−136, https://doi.org/10.1080/13669870802648528.

[12] A. Tibbits, J. Whittaker, Stay and defend or leave early: policy problems and experiences during the 2003 Victorian bushfires, Environ. Hazards 7 (2007) 283−290, https://doi.org/10.1016/j.envhaz.2007.08.001.

[13] N. Krusel, S.N. Petris, A Study of Civilian Deaths In the 1983 Ash Wednesday Bushfire Victoria, CFA Occas. Pap. No 1, 1999, 28p..

[14] G. Worboys, A brief report on the 2003 Australian alps bushfires, Mt. Res. Dev. 23 (2003).

[15] M. Fromm, A. Tupper, D. Rosenfeld, R. Servranckx, R. Mcrae, Violent Pyro-convective Storm Devastates Australia's Capital and Pollutes the Stratosphere, 2006, https://doi.org/10.1029/2005GL025161.

[16] I. White, A. Wade, M. Worthy, N. Mueller, T. Daniell, R. Wasson, The vulnerability of water supply catchments to bushfires: impacts of the January 2003 wildfires on the Australian Capital Territory, Australas. J. Water Resour. 10 (2006) 179−194, https://doi.org/10.1080/13241583.2006.11465291.

[17] M.G. Cruz, A.L. Sullivan, J.S. Gould, N.C. Sims, A.J. Bannister, J.J. Hollis, R.J. Hurley, Anatomy of a catastrophic wildfire: the black Saturday Kilmore east fire in Victoria, Australia, For. Ecol. Manage. 284 (2012) 269−285, https://doi.org/10.1016/j.foreco.2012.02.035.

[18] J.W. Mitchell, Power line failures and catastrophic wildfires under extreme weather conditions, Eng. Fail. Anal. 35 (2013) 726−735, https://doi.org/10.1016/j.engfailanal.2013.07.006.

[19] S.J. O'Neill, J. Handmer, Responding to bushfire risk: the need for transformative adaptation, Environ. Res. Lett. 7 (2012), https://doi.org/10.1088/1748-9326/7/1/014018 doi: Artn 014018.

[20] R.P. Crompton, K.J. Mcaneney, K. Chen, R.A. Pielke, Influence of Location , Population , and Climate on Building Damage and Fatalities Due to Australian Bushfire: 1925–2009, 2010.

[21] R.G. Edmond, E.D.H. Hardin, 2016 Gatlinburg Wildfires Lessons Learned, Argonne National Laboratory – National Public Affairs Academy, 2017.

[22] U.S. Census Bureau, American Community Survey 5-year Estimates. Census Reporter Profile Page for Gatlinburg, 2017. https://censusreporter.org/profiles/16000US4728800-gatlinburg-tn/.

[23] D. Jacobs, Firefighters raced to Gatlinburg, only to find some hydrants were running dry, USA Today Netw. (2017). https://eu.citizen-times.com/story/news/local/2017/02/19/firefighters-raced-gatlinburg-only-find-some-hydrants-were-running-dry/98041968/.

[24] D. Jacobs, Social Media Underused in Gatlinburg Firestorm, Knoxv. News Sentin, 2016. https://eu.knoxnews.com/story/news/local/tennessee/2016/12/27/social-media-underused-gatlinburg-firestorm/95442984.

[25] T. McGee, B. McFarlane, C. Tymstra, Wildfire: a Canadian perspective, in: D. Paton (Ed.), Wildfire Hazards, Risks and Disasters, Elsevier, 2015, pp. 35–58, https://doi.org/10.1016/B978-0-12-410434-1.00003-8.

[26] Fort McMurray Tourism. Five rivers flow through Fort McMurray, (n.d.). http://www.fortmcmurraytourism.com/facts/five-rivers-flow-through-fort-mcmurray.

[27] Statistics Canada, Alberta [Province] and Canada [Country] (Table. Census Profile. 2016 Census. Statistics Canada Catalogue No. 98-316-X2016001, Ottawa, 2017, https://www12.statcan.gc.ca/census-recensement/2016/dp-pd/prof/index.cfm?Lang=E.

[28] M.S. Landis, E.S. Edgerton, E.M. White, G.R. Wentworth, A.P. Sullivan, A.M. Dillner, The impact of the 2016 Fort McMurray Horse River wildfire on ambient air pollution levels in the Athabasca oil sands region, Alberta, Canada, Sci. Total Environ. 618 (2018) 1665–1676, https://doi.org/10.1016/j.scitotenv.2017.10.008.

[29] MNP, A Review of the 2016 Horse River Wildfire Alberta Agriculture and Forestry Preparedness and Response, Office for Official Publications of the European Communities, Luxembourg, 2017.

[30] T.K. McGee, Residents' experiences of the 2016 Fort McMurray wildfire, Alberta, in: Adv. for. Fire Res. 2018, Imprensa da Universidade de Coimbra, Coimbra, 2018, pp. 1155–1159, https://doi.org/10.14195/978-989-26-16-506_129.

[31] M.R.G. Brown, V. Agyapong, A.J. Greenshaw, I. Cribben, P. Brett-MacLean, J. Drolet, C. McDonald-Harker, J. Omeje, M. Mankowsi, S. Noble, D. Kitching, P.H. Silverstone, After the Fort McMurray wildfire there are significant increases in mental health symptoms in grade 7–12 students compared to controls, BMC Psychiatry 19 (2019) 18, https://doi.org/10.1186/s12888-018-2007-1.

Part Two

Extreme Wildfire Events and Disasters: The Root of the Problem

The role of weather and climate conditions on extreme wildfires

Mário G. Pereira [1,2], Joana Parente [1], Malik Amraoui [1], António Oliveira [1], Paulo M. Fernandes [1]
[1]Centre for the Research and Technology of Agro-Environmental and Biological Sciences (CITAB), University of Trás-os-Montes and Alto Douro, Vila Real, Portugal; [2]Instituto Dom Luiz, University of Lisbon, Lisbon, Portugal

3.1 Introduction

Wildfire occurrence and spread require (1) the existence of fuels in adequate quantity, continuity, and condition (i.e., dryness); (2) a source of ignition, such as lightning or a spark/flame from human activities; and (3) conditions conducive to fire spread [1]. All these aspects are strongly dependent on weather or climatic conditions. Weather is defined as the thermohydrodynamic state of the atmosphere, by the values of the climatic elements (e.g., temperature, precipitation, wind, etc.). It is the most important driver of global fire incidence and is a determinant of all stages of a wildfire, by providing the ignition source (i.e., lightning), controlling dead fuel moisture and flammability (precipitation, air humidity, and temperature), favoring fire spread (e.g., wind, atmospheric instability), and facilitating fire suppression (precipitation) [2–4]. Despite the high variability of weather and climatic factors (e.g., latitude, altitude, and distance to sea), regions can be identified where the climatic elements present similar statistical distribution on longer scales, i.e., the same type of climate, which, in turn, defines the existence, type, and cycle of the vegetation [5,6]. Therefore, the high resemblance between global spatial patterns of climate types, vegetation cover/ecosystems/biomes, and fire incidence is not surprising [7–9], especially for large wildfires. Therefore, the main aim of this chapter is to discuss the role of weather and climate (variability and change) on the incidence of extreme wildfire events (EWEs) at global scale.

3.2 The influence of climate

3.2.1 Fire and climate patterns

Biomass burning is unequivocally a global scale and continuous phenomenon. Fire is present in almost all regions and climates of the world, namely, in Australia, in the savanna ring surrounding the tropical forests of central Africa, in the agriculture lands and savanna extending into the Amazon Basin, in the forest of the Mediterranean Basin, in the rangelands and forests of North America and Central Asia, in southern Asia's forests and rice fields, in the grain producing lands, and in the boreal forests

[10,11]. Fire is at an absolute minimum at the extremes of the climatic continuum [12], such as in the Sahara and in the very high latitudes above the Arctic and Antarctic Circles [10,13].

According to the Köppen-Geiger climate classification [8], the world has five large climate types, namely, equatorial (A), arid (B), warm temperate (C), snow (D), and polar (E). On the other hand, a total of 15 terrestrial biomes can be defined [14]. Combining these data sets with the Terra and Aqua combined MCD64A1, version 6, Burned Area dataset [15] allowed us to conclude the high resemblance between fire patterns in the climate and biome basis, with 78% of the burnt area (BA) worldwide occurring in climate types A (50%), B (19%), and C (9%), as well as in just four of the 15 biomes, especially in coincident grasslands, savannas, and scrublands [16].

3.2.2   Fire incidence seasonality

Another important evidence of the influence of climate on fire incidence is the significant seasonality of the global spatial patterns. Global fire peaks occur during the months of July, August, and September, while February is usually the month with the lowest fire incidence, with marked differences between the Northern and Southern hemispheres [11]. In the former, at middle and high latitudes (40°N to 70°N), wildfires are likely to happen from June to September. From 15°N to 40°N, the fire season lasts from February to June, and the fire activity peak month is shifted gradually from summer to spring toward the equator. From 15°N to the Equator, the fire season lasts from November to February. In the Southern hemisphere, from the Equator to 10°S, high fire activity is observed in June and July. From 10°S to 30°S, the fire peak occurs from August to September, and finally, from 30°S to 50°S, fire activity is concentrated in December to February [17].

3.2.3   Existence, life cycle, and type of vegetation cover

Weather and climate exert an indirect influence on the fire regime by determining the characteristics of live and dead fuels in each location. At regional spatial scale and seasonal or interannual temporal scales, the existence, type, and life cycle of vegetation are conditional on climate, while at local and daily scales, weather defines its state [18].

The distribution of the world's vegetation, major biomes and ecoregions—desert, tundra, grasslands, savannas, and forests (tropical, temperate, and boreal)—is determined by climate, broadly predicted from temperature and precipitation patterns and well correlated with water availability [14,19].

Fire is both a natural factor and anthropogenic disturbance influencing the distribution, structure, and functioning of terrestrial ecosystems around the world, including boreal and dry conifer forests, many grasslands (especially those dominated by tall grasses), temperate and eucalypt woodlands, tropical savannas, Mediterranean-type scrublands, and heathlands [12].

Some ecosystems, namely rainforests, are extremely fire intolerant and need protection from fire. Other ecosystems are dominated by species that depend on fire to complete their life cycles, while others are dominated by species that tolerate burning but have no direct dependence on fire. Ecosystems that seldom or never burn, except when disturbed by human activity, contain mixtures of species that fortuitously tolerate burning and species extremely intolerant of fire [20,21].

3.3 The role of weather

3.3.1 Physiological state of the vegetation

The fuel of wildfires encompasses both live (grass, shrubs, trees) and dead vegetation (e.g., leaf litter, down dead woody fuels, standing necromass) aboveground components, and below-ground partially decayed biomass (e.g., humus, peat). Weather and climate have a profound influence on vegetation at different spatial and temporal scales. On the long term, climate shapes the global distribution patterns of vegetation (section 2.3). At interannual timescales, climate variability may modify local to regional vegetation characteristics, especially in arid and semiarid ecosystems, while down to the seasonal scale, the phenology and fuel moisture cycles are the result of the interaction between climate and vegetation [7,22].

The fuel must be dry enough to burn, which is achieved with hot and dry winds. Vegetation flammability also depends on the moisture content of the biomass. Dead fuel has lower moisture content, whereas live leaves will burn more easily if their moisture content is low. The shape, size, and arrangement of plant parts influence vegetation moisture content and flammability. Plants with narrow leaves or thin branches dry rapidly and burn readily. Productive ecosystems that accumulate slow-decomposing litter are highly flammable. Mineral content, including silica, and other components such as oils, fats, waxes, and terpenes are important fuel properties and facilitators of combustion. Volatile substances enhance burning because they are released from leaves, burn fiercely, and thus dry and heat adjacent fuel [22].

Because fire behavior depends on the moisture content of dead and live biomass, the antecedent climatic and weather conditions exert marked influences on the timing of the wildfire. Dead fuels moisture depends solely on environmental conditions, while live fuels moisture depends on the stage of vegetative development and ability of the vegetation to capture the available soil water. However, the length of warm and dry periods needed for fire occurrence and spread depends greatly on vegetation properties. For instance, a few days of hot and dry weather are sufficient to dry tall grasslands enough to sustain a wildfire, whereas months of extremely hot and dry conditions are needed for fire to burn pristine humid tropical forests [22]. On the other hand, large wildfires in arid grassy ecosystems are limited by the availability of fuel and are more common after high rainfall years [12,23].

3.3.2 Fire weather for ignition, spread, and extinction

The combustion triangle links the three fire ingredients: oxidant, fuel, and ignition source. Atmospheric oxygen and vegetation around the Globe provides two of those components, while the general principle of vegetation wildfires is that there is an initial high-temperature heat source, which may be produced by lightning, volcanic activity, a spark from a rock fall, or humans [24].

Lightning strikes are all over the world but can only effectively initiate a wildfire when dissociated from rainfall. The number of wildfires is heavily skewed by the events caused intentionally, accidentally or negligently by humans in agricultural, land management, or fire suppression activities [25]. Fire occurrence and spread are conditioned by fuel structure, nature, and moisture content [24]. The most important fire spread factors are local wind and topography [26]. The major effect of wind is to tilt flames and induce fire spread by convection or flame contact [27]. Wildfires may significantly change the prevailing local winds; the upward convective column of heated gases draws in air from around the wildfire, and the fire front will spread outward into this self-generated wind. Finally, the decrease of wind speed and temperature and increase of humidity but, mainly, rainfall are conditioning factors for the occurrence and spread but help the suppression and extinction of wildfires.

3.3.3 The influence of ridges, blockings, and other synoptic patterns

Weather conditions present high spatial autocorrelation across large geographic areas, and their temporal evolution in a given location is strongly dependent on the dynamics of large-scale space−time atmospheric systems. Therefore, several aspects of the fire regime are highly related to large-scale atmospheric-ocean variability patterns [28,29]. Summer high-pressure systems provide clear sky conditions and dry weather, which maximize biomass drying, permit sufficient convective activity to produce lightning, and therefore favor fire ignition and spread, but long-wave ridging at 500 hPa level is much more persistent than surface high-pressure systems [30].

Catastrophic large wildfires and high burnt area are related with anomalous lower- (e.g., 700−850 hPa), mid- (500 hPa), and upper-tropospheric circulation at middle and high latitudes, including the breakdown of upper level synoptic ridges causing anomalous wind patterns and increased lightning activity in Canada [31,32]; the existence and position of intense short-wave trough in the mid-troposphere in eastern US [33]; and anomalous synoptic patterns at different levels of the atmosphere in the Mediterranean region [29,34−37]. These anomalous circulation patterns lead to warm and dry spells over the affected region, which explain the occurrence of extreme wildfires, especially when they occur during drought periods (DPs) [29,37]. Blocks and extratropical ridges were linked to the occurrence of heatwaves (HWs) in Europe and, consequently, of extreme wildfires [38].

Sea-surface temperature (SST) reflects the upper ocean response to air−sea (mass, energy, and momentum) fluxes [39]. Large-scale SST drives wind patterns [40], anomalous weather, extreme climatic events [41], and the climate controls on fuel structure and moisture content [42]. For example, high SST in the tropical North Atlantic and tropical eastern Pacific lead less precipitation over the Amazon Basin and reduced terrestrial water storage and soil moisture reserves, which limit evapotranspiration during the following dry season but enhance fire spread [43,44]. Almost half of the global interannual BA variability can be forecasted with SST-based ocean-climate indices three or more months before the BA peak [42]. The role of the oceans on wildfires based on ocean-climate indices will be detailed in the following sections.

3.3.4 Wildfire teleconnections

In atmospheric science, a teleconnection is a relationship between climate anomalies at large distances (typically thousands of kilometers). These relationships are also known as large-scale climatic patterns [45]. The most and best-known ocean-atmosphere teleconnection is the ENSO, which includes the El-Niño (EN), defined as the warming of the ocean surface water of the tropical eastern Pacific Ocean around Christmas, and the Southern Oscillation (SO), which is the atmospheric component of the ENSO and reflect the link between the sea-level pressure in East and West equatorial Pacific.

The ENSO has been related to many different weather patterns and, consequently to (high or low) fire activity around the world [46]. Fire—ENSO relationships were studied in many regions, including Southeast Asia [47], regions of the USA [48–52], Tropical Mexico [53], Brazilian Amazon [54,55], Patagonia [54,56,57], and Australia [58,59].

There are other robust ocean-atmosphere and atmospheric patterns of climate variability with impacts on fire activity patterns. These modes of variability are easily translated into indexes; for example, the signal/phase and strength of the SO is assessed by the SO Index (SOI), which can be defined as the difference between the sea-level pressure at Tahiti and Darwin, Australia. Teleconnection time series can be used (alone or in a group of potential factors/predictors) to study the influence of climate variability on the fire regime on a wide range of long timescales (interannual, interdecadal, and multidecadal). There is a long list of studies about the role of these teleconnections on the fire regime in many regions.

3.4 The role of climatic and weather extreme events

High fire incidence is frequently explained by the occurrence of extreme atmospheric conditions that tend to last for periods of a few days to several months, namely HWs and DPs, which lead to significant thermal and/or hydric stress of the vegetation.

3.4.1 Drought period

A DP can be defined as an extended period of moisture deficiency in the land surface, which can last for months or years [60]. Several indices can be used to quantitatively assess and characterize DP, such as the Standardized Precipitation Evapotranspiration Index, Standardized Precipitation Index, Reconnaissance Drought Indicator, and Palmer Drought Severity Index. These indices allow to identify and characterize (duration, severity, and spatial extent) the different types of DP, namely, (1) meteorological [61]; (2) agricultural [62]; and (3) hydrological drought [51].

DPs have major socioeconomic and environmental consequences and are associated to high fire incidence all over the Globe, namely, in the USA [63], Canada [64], Borneo [65], Amazon [20], Israel [66], and Southern Europe [6,67−71]. Conversely, fire activity in Florida decreases during the DP because of the tight control exerted by prescribed burning.

3.4.2 Effects of heatwaves

In general terms, an HW can be defined as an event of abnormally and uncomfortably hot weather over a large area, which usually lasts from a few days to a few weeks, with local thermal conditions recorded above given thresholds [72]. However, there is no unique or universally accepted HW definition. Instead, several definitions have been used [38,73,74]. Regardless of the definition, HWs have social, economic, and environmental catastrophic impacts, which can be experienced well outside of the officially affected region.

In Europe, more than 25,000 wildfires were recorded during the 2003 HW, which burned a total of 650,000 ha of forests, scrublands, and also agricultural areas [37], causing devastating effects on both the natural and built environments [75]. Also, during the 2003 HW and 2003 summer DP, the estimated global financial impact of wildfires in Portugal exceeded one billion euros, while the joint impacts of drought and wildfires over Europe exceeded 13 billion euros [76,77].

For a qualitative and quantitative evaluation, several HW definitions, indices, and statistics have been developed and used to detect and monitor an event in real time [78,79], to quantify the event duration, intensity, peak, and spatial extent [38,80], and to associate these extreme events with the number and BA of medium and large wildfires [38,81].

Owing to the effects of weather, climate, human, and landscape factors, the fire incidence is not a random process, neither in time nor in space, but explains the existence of time, space, and space−time clusters [3,82−84], the increase of fire incidence in the wildland−rural−urban interface [85] and define regions of different fire susceptibility/danger/risk [5,86].

3.5 Fire weather danger and risk rating

Another illustrative and demonstrative aspect of the key role of weather in the fire regime and, in particular, the fire incidence is the development and use of fire danger/hazard/risk indices around the Globe. In the following subsections, several of those indices will be presented, with emphasis on the Canadian Fire Weather Index (FWI), for being widely used in the world as well as their use for operational and research purposes.

3.5.1 Fire danger rating

Several fire danger rating equations and systems have been developed, solely based on meteorological variables or combined with information such as vegetation characteristics. Fire danger is the sum of constant and variable hazard factors that affect fire ignition, behavior, resistance, and control. Available weather/fire danger rating methods range from very simple equations that portray the ease of ignition to complex semiempirical systems that address the various aspects of fire behavior. Some of the most well-known methods are (1) the Nesterov Index (and modified version), developed for the boreal forests of Russia; (2) the Angstrom Index, developed in Sweden; (3) the Baumgartner Index, used in Germany and Slovakia; (4) the Fosberg's Fire Weather Index, developed for California; (5) the McArthur's fire-danger meters widely used in Australia; (6) the US National Fire Danger Rating System; and (7) the FWI, globally available.[1]

Some of these methods are cumulative, i.e., have a memory and take into account values of meteorological variables from previous days, weeks, and months, which enables to track the evolution of fuel moisture content. These indexes can be very simple or aimed to depict the complex processes of fuel moisture variation, such as the FWI [87] and the NFDRS [88]. Several studies have been focused on evaluating and updating these indexes [61,89—92] as well as their comparison in specific regions [93—97]. On the next subsection, the FWI will be described for being the most widely used and for illustrating the role of meteorological conditions in fuel moisture content, different aspects of burning conditions, and rate fire hazard.

3.5.2 The Canadian Fire Weather Index

The FWI was developed as part of the Canadian Forest Fire Danger Rating System [98,99] to account for the effects of weather in fire danger estimation. It was specifically conceived for Canada, but its appropriateness has been shown for other regions [89,97,100—102], making this method the most widely used fire danger rating system in the world [103].

The FWI consists of three fuel moisture codes and three fire behavior indices. Each moisture code represents the moisture content of distinct layers of the forest floor, namely, the Fine Fuel Moisture Code (FFMC) for surface fine dead fuels; the Duff Moisture Code (DMC) for loosely compacted duff of moderate depth; and the Drought Code (DC) for deep and more compact organic material. The system keeps track of fuel wetting and drying, considering the different time lag of each layer and using previous day code values as inputs for the computation. The FWI only requires meteorological information as input data such as air temperature, relative humidity, and wind speed measured at noon local standard time, as well as the precipitation accumulated in the previous 24 hours [104]. The FFMC needs all these variables. The DMC does not use wind, and the DC solely needs air temperature and rainfall. The different indices suit different fire management needs, from public awareness to specialized

---

[1] http://gwis.jrc.ec.europa.eu/static/gwis_current_situation/public/index.html.

applications, e.g., prescribed burning [99,105]. The FFMC is used as an indicator of potential fire occurrence [99,105] as fires start and spread in fine fuels. As ignitions caused by lightning can smolder in the duff present in forest floors [106], the DMC is used as a lightning fire predictor. The potential for the occurrence of deep burning fires and the consequent difficulty of mop-up is related with the DC. The DMC and DC are combined into the Buildup Index (BUI) that indicates fuel availability for combustion. The FFMC together with wind speed is used to estimate initial potential fire-spread rate, through the Initial Spread Index (ISI). Finally, the ISI and BUI are combined as the FWI which is an estimate of the potential intensity of a spreading wildfire. The daily severity rating (DSR) is a power function of the FWI and is used to rate the difficulty of fire suppression; the scaling takes into account the nonlinear increase of difficulty of control as wildfires intensify [107].

3.5.3 Fire danger rating for operational and research purposes

Long-term fire danger information based on seasonal forecasts can anticipate the next fire season trend and support long-term fire preparedness. Short-term products (1−2 weeks) support large-scale decision-making that requires time for coordination, namely rearrangement and transfer of fire suppression resources between different regions, and even countries. Early warning products (1−7 days) provide information to rationalize fire management activities and support strategic positioning of firefighting resources to contain and control new and ongoing wildfires; it is used to define preparedness levels and derive guidelines for fire control and use, including prescribed burning, ignition control, allocation of resources, and fire staffing. Further uses include evacuation planning, insurance estimates, justification of budgets, and modeling fire and climate change impacts [103]. FWIs have been used to assess fire hazard and risk and to support fire management in many locations for current climate conditions [91,108−114], as well as for future climate conditions, which will be discussed in the next section.

3.6 Climate change: The future of extreme wildfires

3.6.1 Climate change projections

The Fifth Assessment Report of the Intergovernmental Panel on Climate Change recently updated the knowledge on the observed, future, global, and regional changes of the climate system [115]. Observed changes since the 1950s with potential impacts on fire activity include warming of the atmosphere and oceans and reduction of ice and snow. At regional scale, it is worth noting the decreasing trends of precipitation in some land regions and the increase of frequency and/or intensity by the end of the 21st century of weather/climate extremes such as DP, windstorms, and HW [38,60,116,117]. The combined effect of increased air temperature and decreased groundwater and air moisture is a key factor in vegetation change, raising (or decreasing) wildfire size and frequency [118], and changing several aspects (e.g., frequency, intensity, seasonality) of the fire regime [119]. In highly anthropized areas and where the fire—vegetation interaction is uncertain, it is difficult to predict future fire activity because of the strong influence of others factors than weather and climate [120,121]. Nevertheless, several studies assessed the impact of climate change on wildfires [17,122—125].

3.6.2 Robust projections for the future

Precipitation and air temperature projections over South America suggest the increase of DP, which combined with land-use change (deforestation, selective logging, and forest fragmentation) will lead to the increase of fire danger in Amazonian forests [118]. Regional warming in the southwestern USA will foster fuel drying, the increase of large-fire frequency, and fire season length [126—129]. Moreover, in densely forested areas, trees will be more sensitive to drought, leading to fuel availability, which will influence future fire behavior [127].

Future projections suggest an increase: in fire season days by 2050 [130] and in days with potential for unmanageable fire in Canada [131]; in total BA, between 66% and 140%, by the end of 21st century in European Mediterranean countries [132]; of fire risk and likelihood of much larger BAs, under future climate conditions in Portugal [133—135]; of annual critical fire risk days [136], and of larger wildfires that resist initial fire suppression in Greece [137]; of total BA and a decrease of effective fire suppression future opportunities in Spain, especially if no improvements in fire management are introduced [21,138]; and of BA and fire season extent in Italy and France [139].

In Australia, future changes in climate and atmospheric CO_2 will increase fuel load more significantly in grasslands than in forests, and therefore, fire hazard could rise much more significantly in areas where fuel amount tends to limit fire incidence than in areas where it is limited more by weather conditions [140,141]. Future climate projections for West African savannas unveil that without anthropized land use changes, these areas will shift to wood dominated vegetation because of CO_2 fertilization effects, increased water use efficiency, and decreased fire activity [142].

3.7 Concluding remarks

This chapter was devoted to the role of weather, climate, climate variability, and climate change on wildfires, in general, and extreme wildfires, in particular. In fact, a significant part of the studies cited in this chapter relate weather/climate conditions and high values of fire incidence, especially large burnt area. It is well known that the fire size distribution is right skewed, meaning that a few number of extreme wildfires are responsible for the majority of the total burnt area. For small wildfires to become extreme wildfires, it is necessary to have extreme weather (heatwaves) and climatic (droughts) conditions. It was precisely these conditions and essentially extreme wildfires that were analyzed in this chapter although it has not always been made explicit in the cited studies.

Essentially, weather and climate are the same fire driver (atmospheric conditions) acting at different timescales, ranging from a place and periods from hours to days (weather) to the whole Globe or a continent and periods of years or decades (climate). Atmospheric conditions are involved in all fire stages and influence all aspects of the fire regime.

Climate determines the existence, type, and life cycle of the vegetation, while weather and climate variability determine the state of the fuels. Fire climate patterns reflect this complex interaction, which becomes evident in the intraannual variability (seasonality) of the fire incidence, during short-term extreme weather events (e.g., heatwaves) and/or medium-term extreme climatic patterns of variability (e.g., drought period) as well as large-scale ocean-atmosphere teleconnection patterns, all causing high fire activity including extreme wildfires. Finally, fire danger rating indices aiming to proxy the role of weather and climate on wildfire as well as potential future climate change impacts on fire incidence and regime are also addressed.

References

[1] C. Whitlock, P.E. Higuera, D.B. McWethy, C.E. Briles, Paleoecological perspectives on fire ecology: revisiting the fire-regime concept ~ !2009-09-02 ~ !2009-11-09 ~ !2010-03-05 ~ !, Open Ecol. J. 3 (2010) 6—23.

[2] E.A. Johnson, Fire and Vegetation Dynamics: Studies from the North American Boreal Forest, Cambridge University Press, 1996.

[3] J. Parente, M.G. Pereira, M. Tonini, Space-time clustering analysis of wildfires: the influence of dataset characteristics, fire prevention policy decisions, weather and climate, Sci. Total Environ. 559 (2016) 151—165.

[4] F. Tedim, V. Leone, M. Amraoui, C. Bouillon, M. Coughlan, G. Delogu, P. Fernandes, C. Ferreira, S. McCaffrey, T. McGee, J. Parente, D. Paton, M. Pereira, L. Ribeiro, D. Viegas, G. Xanthopoulos, Defining extreme wildfire events: difficulties, challenges, and impacts, Fire 1 (2018) 9.

[5] J. Parente, M.G. Pereira, Structural fire risk: the case of Portugal, Sci. Total Environ. 573 (2016) 883—893.

[6] I.R. Urbieta, G. Zavala, J. Bedia, J.M. Gutiérrez, J.S. Miguel-Ayanz, A. Camia, J.E. Keeley, J.M. Moreno, Fire activity as a function of fire—weather seasonal severity and antecedent climate across spatial scales in southern Europe and Pacific western USA, Environ. Res. Lett. 10 (2015) 114013.

[7] R.M.B. Harris, T.A. Remenyi, G.J. Williamson, N.L. Bindoff, M.J. David, Climate-vegetation-fire interactions and feedbacks: trivial detail or major barrier to projecting the future of the Earth system? Wiley Interdiscip. Rev. Clim. Change. 7 (2016) 910—931.

[8] M. Kottek, J. Grieser, C. Beck, B. Rudolf, F. Rubel, World Map of the Köppen-Geiger climate classification updated, Meteorol. Z. 15 (2006) 259—263.

[9] M.A. Krawchuk, M.A. Moritz, M.-A. Parisien, J. Van Dorn, K. Hayhoe, Global pyrogeography: the current and future distribution of wildfire, PLoS One 4 (2009) e5102.

[10] C. Carmona-Moreno, A. Belward, J.-P. Malingreau, A. Hartley, M. Garcia-Alegre, M. Antonovskiy, V. Buchshtaber, V. Pivovarov, Characterizing interannual variations in global fire calendar using data from Earth observing satellites, Glob. Chang. Biol. 11 (2005) 1537−1555.

[11] L. Giglio, I. Csiszar, C.O. Justice, Global distribution and seasonality of active fires as observed with the Terra and Aqua moderate resolution imaging spectroradiometer (MODIS) sensors, J. Geophys. Res.: Biogeosciences 111 (2006).

[12] W. Bond, Fires, ecological effects of, in: Encyclopedia of Biodiversity, 2001, pp. 745−753.

[13] M.A. Cochrane, K.C. Ryan, Fire and fire ecology: concepts and principles, in: Tropical Fire Ecology, 2009, pp. 25−62.

[14] E. Dinerstein, D. Olson, A. Joshi, C. Vynne, N.D. Burgess, E. Wikramanayake, N. Hahn, S. Palminteri, P. Hedao, R. Noss, M. Hansen, H. Locke, E.C. Ellis, B. Jones, C.V. Barber, R. Hayes, C. Kormos, V. Martin, E. Crist, W. Sechrest, L. Price, J.E.M. Baillie, D. Weeden, K. Suckling, C. Davis, N. Sizer, R. Moore, D. Thau, T. Birch, P. Potapov, S. Turubanova, A. Tyukavina, N. de Souza, L. Pintea, J.C. Brito, O.A. Llewellyn, A.G. Miller, A. Patzelt, S.A. Ghazanfar, J. Timberlake, H. Klöser, Y. Shennan-Farpón, R. Kindt, J.-P.B. Lillesø, P. van Breugel, L. Graudal, M. Voge, K.F. Al-Shammari, M. Saleem, An Ecoregion-based approach to protecting half the terrestrial realm, Bioscience 67 (2017) 534−545.

[15] L. Giglio, C. Justice, L. Boschetti, D. Roy, MCD64A1 MODIS/Terra+Aqua Burned Area Monthly L3 Global 500m SIN Grid V006 [Data Set], NASA EOSDIS Land Processes DAAC, 2015.

[16] A. Oliveira, J. Parente, M. Amraoui, M.G. Pereira, P. Fernandes, Global-scale Analysis of Wildfires, in: n.d.

[17] J. Yang, H. Tian, B. Tao, W. Ren, J. Kush, Y. Liu, Y. Wang, Spatial and temporal patterns of global burned area in response to anthropogenic and environmental factors: reconstructing global fire history for the 20th and early 21st centuries, J. Geophys. Res.: Biogeosciences 119 (2014) 249−263.

[18] M.D. Flannigan, B.M. Wotton, G.A. Marshall, W.J. de Groot, J. Johnston, N. Jurko, A.S. Cantin, Fuel moisture sensitivity to temperature and precipitation: climate change implications, Clim. Change 134 (2015) 59−71.

[19] F.I. Woodward, Climate and Plant Distribution, Cambridge University Press, 1987.

[20] P.M. Brando, J.K. Balch, D.C. Nepstad, D.C. Morton, F.E. Putz, M.T. Coe, D. Silvério, M.N. Macedo, E.A. Davidson, C.C. Nóbrega, A. Alencar, B.S. Soares-Filho, Abrupt increases in Amazonian tree mortality due to drought-fire interactions, Proc. Natl. Acad. Sci. U.S.A. 111 (2014) 6347−6352.

[21] L. Brotons, N. Aquilué, M. de Cáceres, M.-J. Fortin, A. Fall, How fire history, fire suppression practices and climate change affect wildfire regimes in Mediterranean landscapes, PLoS One 8 (2013) e62392.

[22] M.A. Cochrane, Fire science for rainforests, Nature 421 (2003) 913−919.

[23] W.J. Bond, R.E. Keane, Fires, ecological effects of ☆, in: Reference Module in Life Sciences, 2017.

[24] J.H.A.H. van Konijnenburg-van Cittert, No. of pages: xix 413. Price: UK£39.95, in: M.S. David, J. Bowman, W.J. Bond, S.J. Pyne, M.E. Alexander (Eds.), Fire on Earth: An Introduction by Andrew C. Scott, Wiley-Blackwell, Chichester, 2014 (paperback), Geol. J. 49 (2014) 656−657 ISBN: 978-1-119-952356.

[25] T.T. Veblen, T. Kitzberger, J. Donnegan, Climatic and human influences on fire regimes in ponderosa pine forests in the Colorado front range, Ecol. Appl. 10 (2000) 1178.

[26] P.G. Baines, Physical mechanisms for the propagation of surface fires, Math. Comput. Model. 13 (1990) 83–94.

[27] M.A. Finney, J.D. Cohen, J.M. Forthofer, S.S. McAllister, M.J. Gollner, D.J. Gorham, K. Saito, N.K. Akafuah, B.A. Adam, J.D. English, Role of buoyant flame dynamics in wildfire spread, Proc. Natl. Acad. Sci. U.S.A. 112 (2015) 9833–9838.

[28] M.D. Flannigan, B.M. Wotton, Climate, weather, and area burned, in: Forest Fires, 2001, pp. 351–373.

[29] M.G. Pereira, R.M. Trigo, C.C. da Camara, J.M.C. Pereira, S.M. Leite, Synoptic patterns associated with large summer forest fires in Portugal, Agric. For. Meteorol. 129 (2005) 11–25.

[30] M.J. Newark, The relationship between forest fire occurrence and 500 mb longwave ridging, Atmosphere 1 (1975) 26–33.

[31] N. Nimchuk, Alberta. Forest Service, Wildfire Behavior Associated with Upper Ridge Breakdown, 1983.

[32] W.R. Skinner, M.D. Flannigan, B.J. Stocks, D.L. Martell, B.M. Wotton, J.B. Todd, J.A. Mason, K.A. Logan, E.M. Bosch, A 500 hPa synoptic wildland fire climatology for large Canadian forest fires, 1959–1996, Theor, Appl. Climatol. 71 (2002) 157–169.

[33] E.A. Brotak, W.E. Reifsnyder, An investigation of the synoptic situations associated with major wildland fires, J. Appl. Meteorol. 16 (1977) 867–870.

[34] M. Amraoui, M.L.R. Liberato, T.J. Calado, C.C. DaCamara, L.P. Coelho, R.M. Trigo, C.M. Gouveia, Fire activity over mediterranean Europe based on information from meteosat-8, For. Ecol. Manage. 294 (2013) 62–75.

[35] M. Amraoui, M.G. Pereira, C.C. DaCamara, T.J. Calado, Atmospheric conditions associated with extreme fire activity in the Western Mediterranean region, Sci. Total Environ. 524–525 (2015) 32–39.

[36] J. Ruffault, V. Moron, R.M. Trigo, T. Curt, Daily synoptic conditions associated with large fire occurrence in Mediterranean France: evidence for a wind-driven fire regime, Int. J. Climatol. 37 (2016) 524–533.

[37] R.M. Trigo, J.M.C. Pereira, M.G. Pereira, B. Mota, T.J. Calado, C.C. Dacamara, F.E. Santo, Atmospheric conditions associated with the exceptional fire season of 2003 in Portugal, Int. J. Climatol. 26 (2006) 1741–1757.

[38] J. Parente, M.G. Pereira, M. Amraoui, E.M. Fischer, Heat waves in Portugal: current regime, changes in future climate and impacts on extreme wildfires, Sci. Total Environ. 631–632 (2018) 534–549.

[39] C. Frankignoul, Sea surface temperature anomalies, planetary waves, and air-sea feedback in the middle latitudes, Rev. Geophys. 23 (1985) 357.

[40] S. Sagarika, A. Kalra, S. Ahmad, Pacific ocean SST and Z500climate variability and western U.S. seasonal streamflow, Int. J. Climatol. 36 (2015) 1515–1533.

[41] G.D. McCarthy, I.D. Haigh, J.J.-M. Hirschi, J.P. Grist, D.A. Smeed, Ocean impact on decadal Atlantic climate variability revealed by sea-level observations, Nature 521 (2015) 508–510.

[42] Y. Chen, D.C. Morton, N. Andela, L. Giglio, J.T. Randerson, How much global burned area can be forecast on seasonal time scales using sea surface temperatures? Environ. Res. Lett. 11 (2016) 045001.

[43] Y. Chen, I. Velicogna, J.S. Famiglietti, J.T. Randerson, Satellite observations of terrestrial water storage provide early warning information about drought and fire season severity in the Amazon, J. Geophys. Res.: Biogeosciences 118 (2013) 495–504.

[44] D.C. Morton, Y. Le Page, R. DeFries, G.J. Collatz, G.C. Hurtt, Understorey fire frequency and the fate of burned forests in southern Amazonia, Philos. Trans. R. Soc. Lond. B Biol. Sci. 368 (2013) 20120163.

[45] M. Macias Fauria, M.M. Fauria, E.A. Johnson, Large-scale climatic patterns control large lightning fire occurrence in Canada and Alaska forest regions, J. Geophys. Res.: Biogeosciences 111 (2006).

[46] Y. Le Page, J.M.C. Pereira, R. Trigo, C. da Camara, D. Oom, B. Mota, Global fire activity patterns (1996−2006) and climatic influence: an analysis using the World Fire Atlas, Atmos. Chem. Phys. Discuss. 7 (2007) 17299−17338.

[47] D.O. Fuller, K. Murphy, The Enso-fire dynamic in insular Southeast Asia, Clim. Change 74 (2006) 435−455.

[48] R.M. Beaty, R. Matthew Beaty, A.H. Taylor, Fire history and the structure and dynamics of a mixed conifer forest landscape in the northern Sierra Nevada, Lake Tahoe Basin, California, USA, For. Ecol. Manage. 255 (2008) 707−719.

[49] P.G. Dixon, P. Grady Dixon, G.B. Goodrich, W.H. Cooke, Using teleconnections to predict wildfires in Mississippi, Mon. Weather Rev. 136 (2008) 2804−2811.

[50] T. Schoennagel, T.T. Veblen, W.H. Romme, J.S. Sibold, E.R. Cook, ENSO and PDO variability affect drought-induced fire occurrence in Rocky Mountain Subalpine forests, Ecol. Appl. 15 (2005) 2000−2014.

[51] T.W. Swetnam, C.H. Baisan, Tree-ring reconstructions of fire and climate history in the sierra Nevada and southwestern United States, in: Ecological Studies, n.d.: pp. 158−195.

[52] A.H. Taylor, R.M. Beaty, Climatic influences on fire regimes in the northern Sierra Nevada mountains, Lake Tahoe Basin, Nevada, USA, J. Biogeogr. 32 (2005) 425−438.

[53] R.M. Román-Cuesta, M. Gracia, J. Retana, Environmental and human factors influencing fire trends in ENSO and non-ENSO years in tropical Mexico, Ecol. Appl. 13 (2003) 1177−1192.

[54] A. Alencar, D. Nepstad, M.C.V. Diaz, Forest understory fire in the Brazilian Amazon in ENSO and non-ENSO years: area burned and committed carbon Emissions, Earth Interact. 10 (2006) 1−17.

[55] D.C. Nepstad, A. Verssimo, A. Alencar, C. Nobre, E. Lima, P. Lefebvre, P. Schlesinger, C. Potter, P. Moutinho, E. Mendoza, M. Cochrane, V. Brooks, Large-scale impoverishment of Amazonian forests by logging and fire, Nature 398 (1999) 505−508.

[56] T. Kitzberger, ENSO as a forewarning tool of regional fire occurrence in northern Patagonia, Argentina, Int. J. Wildland Fire 11 (2002) 33.

[57] T. Kitzberger, T.T. Veblen, Influences of humans and ENSO on fire history of Austrocedrus chilensiswoodlands in northern Patagonia, Argentina, Écoscience 4 (1997) 508−520.

[58] S. Harris, N. Nicholls, N. Tapper, Forecasting fire activity in Victoria, Australia, using antecedent climate variables and ENSO indices, Int. J. Wildland Fire 23 (2014) 173.

[59] M. Mariani, M.-S. Fletcher, A. Holz, P. Nyman, ENSO controls interannual fire activity in southeast Australia, Geophys. Res. Lett. 43 (2016) 10,891−10,900.

[60] J. Spinoni, J.V. Vogt, G. Naumann, P. Barbosa, A. Dosio, Will drought events become more frequent and severe in Europe? Int. J. Climatol. 38 (2017) 1718−1736.

[61] R. Romero, A. Mestre, R. Botey, A new calibration for fire weather index in Spain (AEMET), in: Advances in Forest Fire Research, 2014, pp. 1044−1053.

[62] N.O. Agutu, J.L. Awange, A. Zerihun, C.E. Ndehedehe, M. Kuhn, Y. Fukuda, Assessing multi-satellite remote sensing, reanalysis, and land surface models' products in characterizing agricultural drought in East Africa, Remote Sens. Environ. 194 (2017) 287−302.

[63] A.E. Hessl, D. McKenzie, R. Schellhaas, Drought and pacific decadal oscillation linked to fire occurrence in the inland pacific northwest, Ecol. Appl. 14 (2004) 425−442.

[64] M.P. Girardin, B. Mike Wotton, Summer moisture and wildfire risks across Canada, J. Appl. Meteorol. Climatol. 48 (2009) 517−533.

[65] M. Taufik, J.J. Paul, R. Uijlenhoet, P.D. Jones, D. Murdiyarso, H.A.J. Van Lanen, Amplification of wildfire area burnt by hydrological drought in the humid tropics, Nat. Clim. Chang. 7 (2017) 428–431.

[66] M. Turco, N. Levin, N. Tessler, H. Saaroni, Recent changes and relations among drought, vegetation and wildfires in the Eastern Mediterranean: the case of Israel, Glob. Planet. Chang. 151 (2017) 28–35.

[67] A.P. Dimitrakopoulos, M. Vlahou, C.G. Anagnostopoulou, I.D. Mitsopoulos, Impact of drought on wildland fires in Greece: implications of climatic change? Clim. Change 109 (2011) 331–347.

[68] L. Gudmundsson, F.C. Rego, M. Rocha, S.I. Seneviratne, Predicting above normal wildfire activity in southern Europe as a function of meteorological drought, Environ. Res. Lett. 9 (2014) 084008.

[69] R. Marcos, M. Turco, J. Bedia, M.C. Llasat, A. Provenzale, Seasonal predictability of summer fires in a Mediterranean environment, Int. J. Wildland Fire 24 (2015) 1076.

[70] J. Ruffault, T. Curt, N.K. Martin St-Paul, V. Moron, R.M. Trigo, Extreme wildfire events are linked to global-change-type droughts in the Northern Mediterranean, Nat. Hazards Earth Syst. Sci. 18 (2018) 847–856.

[71] A. Russo, C.M. Gouveia, P. Páscoa, C.C. DaCamara, P.M. Sousa, R.M. Trigo, Assessing the role of drought events on wildfires in the Iberian Peninsula, Agric. For. Meteorol. 237–238 (2017) 50–59.

[72] WMO, Guidelines on the Definition and Monitoring of Extreme Weather and Climate Events: Draft Version- First Review by TT-Dewce, 2016.

[73] D. Jacob, J. Petersen, B. Eggert, A. Alias, O.B. Christensen, L.M. Bouwer, E. Georgopoulou, EURO-CORDEX: new high-resolution climate change projections for European impact research, Reg. Environ. Chang. 14 (2014) 563–578.

[74] P.J. Robinson, On the definition of a heat wave, J. Appl. Meteorol. 40 (2001) 762–775.

[75] S.E. Perkins, L.V. Alexander, J.R. Nairn, Increasing frequency, intensity and duration of observed global heatwaves and warm spells, Geophys. Res. Lett. 39 (2012).

[76] C. Cogeca, Assessment of the Impact of the Heat Wave and Drought of the Summer 2003 on Agriculture and Forestry, vol. 15, Committee of Agricultural Organisations in the European Union and General Committee for Agricultural Cooperation in the European Union, Brussels, 2003.

[77] A. De Bono, P. Peduzzi, S. Kluser, G. Giuliani, Impacts of Summer 2003 Heat Wave in Europe, Environment Alert Bulletin, 2004.

[78] A. Ghulam, Q. Qin, T. Teyip, Z.-L. Li, Modified perpendicular drought index (MPDI): a real-time drought monitoring method, ISPRS J. Photogrammetry Remote Sens. 62 (2007) 150–164.

[79] A.F. Van Loon, R. Kumar, V. Mishra, Testing the use of standardised indices and GRACE satellite data to estimate the European 2015 groundwater drought in near-real time, Hydrol. Earth Syst. Sci. 21 (2017) 1947–1971.

[80] O. Lhotka, J. Kyselý, Characterizing joint effects of spatial extent, temperature magnitude and duration of heat waves and cold spells over Central Europe, Int. J. Climatol. 35 (2014) 1232–1244.

[81] A. Cardil, C.S. Eastaugh, D.M. Molina, Extreme temperature conditions and wildland fires in Spain, Theor. Appl. Climatol. 122 (2014) 219–228.

[82] M. Kanevski, M.G. Pereira, Local fractality: the case of forest fires in Portugal, Phys. A Stat. Mech. Appl. 479 (2017) 400–410.

[83] M.G. Pereira, L. Caramelo, C.V. Orozco, R. Costa, M. Tonini, Space-time clustering analysis performance of an aggregated dataset: the case of wildfires in Portugal, Environ. Model. Softw 72 (2015) 239–249.

[84] M. Tonini, M.G. Pereira, J. Parente, C.V. Orozco, Evolution of forest fires in Portugal: from spatio-temporal point events to smoothed density maps, Nat. Hazards 85 (3) (2017) 1489–1510.

[85] M. Tonini, J. Parente, M.G. Pereira, Global assessment of land cover changes and rural-urban interface in Portugal, Nat. Hazards Earth Syst. Sci. 18 (2018) 1647–1664.

[86] M. Leuenberger, J. Parente, M. Tonini, M.G. Pereira, M. Kanevski, Wildfire susceptibility mapping: deterministic vs. stochastic approaches, Environ. Model. Softw 101 (2018) 194–203.

[87] N. Hamadeh, B. Daya, A. Hilal, P. Chauvet, An analytical review on the most widely used meteorological models in forest fire prediction, in: 2015 Third International Conference on Technological Advances in Electrical, Electronics and Computer Engineering, TAEECE, 2015.

[88] J.D. Cohen, J.E. Deeming, The National Fire-Danger Rating System: Basic Equations, 1985.

[89] A.P. Dimitrakopoulos, A.M. Bemmerzouk, I.D. Mitsopoulos, Evaluation of the Canadian fire weather index system in an eastern Mediterranean environment, Meteorol. Appl. 18 (2011) 83–93.

[90] M. Ertuğrul, Evaluation of fire activity in some regions of aegean coasts of Turkey via Canadian forest fire weather index system (CFFWIS), Appl. Ecol. Environ. Res. 14 (2016) 93–105.

[91] H. Tatli, M. Türkeş, Climatological evaluation of Haines forest fire weather index over the Mediterranean Basin, Meteorol. Appl. 21 (2013) 545–552.

[92] J.M. Waddington, D.K. Thompson, M. Wotton, W.L. Quinton, M.D. Flannigan, B.W. Benscoter, S.A. Baisley, M.R. Turetsky, Examining the utility of the Canadian forest fire weather index system in boreal peatlands, Can. J. For. Res. 42 (2012) 47–58.

[93] P.L. Andrews, D.O. Loftsgaarden, L.S. Bradshaw, Evaluation of fire danger rating indexes using logistic regression and percentile analysis, Int. J. Wildland Fire 12 (2003) 213.

[94] A. Arpaci, C.S. Eastaugh, H. Vacik, Selecting the best performing fire weather indices for Austrian ecoregions, Theor. Appl. Climatol. 114 (2013) 393–406.

[95] A.J. Dowdy, G.A. Mills, K. Finkele, W. de Groot, Index sensitivity analysis applied to the Canadian forest fire weather index and the McArthur forest fire danger index, Meteorol. Appl. (2009).

[96] C.C. Simpson, H. Grant Pearce, A.P. Sturman, P. Zawar-Reza, Behaviour of fire weather indices in the 2009–10 New Zealand wildland fire season, Int. J. Wildland Fire 23 (2014) 1147.

[97] D.X. Viegas, D. Xavier Viegas, G. Bovio, A. Ferreira, A. Nosenzo, B. Sol, Comparative study of various methods of fire danger evaluation in southern Europe, Int. J. Wildland Fire 9 (1999) 235.

[98] C.E. Van Wagner, Canadian Forestry Service, Development and Structure of the Canadian Forest Fire Weather Index System, 1987.

[99] B.M. Wotton, Interpreting and using outputs from the Canadian forest fire danger rating system in research applications, Environ, Ecol. Stat. 16 (2009) 107–131.

[100] W.J. de Groot, R.D. Field, M.A. Brady, O. Roswintiarti, M. Mohamad, Development of the Indonesian and Malaysian fire danger rating systems, mitigation and adaptation strategies for global change 12 (2006) 165–180.

[101] P.M. Palheiro, P. Fernandes, M.G. Cruz, A fire behaviour-based fire danger classification for maritime pine stands: comparison of two approaches, For. Ecol. Manage. 234 (2006) S54.

[102] S. Taylor, M. Alexander, Science, technology, and human factors in fire danger rating: the Canadian experience, Int. J. Wildland Fire 15 (2006) 121−135.

[103] W.J. de Groot, M.D. Flannigan, Climate change and early warning systems for wildland fire, in: Reducing Disaster: Early Warning Systems for Climate Change, 2014, pp. 127−151.

[104] B.J. Stocks, T.J. Lynham, B.D. Lawson, M.E. Alexander, C.E. Van Wagner, R.S. McAlpine, D.E. Dubé, Canadian forest fire danger rating system: an overview, For. Chron. 65 (1989) 258−265.

[105] W.J. de Groot, Wardati, Y. Wang, Calibrating the fine fuel moisture code for grass ignition potential in Sumatra, Indonesia, Int. J. Wildland Fire 14 (2005) 161.

[106] B.M. Wotton, D.L. Martell, A lightning fire occurrence model for Ontario, Can. J. For. Res. 35 (2005) 1389−1401.

[107] C.E. van Wagner, Conversion of Williams' Severity Rating for Use with the Fire Weather Index, Canadian Forestry Service, 1970.

[108] A. Carvalho, M.D. Flannigan, K. Logan, A.I. Miranda, C. Borrego, Fire activity in Portugal and its relationship to weather and the Canadian fire weather index system, Int. J. Wildland Fire 17 (2008) 328.

[109] H. Clarke, C. Lucas, P. Smith, Changes in Australian fire weather between 1973 and 2010, Int. J. Climatol. 33 (2012) 931−944.

[110] M. Flannigan, B. Stocks, M. Turetsky, M. Wotton, Impacts of climate change on fire activity and fire management in the circumboreal forest, Glob. Chang. Biol. 15 (2009) 549−560.

[111] S.L. Goodrick, Modification of the Fosberg fire weather index to include drought, Int. J. Wildland Fire 11 (2002) 205.

[112] H.D. Kambezidis, G.K. Kalliampakos, Fire-risk assessment in northern Greece using a modified Fosberg fire-weather index that includes forest coverage, Int. J. Atmospheric Sci. (2016) (2016) 1−8.

[113] J.J. Keetch, G.M. Byram, A Drought Index for Forest Fire Control, 1968.

[114] B.S. Lee, M.E. Alexander, B.C. Hawkes, T.J. Lynham, B.J. Stocks, P. Englefield, Information systems in support of wildland fire management decision making in Canada, Comput. Electron. Agric. 37 (2002) 185−198.

[115] Climate Change 2013: The Physical Science Basis: Summary for Policymakers: Working Group I Contribution to the IPCC Fifth Assessment Report, 2013.

[116] E.-S. Im, J.S. Pal, E.A.B. Eltahir, Deadly heat waves projected in the densely populated agricultural regions of South Asia, Sci Adv 3 (2017) e1603322.

[117] IPCC, in: R.K. Pachauri, L.A. Meyer (Eds.), Climate Change 2014: Synthesis Report. Contribution of Working Groups I, II and III to the Fifth Assessment Report of the Intergovernmental Panel on Climate Change [Core Writing Team, 2014, p. 151.

[118] B.C. Bates, Z.W. Kundzewicz, S. Wu, J.P. Palutikof (Eds.), Climate Change and Water. Technical Paper of the Intergovernmental Panel on Climate Change.Technical Paper of the Intergovernmental Panel on Climate Change, IPCC Secretariat, Geneva, 2008, p. 210.

[119] V.H. Dale, L.A. Joyce, S. Mcnulty, R.P. Neilson, M.P. Ayres, M.D. Flannigan, P.J. Hanson, L.C. Irland, A.E. Lugo, C.J. Peterson, D. Simberloff, F.J. Swanson, B.J. Stocks, B. Michael Wotton, Climate change and forest disturbances, Bioscience 51 (2001) 723.

[120] T. Fréjaville, T. Curt, Spatiotemporal patterns of changes in fire regime and climate: defining the pyroclimates of south-eastern France (Mediterranean Basin), Clim. Change 129 (2015) 239–251.

[121] M. Wu, W. Knorr, K. Thonicke, G. Schurgers, A. Camia, A. Arneth, Sensitivity of burned area in Europe to climate change, atmospheric CO_2 levels, and demography: a comparison of two fire-vegetation models, J. Geophys. Res.: Biogeosciences 120 (2015) 2256–2272.

[122] J. Bedia, S. Herrera, J.M. Gutiérrez, A. Benali, S. Brands, B. Mota, J.M. Moreno, Global patterns in the sensitivity of burned area to fire-weather: implications for climate change, Agric. For. Meteorol. 214–215 (2015) 369–379.

[123] M.D. Flannigan, M.A. Krawchuk, W.J. de Groot, B. Mike Wotton, L.M. Gowman, Implications of changing climate for global wildland fire, Int. J. Wildland Fire 18 (2009) 483.

[124] Y. Liu, J. Stanturf, S. Goodrick, Trends in global wildfire potential in a changing climate, For. Ecol. Manage. 259 (2010) 685–697.

[125] M.A. Moritz, M.-A. Parisien, E. Batllori, M.A. Krawchuk, J. Van Dorn, D.J. Ganz, K. Hayhoe, Climate change and disruptions to global fire activity, Ecosphere 3 (2012) art49.

[126] R. Barbero, J.T. Abatzoglou, N.K. Larkin, C.A. Kolden, B. Stocks, Climate change presents increased potential for very large fires in the contiguous United States, Int. J. Wildland Fire (2015).

[127] M.D. Hurteau, J.B. Bradford, P.Z. Fulé, A.H. Taylor, K.L. Martin, Climate change, fire management, and ecological services in the southwestern US, For. Ecol. Manage. 327 (2014) 280–289.

[128] A.L. Westerling, Increasing western US forest wildfire activity: sensitivity to changes in the timing of spring, Philos. Trans. R. Soc. Lond. B Biol. Sci. 371 (2016).

[129] A.L. Westerling, B.P. Bryant, Climate change and wildfire in California, Clim. Change 87 (2007) 231–249.

[130] X. Wang, D.K. Thompson, G.A. Marshall, C. Tymstra, R. Carr, M.D. Flannigan, Increasing frequency of extreme fire weather in Canada with climate change, Clim. Change 130 (2015) 573–586.

[131] B.M. Wotton, M.D. Flannigan, G.A. Marshall, Potential climate change impacts on fire intensity and key wildfire suppression thresholds in Canada, Environ. Res. Lett. 12 (2017) 095003.

[132] G. Amatulli, A. Camia, J. San-Miguel-Ayanz, Estimating future burned areas under changing climate in the EU-Mediterranean countries, Sci. Total Environ. 450–451 (2013) 209–222.

[133] A. Carvalho, Climate change, forest fires and air quality in Portugal in the 21st century, in: Climate Change and Variability, 2010.

[134] A. Carvalho, M.D. Flannigan, K.A. Logan, L.M. Gowman, A.I. Miranda, C. Borrego, The impact of spatial resolution on area burned and fire occurrence projections in Portugal under climate change, Clim. Change 98 (2009) 177–197.

[135] M.G. Pereira, T.J. Calado, C.C. DaCamara, T. Calheiros, Effects of regional climate change on rural fires in Portugal, Clim. Res. 57 (2013) 187–200.

[136] A. Karali, M. Hatzaki, C. Giannakopoulos, A. Roussos, G. Xanthopoulos, V. Tenentes, Sensitivity and evaluation of current fire risk and future projections due to climate change: the case study of Greece, Nat. Hazards Earth Syst. Sci. Discuss. 1 (2013) 4777–4800.

[137] K. Kalabokidis, P. Palaiologou, E. Gerasopoulos, C. Giannakopoulos, E. Kostopoulou, C. Zerefos, Effect of climate change projections on forest fire behavior and values-at-risk in southwestern Greece, For. Trees Livelihoods 6 (2015) 2214−2240.

[138] M. Turco, M.-C. Llasat, J. von Hardenberg, A. Provenzale, Climate change impacts on wildfires in a Mediterranean environment, Clim. Change 125 (2014) 369−380.

[139] O.M. Lozano, M. Salis, A.A. Ager, B. Arca, F.J. Alcasena, A.T. Monteiro, M.A. Finney, L. Del Giudice, E. Scoccimarro, D. Spano, Assessing climate change impacts on wildfire Exposure in mediterranean areas, Risk Anal. 37 (2017) 1898−1916.

[140] M.M. Boer, M.J. David, B.P. Murphy, G.J. Cary, M.A. Cochrane, R.J. Fensham, M.A. Krawchuk, O.F. Price, V.R. De Dios, R.J. Williams, R.A. Bradstock, Future changes in climatic water balance determine potential for transformational shifts in Australian fire regimes, Environ. Res. Lett. 11 (2016) 065002.

[141] H. Clarke, A.J. Pitman, J. Kala, C. Carouge, V. Haverd, J.P. Evans, An investigation of future fuel load and fire weather in Australia, Clim. Change 139 (2016) 591−605.

[142] S. Scheiter, P. Savadogo, Ecosystem management can mitigate vegetation shifts induced by climate change in West Africa, Ecol. Model. 332 (2016) 19−27.

The relation of landscape characteristics, human settlements, spatial planning, and fuel management with extreme wildfires

4

Christophe Bouillon[1], Michael Coughlan[2], José Rio Fernandes[3], Malik Amraoui[4], Pedro Chamusca[3], Helena Madureira[3], Joana Parente[4], Mário G. Pereira[4,5]

[1]National Research Institute of Science and Technology for Environment and Agriculture (IRSTEA), Risks Ecosystems Environment Vulnerability Resilience (RECOVER) research unit, Aix-en-Provence, France; [2]Institute for a Sustainable Environment, University of Oregon, Eugene, OR, United States; [3]Centre of Studies of Geography and Spatial Planning (CEGOT), University of Porto, Portugal; [4]Centre for the Research and Technology of Agro-Environmental and Biological Sciences (CITAB), University of Trás-os-Montes and Alto Douro, Vila Real, Portugal; [5]Instituto Dom Luiz, University of Lisbon, Lisbon, Portugal

4.1 Introduction

It is difficult to comprehend extreme wildfire events (EWEs) outside of the geographical contexts within which they occur. National- to regional-level differences in landscape characteristics, patterning, and processes of human settlement, as well as variation in fire-related spatial planning and fuel management, each differentially contributes to the character and severity of EWEs. In this chapter we describe various geographical and historical circumstances of three diverse territories threatened by EWEs: France, Portugal, and the United States of America (USA), showing their unique context and challenges for confronting this increasingly common and disastrous phenomenon.

In France, rural land abandonment, urban growth, and changes in the rules of land use and development have created a volatile "*firescape*" where official policy places the onus of fire prevention upon individual landowners. In Portugal, where most of the lands are privately owned, rural land abandonment combined with the splitting of land ownership creates a complex landscape mosaic of neglected shrublands and small- to large-scale forest plantations. The resulting peri-urban and rural firescape presents a considerable challenge to fire prevention and suppression activities, considering that the casual use of fire by the populations continues to represent a major source of wildfire ignitions. In the USA, where much of the landscape is publicly

Extreme Wildfire Events and Disasters. https://doi.org/10.1016/B978-0-12-815721-3.00004-7

owned and managed by Federal and State Governments, highly fire-prone forests and shrublands are to a large degree a testament to a century of successful top-down execution of fire exclusion and suppression policies. At the same time, the regional diversity in fuel types, land use and settlement patterns, and the highly decentralized context of spatial planning specifically in relation to wildfire management and prevention have rendered vulnerable homes in the wildland−urban interface (WUI).

In spite of these different geographical and historically contingent contexts, EWEs represent a formidable threat in all three countries. In the following sections, we expand on the details of each of these country-specific contexts.

4.2 France

4.2.1 Landscape and property

As in other countries with a long and rich history, French landscapes have been shaped through different epochs and different property laws.

After the French Revolution (1789−93), the creation of 44,000 municipalities and private properties triggered off the splitting up of great estates [1]. The main consequence was that forest became propriety of the State, with most of the commons disappearing and properties being fenced [1].

Since the end of World War I, forest areas have increased: from 9.9 to 15.6 million hectares (+58%) in just a century [2]. Forest property is shared between State, territorial authorities (*Régions, Départements, and Municipalités*) (24%) and private (76%). Private forests are mainly present in the southern half of France and split into numerous small owners. Great estates are present especially in the northern half, except for the *Département des Landes* (in the extreme SW of the country) [3]. Today 10% of the French territory is still composed of "commons" [4].

The rural exodus and the decrease in the demand of forest products have made forests less profitable [5], especially in the south [6]. Land abandonment concerns mainly isolated rural areas, especially in the Mediterranean hinterland and the piedmonts of the mountains (*Alpes, Massif Central et Pyrénées*). Land is also being abandoned in areas around cities and along the Mediterranean coast. That, together with changing agricultural techniques, has triggered significant landscape changes as some agricultural lands were converted to woodlands [7] and poorer soils and steeper slopes and the more remote areas are gradually abandoned or converted into pasture and wasteland [7]. As a consequence, landscape is more undefined with large more fire-prone areas and create large areas likable to undergo a great event [8], including extreme wildfire.

4.2.2 Property and land management

Landowners' strategies vary according place, value of the land, and public policies. In the south of the Alps [7], several landowners are expecting a profit because of the change of zoning, as some agricultural areas may change into building areas. In some other cases landowners can benefit from grants given to convert agricultural lands into woods. In the the case of *Languedoc-Roussillon*, the uprooting of vineyards (funded by the State) has resulted in the disappearance of half of the vineyards (80,000 ha) since 1988 [9]. An important part of these lands, from 40% to 80% in some municipalities [10], are no longer tended, increasing the number and area of wastelands, shrublands, and also the number of fires [6,9], sometimes very near to the city limit.

If rural migration was an ancient and continuous phenomenon at least from the end of the 19th century onward, a return to the rural areas occurred in the seventies of the 20th century. Henri Mendras talked of *"the end of the farmers"* [11], heralding the progressive collapsing of agrarian society, and the emergence of neo-ruralization [12] with the changing of rural land by town citizens and the increase of the use of land now called "natural" [13,14]. In the south, the Mediterranean forest is more and more perceived as a "natural space" [15], creating tensions between traditional farmers and the neo-rurals [14,16].

4.2.3   Spatial planning

The State was responsible for the implementation of urban rules until the decentralization laws of 1982 [17] that gave municipalities the responsibility for local town planning, with the most important document being the *Plan d'occupation des sols* (POS), very popular between 1967 and 2015. The plans had a zoning area allowing buildings on big plots (from 2000 to 20, 000 square meters) in natural areas, which created a scattering of houses in forest or nearby [18]. The POS was replaced by the *Plan local d'urbanisme* (PLU), with local urban planning rules to be implemented in the areas.

In France, each year, a total area of 25,000 ha is urbanized at the expense of agricultural, natural, or forest areas. After the slowdown after the 2008 crisis, since 2015, there has been again an increase of urbanization, including on risky areas as in the *Provence-Alpes-Côte d'Azur region* [19], with peri-urban areas having a particular important increase in population, also along the Atlantic coast [20].

If municipalities are in favor of fire protection, they find it difficult to resist the demand for urban land because of population increase, and the development of infrastructure, especially on the Mediterranean coast (for example, in the *Département du Var*, the population increased by 43% from 1982 to 2011) [19]. As a result, urban areas are spreading and the WUI increases by 1% a year [20], although measures taken recently to curb that trend, namely, the ALUR Law [21], for urban spreading control and the PLU establish economical space-consumption limits based on the analysis of potential density.

4.2.4   Uses of fire and fire regulations

Landscapes were tended for a very long time with traditional practices including a large use of fire. Many historical texts give evidence of the use of fire by local people, although that has been often criticized and the source of tensions between various persons, shepherds, and foresters.

The pastoral practice of burnover is especially controversial [13,22] since a long time, as we may conclude from the fact that training of foresters in the Napoleonic schools excluded the use of fire, considered as primitive [23] or the fact that still today the law banishes fire except when it is used by the "right people": the technical and scientific people in charge [24,25].

Some researchers have changed the image of pastoral fires in the *Pyrénées* [22,26], in the *Cévennes* [27], or in *Auvergne* [28] as they have shown that fire is not evil and that it can be an efficient agricultural technique without any particular damage [13]. Some studies in the *Pyrénées* [22,29] show that pastoral fire is used to improve pasture grasses [30]. In dryer regions, pastoral fire is still considered destructive, whereas the controlled burning corresponds to *"an expert and allowed application of fire"* in a defined area with the adoption of authorized prescriptions and procedures on a precisely targeted parcel of land [31].

In view of strict regulations, the current use of fire is technically banned by the *Code Forestier* because of the minimum distance of 200 m from the forest boundary. But, depending on forest-fire risk, fire regulation differs from region to region. In regions with a live pastoral fire tradition, the *Commission locale d'écobuage* [32,33], under departmental authority rules, manages pastoral fires with farmers and shepherds [34]. By contrast, in risky areas, only some experts have the skills and the competency to make pastoral fire and controlled burning. However, with climate change, the risk of fire is increasing even in traditional pastoral areas, with 5000 ha of forest burned only in 2000 with five fatalities [30] *(Commune d'Esterençuby, February 10, 2000)*.

Laws and rules about the forest have a long history, dating back to King Philippe le Bel (1110). The rules were collected under the reign of Louis XIV from the 1669 *"Ordonnance de Louis XIV Roi de France et de Navarre sur le fait des Eaux et Forets"* [35], and some articles concern restriction in the use of fire. A great part of the first Forest Code (May 21st, 1827) was inspired by the 1669 *Ordonnance* [36,37].

Today the rules for the protection of forests against fire are contained in the Forest Code [38], to be applied in the regions with a higher fire risk: *Corse, Provence-Alpes-Côte-d'Azur, Occitanie, Aquitaine* and *Poitou-Charentes* and the Départements of *Drôme* and *Ardèche*. It creates the legal obligation of brush clearing for properties, allowing only the landowner or the occupant to light a fire at a 200 m distance away from forests and woods. He is subject to periods decided by the *Préfet* of the concerned *Département* when fire is strictly forbidden [38,39] (Art 131-2 II). Pastoral fire or controlled burning is well supervised [32,34,38] (Art 131-9) and can be lit with the written or tacit agreement of the landowners because of the measures of fire prevention.

The national policy of forest-fire prevention is mainly based on texts from the Forest Code and the Environment Code. The Plan of Prevention for Forest-Fire Risks [40] (Law 1995) whose main aim is to protect people and assets has the purpose to delimit the fire-prone zones by taking into account the type and intensity of the risk. In these zones, buildings or development can be forbidden or allowed only with specific requirements, so as not to increase the danger for people.

The CCFF (*Comité Communal Feux de Forêts*), communal forest fire commitees were created in 1984 [41] and are under the mayors' authority. They are directed to prevent fires (by patrols, on watch) and to assure the diffusion of the environment values and make people aware of danger. They can intervene on incipient stages of fires and guiding firefighters' action.

The DICRIM *(Document d'Information Communal sur les Risques Majeurs)*, a communal information file on major risks [42] informs people about the risks and the protection means and has a chapter on forest fires.

The legal obligation of brush clearing is a key point in the prevention policy. The L134-6 article of the Forest Code [38] defines the implementation, obligation of brush clearing, and maintenance of barren land which can be applied to plots of land located 200 m away from woods and forests in each of the following situations: (1) in the area around buildings, construction sites, and installations of all kinds, a 50-m-wide perimetral strip (the mayor can decide a 100-m width); (2) in the area around private roads to buildings, constructions sites, and installations of all kinds, a 10-m-wide strip on both sides of the road.

At a landscape dimension, the Land Management Contract [43] (financed by European Union) is a contract between the State and the farmer, where different measures can have an impact, including maintenance of fire risk areas, planting of trees on banks to reduce flood risk, restraint of brushwood, reopening of pastoral areas, creation and maintenance of hedges and terraces, and protection of meadow plants [33].

4.3 Portugal

4.3.1 Landscape and property

In Portugal the vast majority of the country's forest areas is privately owned (about 85.5%), and approximately 5%.7% of them belong to big enterprises that usually work with pulp and paper production. Local communities own around 11.8% of the country's forest areas, leaving only 2.7% to the State [44].

Most of the country's rural properties are very small. About 71% of the farms have less than 5 ha of utilized agricultural area (UAA), corresponding to a mere 9% of the total UAA. On the other hand, 67% of the UAA is concentrated in 4% of the farms which are greater than 50 ha [45]. As a result of both biophysical conditions and historical, social, and economic processes, there are marked differences in agroforestry structures. The extreme fragmentation of the rural property, especially north of the Tagus River, is due also due to the inheritance laws that have been in force in Portugal throughout history, alternately favoring conditions for the fragmentation or the merging of the rural property. One particular note is the effects of inheritance laws over the last two centuries, which have tended to encourage fragmentation of ownership. As a consequence, the fragmentation of the rural property has been a concern for governments since the early 20th century, introducing the principle of indivisibility and coownership among the coheirs for small properties. However, the problem persists until today, and moreover with the spread of small and unmanageable "indivisible" plots that are divided among increasing numbers of coowners [46].

During the 1960s Portugal went through deep economic and social changes. There was a great movement of rural population into the cities (in Portugal and abroad), attracted by better job opportunities in the growing industrial and services sector, leaving behind depopulated and aged population a tendency toward dissociation between agriculture and forest management occurred, as at the same time that the forest market became more competitive.

These changes affected the country differently. In the South, the prevalence of large properties enhanced mechanization, allowing agricultural activity to remain dynamic and able to access communitarian funds. In the North and Center, the exploitation of small rural properties became economic unviable, so most of them were abandoned and covered by shrubs or converted to plantation of pine and eucalyptus.

4.3.2   Spatial planning

In the late decades spatial planning went through several transformations, associated with major changes, which can be separated in three fields: constitutional law, legal order, and institutional agreements. At the constitutional plan, the changes point out to strengthen the role of the State, which must "ensure a correct spatial planning" (1989) or more recently in its association to the "promotion of the well-being and quality of life of citizens" the "balanced promotion of all sectors and regions" or "phasing-out of the economic and social differences between the city and the countryside" [47]. The legal system is essentially associated with the fundamentals of the Law of Bases of Spatial Planning and Urban Planning[1] [48]. Finally, due to the need to maximize the use of communitarian funds made available and to increase GDP associated with a bad relationship between the legal system and the tax system, planning was relegated to a secondary position with a prevailing bureaucratic perspective. In fact, despite the conceptual and legislative changes, the territorial planning (especially local and regional) is too much time-consuming, badly treated by the central administration and not really related with spatial planning strategies, which reinforces the idea of a "nonplanned" country.

This context is relevant to understand the case of forest planning. The Law of Bases for Forest Policy[2] [49] establishes, in its article 2, that "the forest, for its diversity and the nature of the goods and services it provides (…), is essential to the maintenance of all forms of life, and all citizens are responsible for its conservation and protection" defining as a compulsory measure of forest planning and management, the elaboration of regional plans of forest planning. Despite this legal framework, the planning and the actions taken over the last few years are clearly insufficient. To this end, three factors compete:

a) The excessive fragmentation of property. Throughout the 20th century and to the present, several measures have been implemented (or at least tried) to solve this problem, through

---

[1] Law nr. 48/98 of 11/8.
[2] Law nr. 33/96 of 17/8.

the institution of culture units and rustic soil reparcelling policies, or through the introduction of Forest Intervention Areas[3] (ZIF), with no significant impact;

b) The lack of an updated land/property cadaster. There are only a few Portuguese towns or municipalities that have developed initiatives to have a complete and trustful ownership registration and limits identification. That poses several problems, in particular in the management of the forest, cleaning, and fire prevention, as well as general spatial planning. It is estimated that the properties without an owner or without a known owner represent between 10% and 20% of the Portuguese territory [50], while the ENF[4] [44] (2015) considers that 50% of the forest spaces are located in areas with no cadaster;

c) The disarticulation between the State bodies (central and local) the policy instruments of spatial planning and the sectoral interventions [51,52]. Portugal is also characterized by excessive centralism [53]. Finally, the Portuguese framework of forest/fire governance is very complex, considering different structures (National System of Civil Protection [SNPC], Integrated System of Protection and Assistance Operations [SIOPS], and Special Forest Fire-fighting Device [DECIF]), and multiple principles, concepts, stakeholders, resources, and levels of action. This, of course, poses several problems and claims for governance change, to increase the efficiency of planning and a better balance between plans, prevention, and combat, which takes much effort in view of the solid interests related with property, fire combat, and forest production.

The stakeholders (politicians, technicians, and scientific advisers/researchers) agree that there is a lot to be done. A key idea, especially after the tragic effects of forest fires in 2017 and 2018, is that effective governance and spatial planning play a key role to prevent fire occurrence, to facilitate suppression, and to minimize impacts on people and land.

Ongoing/planned changes include the creation of a public bank of agricultural land, a new legal framework for afforestation, reforestation or forest densification, and a limit to monocultures and invasive species. Also the capacitation of a national system for forest defense against fires, with obligation of land cleaning by the owners, the implementation of a simplified cadastral information system and tax benefits to forest management entities.

4.3.3 Uses of fire and fire regulations

The use of fire dates back many thousands of years in the struggle for security, food, and shelter [54] and is linked to activities such as hunting and gathering and agriculture through the management of game and plant resources, pests, and diseases. It was also used for military purposes and to claim land [55,56]. Some of these uses remain actual, for example, fire is still used as a tool for clearing and preparing agricultural land.

The use of fire is an official forest fire cause for the Portuguese Institute for the Conservation of Nature and Forests (ICNF), which comprises burnings of trash, agricultural and forestry materials, the use of fireworks, bonfires, smoking (throwing of

---

[3] ZIF is a continuous and bounded territorial area and bounded, constituted mostly by forestry spaces, subject to a Forest Management Plan (PGF), that meets the requirements of the Municipal Plans for the Forest Defense against Fires, and that is administered by a single entity.

[4] National Strategy for Forests.

cigarette butts and incandescent matches), fire use by beekeepers, dispersion of sparks, or other type of incandescent material from industrial chimneys, machines, and dwellings or agricultural installations. Based on the Portuguese Rural Fire database [57], the use of fire was the cause of 20% of forest fires with known causes in the 1981−2014 period and 63% in the more recent 2001−13 period.

Despite fire prevention actions, the reckless use of fire for various land management purposes persists and the lack of awareness of fire risk is generalized, especially in peri-urban areas where, for example, pastoral burning increase fire activity in autumn−winter months [58].

A succinct summary of fire regulation [54,59] is as follows. References to the use of fire date back to the 19th century [60], but legislation began in 1970 (Decree-Law n. 488/70), in a context of growing concern about fires, advocating the concentration of the various entities (administrative authorities, forest services, and private owners) and attributing to municipalities the possibility of determining the places and times in which the use of fire may be prohibited or conditioned, also stipulating penalties in case of infraction. In 1980s, the State recognizes humans as the main cause of fires, opts for strengthening the firefighting means, defines different types of fire (including prescribed fire), and prohibits the use of fire during the forest-fire season. In 1996, the Forest Policy Framework Law was adopted, intended to ensure the protection of forests against fire.

After the catastrophic fires of 2003, the Structural Reform for the Forest Sector is approved to revise legislation, rehabilitate the state forest structure, create a fiscal framework, facilitating simple mechanisms of active management of the territory, and solve conjunctural problems of the National Forest Fire Protection System (NFFPS), through a balanced distribution of responsibilities, means, and assignments (Decree-Law n.94/2004 and Decree-Law n.156/2004). During recent decades, the national, regional, and subregional structures with planning, coordination, and execution functions of forest fire prevention, detection, and control activities underwent several changes. Finally, also after the large fires of 2017, the State created a Municipal Emergency Fund for the affected territories (*Resolução do Conselho de Ministros n.101/2017 and n.148/2017*) and proceeded to the fifth amendment of the NFFPS (Law n.76/2017).

Wildfire regulation history is linked to the occurrence of extreme wildfires (e.g., 2003 and 2017), its catastrophic consequences, media exposure, and society demands for better management, and more firefighting resources. This regulation had positive impacts on fire incidence, including the reduction of the number and duration of large fires (burnt area of more than 100 ha) between 2007 and 2013 [61].

The most recent wildfire regulations can be found mainly in the NFFPS (Decree-Law n.124/2006), in the National Plan for the Defense of Forest Against Fires (NPDFAF) (*Resolução do Conselho de Ministros n. 65/2006*), and in the decree-laws that have adjusted them (Decree-Law n.17/2009, Law n.76/2017, and Resolução do Conselho de Ministros n. 58/2017). The NFFPS defines how various types of fire can be used and authorized, such as (1) controlled/prescribed fire; (2) burnings; (3) bonfires; and (4) firecrackers and other forms of fire. This decree was adjusted in Decree-Law n.17/2009 to regulate also the technical and suppression fire. The

NPDFAF is aimed to (1) increase the territory resilience; (2) reduce the fire incidence; (3) improve the efficiency of fire attack and management; (4) recover and rehabilitate ecosystems; (5) adapt and improve organizational and functional structure; and (6) reduce burnt area to average values of the other Southern European and the number of small fires (<1 ha), rekindles and large fires.

4.3.4 Prevention and fuel management

Some studies suggest that, if not combined with fuel management, fire suppression policies in Portugal and other Mediterranean countries will only be effective at reducing small fires and not large ones [62]. However, the impressive afforestation that was carried out in public and common land during the 20th century was not followed by proper forest management and fuel hazard reduction [58].

State and local authorities were responsible for the planning, installation, and maintenance of primary fuel management networks (Decree-Law n.17/2009), which must be efficient and secure for firefighters, taking into account topography, landscape, and socioeconomic characteristics, and large fire history and behavior during high fire weather risk. The NPDFAF highlighted the urban–forest interface and suggested to create and maintain an external buffer around population clusters with the highest fire vulnerability, as well as around parks, industrial polygons, landfills, warehouses, and other buildings. The buffer size should be of 100 m around population clusters, 50 m around houses, and 10 m for each side of a road, with landowners forced to leave at least a 25-m strip for each side of the circulation road (Law n.76/2017).

In the Mediterranean forests, the usual postfire management procedure starts with the cleaning and fell of burned trees and ends with the plantation or direct seeding [63].

4.4 The United States of America

The United States of America (US) is a vast country composed of numerous and diverse ecoregions [64]. Each of these ecoregions is cross-cut and divided by semiautonomous governmental units (States) characterized by unique histories of settlement and land use and divergent trajectories of planning and management. As a consequence, relationships between landscapes, their human attributes, and extreme wildfires differ between somewhat arbitrary political boundaries on the ecoregional type shifts they arbitrarily divide. For the purposes of this chapter section, I contrast the history and trajectories of settlement, planning, and management broadly within the conterminous Eastern and Western portions of the country.

4.4.1 Landscape and property

With the exception of relatively small-scale Native American settlement and land use, significant and persistent landscape and property regime characteristics for the majority of US landscapes began to materialize in earnest during the 17th to 19th centuries

[65,66]. The Eastern US from the Appalachian Mountains to the Atlantic as well as parts of the Mississippi river corridor were settled during Colonial times (before ca. 1776) under British, Dutch, French, and Spanish authorities. Legacy property regimes from these predominantly follow the British metes and bounds system, whereby private landholdings are defined somewhat organically according to physical landscape features (streams, roads, rock walls, hedgerows, etc.) or strait lines between such features (usually trees), and previously claimed neighboring lands surrounding the parcel [67,68]. The resulting landscape mosaic is characterized by an extremely fine-grained and heterogeneous distribution of agricultural and industrial land uses and associated fuel types and quantities.

In the Western US (e.g., west of the Appalachian Mountain states), Euro-American settlements were laid out on a rectangular "lot and block" system directly influenced by government planning under the Public Lands Survey System [69]. Significant exceptions to this rule include areas originally colonized by the Spanish and French (e.g., predominantly Louisiana and Texas, but also large parts of New Mexico, Arizona, and California) which were later purchased or annexed into the US. Legacies of Native American settlements also effect a significant portion of the Western landscape, particularly in New Mexico and Arizona where there are large semiautonomous tribal holdings. Vast tracts of grasslands, mountains, forests, and deserts of the Western US were not well suited to traditional Euro-American agricultural land uses, and so much of this land has remained undeveloped and in the public domain. Consequently, federally permitted livestock grazing, mining (especially gas and oil well drilling), and forestry remain the dominant land uses across the public lands in the US West.

Forestry practices in both the East and West were extremely wasteful, and no forethought was given to the reproduction of timber or the sustainability of ecological services more generally [70]. By the early 20th century, with growing urban and industrial demand for ecological services largely in the form of water and timber, there was a clear need for forest conservation and watershed management. In the Eastern US, destructive agricultural and forestry practices left behind degraded landscapes with many forests subject to frequent wildfire and soil loss. Wildfires were frequently caused by locomotive sparks, other machinery, arsonists, and careless forest workers such that human-caused wildfires were haphazard and, in some cases, disastrous. Debris-strewn logging grounds and agricultural abandonment fueled the fires [71]. Although landscape degradation occurred slightly earlier in the Eastern US, thanks to the construction of railroads, the late 19th century saw the rapid spread of destructive logging and mining operations across the continent.

Conservation legislation at both Federal and State levels resulted in the creation of professional forest and soil conservation agencies, charged with the management of public lands. In the East, where public lands were in short supply, state and national forests and parks were acquired and consolidated from private hands. Cut-over timber lands in the East's mountainous regions were the first to be purchased and consolidated under state and federal management (ca. 1880—1920s), followed by a second wave that created new forest on degraded agricultural lands (ca. 1920—80s) [70,72]. Farm houses were demolished and fields reclaimed, such that many rural towns and remaining farmlands are now surrounded by patchworks of private and publicly managed forests and woodlands.

4.4.2 Spatial planning

Spatial planning in the US is predominantly regulatory-based and controlled at local levels such as counties, municipalities, and even neighborhoods. For this reason, planning and zoning regulations remain highly variable down to the local level. Wildfire-related planning is no exception to this rule. Demand for exurban housing development in fire-prone landscapes is on the increase because of the attraction of many nature-based amenities those landscapes have to offer [73]. As a result, there is a great need for wildfire-relevant planning efforts, especially in the US West, but in key Eastern states as well, such as Florida.

Although wildfire in the WUI has received increasing attention, policymakers and land use planners are only just beginning to address these concerns [74—77]. Indeed, most of the focus on wildfire mitigation in the WUI has been postconstruction, on fuels reduction and homeowner education [78]. Planning for wildfire mitigation and prevention at the individual household-level is currently voluntary "firewise" activities [79], but there does appear to be increasing support from governmental and nongovernmental organizations through a variety of incentives [80].

In addition to educational efforts, the Federal government also focuses on reducing fire hazard in the WUI through fuel-thinning projects on federally managed lands. The National Fire Policy (2001), the Healthy Forest Initiative (2002), and the Healthy Forest Restoration Act (2003), for example, dedicated funding to reduce hazardous fuels in the WUI [73]. However, one of the greatest challenges for these programs, and for wildfire hazard—based spatial planning in general, concerns the fact that ecological conditions in the United States are extremely variable and there can be no "one-size-fits-all" solution [81].

National- to regional-level regulations on land use and development of private lands in the US are virtually nonexistent, unless federal permitting or funding is involved. In the latter case, the National Environmental Policy Act [82] often requires the developer to undertake an environmental impact assessment. In theory, the assessment process could force a private developer to account for wildfire hazard in his development plans. In lieu of national- and regional-level policy, spatial planning falls on the responsibility of local communities which lack appropriate resources and often perceive a need for housing as well as the economic benefits such projects bring. Designing regulatory plans (e.g., zoning) that mitigate for wildfire hazard may offer high economic opportunity costs with respect to perceived or even scientifically forecasted wildfire potential.

4.4.3 Uses of fire and fire regulations

The use of broadcast fire as a land management tool has a long history in the US [71]. The historical use of fire by Native Americans has been documented in a large number of contexts [83,84]. In general, burning by Native Americans was associated with agriculture, vegetation management, and hunting. Most of the Native American burning practices in the Eastern US were lost through the displacement and genocide wrought by European colonization and settlement. Europeans quickly replaced Native

American burning practices with their own forms of fire use [71,85]. In fact, fire-use practices were so integral to Euro-American land use practices that fire frequency increases observed in dendrochronological archives in the central Pennsylvania region reflect the settlement process as a "wave" of fire across the landscape [86].

This tight, but positive, relationship between fire and land use in the Eastern US lasted until the early 20th century [71] when official governmental attitudes toward fire turned negative. Indeed, federal wildfire management has been mostly adversarial since the creation of federally managed parks and forest reserves in the late 19th century [87,88]. Nonetheless, prescribed fire in the Eastern US has roots that extend back to traditional woodland and range management practices [89]. Although official policy strictly prohibited the use of fire during the early 20th century throughout the US, prescribed fire practices survived on private game reserves in the Southeast [90,91] and also as an informal clandestine activity [85,92]. One consequence of this legacy is that the Southeastern US today has the largest and most well-established prescribed fire program in the country [93].

In the US West, Spain was a relatively weak colonial power and much of the area remained under local, indigenous control well into the 19th century. As a consequence, Native burning practices persisted nearly up until fire suppression policies were put into place by the US government. Indeed, some of the first attempts at prescribed fire by professional foresters were derogatorily labeled "Paiute Forestry" in reference to the Native American ethnolinguistic identity, "Paiute" [94]. However, the relative importance of Native American fire regimes to the continent's ecology remains debated [95].

Following the lead of fire managers in the Southeastern US, prescribed fire programs for the Western US were eventually accepted and initiated in the 1960s [96,97]. Prescribed fires now form a significant, if secondary, wildfire management tactic in the US, broadly [87]. These fires are not simply conducted to reduce fuels for wildfire prevention but are also the main tool being used to restore landscapes that have been damaged by intensive land use activities [98].

Fire-use regulations in the US vary from federal land management agency to agency, state to state, and from municipality to municipality [99,100]. Outside of a state-managed or federally managed prescribed fire context, private citizens may burn given that they obtain a proper permit. The legal acceptability of fire use can be understood as the degree of liability assigned to the fire setter for any costs that result from the fire [100] and found that the degree and type of liability varied by state and is statistically tied to the amount of private forest holdings such that the more private the forest holdings in a state, the more relaxed its prescribed burning liability laws. Unsurprisingly, these are Southeastern states. Conversely, Western states with large public forestlands have strong liability laws, thus increasing the potential cost of fire use by individuals.

Given the geographic heterogeneity in the fire-use regulatory environment, there is surprising agreement between federal and state wildfire management policies in that they all promote fire suppression and exclusion. Indeed, these policies have the staggering effect of suppressing of 99% of all wildfires on public lands [101]. State and national fire suppression policies took shape in the wake of the 1910 fire season, which was particularly disastrous, and during the Great Depression with the New Deal formation of the Civilian Conservation Corps [72]. One outcome of this was a policy of fire exclusion from public lands that some have linked to an increase in extreme wildfire and a crisis in public lands management [102,103].

4.5 Conclusion

Fire-prone territories share some common aspects. France and Portugal had a common transition from intensive agricultural land use to abandonment and afforestation. Although the specific contexts of landscape transition are different in the US, depending on the region, land use changes are clearly implicated in the increase of fire hazard. In all three places, fuel accumulation has increased and human settlements have expanded, leaving little or no buffer zone between populated areas and wildlands. Thus the WUI in each of these countries is a major concern for wildfire management. With climate change increasing the probability of EWE occurrence, policies to ensure public safety must account with the increasing relevance of the relationships between landscape characteristics, human settlements, spatial planning, and fuel management.

References

[1] D. Aubin, S. Nahrath, F. Varone, Paysage et propriété: patrimonialisation, communautarisation ou pluridomanialisation, 2004. Cahier de l'IDHEAP (219).

[2] Institut Géographique National, Un siécle d'expansion des forêts françaises. de la statistique Daubrée à l'inventaire forestier de l'IGN, 2013. L'If (31), Saint-Mandé.

[3] AGRESTE, Les Dossiers, N30, La forêt privée française en France métropolitaine : structure, propriétaires et potentiel de production, Ministère de l'Agriculture, Paris, 2015.

[4] N. Vivier, Les biens communaux en France de 1750 à 1914. État, notables et paysans face à la modernisation de l'agriculture, 1998. Ruralia (02).

[5] C. Vigneron, Les moteurs d'évolution des paysages méditerranéens, Forêt Méditerranéenne XV-3 (1994) 321–324.

[6] Conseil général de l'environnement et du développement durable, IGA, Mission d'évaluation relative à la défense de la forêt contre l'incendie, 2015. Paris).

[7] G. Trie, G.C. Aspe, E. Maillé, C. Bouillon, Déprise agricole et stratégies locales des acteurs sur le canton de Banon -Alpes de Haute-Provence, 1995. Irstea (1995).

[8] E. Maillé, C. Bouillon, Consommation des espaces agricoles et naturels par le processus d'urbanisation discontinue, Diagnostic comparé de l'évolution de l'occupation de l'espace des territoires du Pays d'Aix, Cemagref, Aix en Provence, 2005.

[9] C. Arnal, L. Laurens, C. Soulard, The Landscape Mutations Generated by the Uprooting of Vineyard, a Vector for the Mobilization of Local Actors in the Herault Department Méditerranée, vol. 120-1, 2013, pp. 49–58.

[10] FranceAgriMer, Dyopta, Préfet de la région Languedoc Roussillon, Que deviennent les anciennes parcelles de vigne suite aux arrachages avec primes ? Résultats d'une enquête menée en 2012, 2013.

[11] H. Mendras, La fin des paysans. Suivi de : Une réflexion sur la fin des paysans, vingt ans après, France, Actes Sud-réédition 1984, 1967.

[12] G. Martinotti Guido, in: The New Social Morphology of Cities, Conference UNESCO/ MOST Wien, 10−12 February, 1994.

[13] P. Vilain-Carlotti, Perceptions et représentations du risque d'incendie de forêt en territoires méditerranéens La construction socio-spatiale du risque en Corse et en Sardaigne, Thèse de doctorat, Université Paris, 2016, p. 8.

[14] C. Aspe, La naturalité entre profunda scientia et esthétique verte: l'exemple de la forêt, Forêt Méditerranéenne XXIX-4 (2008) 517−524.

[15] J.P. Hethier, Forêt méditerranéenne : vivre avec le feu ? Éléments pour une gestion patrimoniale des écosystèmes forestiers littoraux, Les cahiers du Conservatoire du littoral 2-1993, 1993.

[16] M. Chevalier, Neo-rural phenomena, in: Espace Géographique. Espaces, Modes D'emploi, Two decades of l'Espace géographique, an anthology, 1993.

[17] Loi n° 82-213 du 02/03/1982 relative aux droits et libertés des communes, des départements et des régions.

[18] Ministère de la cohésion des territoires, La caducité des Plans d'occupation des sols (Loi Alur), 2016.

[19] M. Bocquet, La consommation d'espaces et ses déterminants d'après les Fichiers fonciers de la DGFiP, Analyse et état des lieux au 1er janvier 2015, CEREMA direction territoriale Nord-Picardie, 2016 (Lille).

[20] M. Long-Fournel, D. Morge, C. Bouillon, M. Jappiot, Interfaces habitat-forêt dans le département des Bouches-du-Rhône. Phase 2 : Analyse diachronique, convention Dreal-Paca, 2012.

[21] Loi ALUR n°2014-366 du 24 mars 2014 JORF n°0072 du 26/03/2014, 5809, 2014.

[22] M.R. Coughlan, Errakina: pastoral fire use and landscape memory in the Basque region of the French Western Pyrenees, J. Ethnobiol. 33 (1) (2013) 86−104.

[23] G.M. Delogu, Dalla parte del fuoco. Ovvero il paradosso di Bambi, Edizione Il Maestrale, Nuoro, 2013.

[24] N. Ribet, Les savoirs traditionnels en matière de brûlage, XVIIès Rencontres du réseau des Equipes de Brûlage Dirigé, Cardère éditeur, 2006, pp. 30−32.

[25] E. Rigolot, Construire une culture du feu : fire Paradox, in: Communication lors des 17e rencontres des équipes de brûlage dirigé, Rosans (Hautes Alpes) 7−9 juin, 2006.

[26] J.P. Métailié, J. Faerber, Quinze années de gestion des feux pastoraux dans les Pyrénées : du blocage à la concertation, in: In Sud-Ouest Européen, Tome (16-2003), Pastoralisme 37-51, 2003.

[27] D. Richard, feu, savoirs et pratique en Cévennes, Quae NSS, Versailles, 2010.

[28] N. Ribet, Les parcours du feu, in: Techniques de brûlage à feu courant et socialisation de la nature dans les Monts d'Auvergne et les Pyrénées centrales, Thèse en anthropologie sociale et ethnologie, EHESS-Paris, Bourse de Thèse Environnement de la Région Auvergne, 2009.

[29] M.R. Coughlan, Farmers, flames, and forests: historical ecology of pastoral fire use and landscape change in the French Western Pyrenees, 1830−2011, For. Ecol. Manag. 312−2014 (2014) 55−66.

[30] Commissariat général au développement durable, Le risque de feux de forêts en France, Service de l'observation et des statistiques, Études & documents, 2011 n°45, Nancy, France.

[31] G. Bovio, D. Ascoli, Fuoco prescritto: stato dell'arte della normativa italiana, L'Italia Forestale e Montana 67 (4) (2012) 347−358, 2012.

[32] Ville d''Oloron Sainte-Marie, Commission Locale Écobuage, Réglementation, Composition, Sécurité, 2017.

[33] F. Novellas, P. Blot, Mise en place d'une commission locale d'écobuage, Préfecture de l'Ariège, Service environnement, risques, unité biodiversité — forêt, DDTM de l'Ariege, 2016.

[34] Prefet des Pyrénées-Atlantiques, DDTM, Arrêté n° 2012296-0004 portant réglementation des incinérations, 2012.

[35] De Gallon, Le Gras, in: Conférence de l'ordonnance de Louis XIV du mois d'août 1669, sur le fait des eaux et forêts, Original provenant de la Bibliothèque municipale de Lyon, tome 1, 1725.

[36] F. Meyer, Législation et réglementation, Revue forestière française, numéros spécial incendies de forêt, 1974, pp. 249—257.

[37] K. Matteson, Forests in Revolutionary France: Conservation, Community, and Conflict, 1669—1848, Cambridge University Press, 2015.

[38] Legifrancegouvfr, Code forestier nouveau, version consolidée au 9 avril 2018, 2018. https://www.legifrance.gouv.fr/affichCode.do?cidTexte=LEGITEXT000025244092.

[39] Préfet des Bouches du Rhône, arrêté préfectoral n°2013354-0004 du 20/12/2013.

[40] Loi n° 95-101 du 2 février 1995 relative au renforcement de la protection de l'environnement, Journal officiel de la république française, n°29 du 3 février 1995.

[41] Préfet, des Bouches-du-Rhône, arrete 850 du 4 mars 1996 : creation des Comité communaux feux de forêt organisation, missions, fonctionnement.

[42] Décret 90-918 du 11 octobre 1990 introduit le document d'information communal sur les risques majeurs (DICRIM) Journal officiel de la République Française.

[43] F. Alavoine, G. Giraud, C. Bouillon, L'enjeu paysager dans les contrats territoriaux d'exploitation, Colloque L'évaluation du paysage, une utopie nécessaire?, 2004 (Montpellier).

[44] Estratégia Nacional para as Florestas (ENF), através da Resolução do Conselho de Ministros n.º 6-B/2015 - Diário da República n.º 24/2015, 1º Suplemento, Série I de 2015-02-04.

[45] INE, Inquerito a Estrutura das Exploracões Agricolas 2016, Instituto Nacional de Estatística, Lisboa, 2016.

[46] P. Bringe, Ordenamento florestal ou ordenamento territorial? in: P.C. Ferreira (Ed.), Economia da floresta e ordenamento do territorio Conselho Economico e Social, Lisboa, 2017, pp. 55—66.

[47] Portuguese Republic Constitution, 2005, p. 4655.

[48] Assembleia da República, Lei de Bases de Ordenamento do Território e de Urbanismo (LBOTDU). Lei n.º 48/98, de 11 de Agosto: Estabelece as bases da política de ordenamento do território e do urbanismo. Diário da República n.º 184/1998, Série I-A de 1998-08-11, 1998.

[49] Assembleia da República, Lei de Bases da Política Florestal (LBPF). Lei n.º 33/96, de 17 de agosto: define as bases da Política Florestal nacional. Diário da República — I Série-A N.o 190 17-8-1996, 1996.

[50] R.S. Beires, J. Amaral, P. Ribeiro, O Cadastro e a Propriedade Rústica em Portugal, Fundação Francisco Manuel dos Santos, Lisboa, 2013.

[51] P. Chamusca, Governança e regeneração urbana: entre a teoria e algumas práticas, FLUP. Porto. 400pp. Tese de doutoramento, 2012.

[52] J. Ferrão, Governança e ordenamento do território: reflexões para uma governança territorial eficiente, justa e democrática, in: Prospectiva e planeamento, vol. 17, 2010, pp. 129—139.

[53] J.A. Rio Fernandes, Reestruturação da administração territorial portuguesa : o duplo centralismo em busca de escalas intermédias. [Santiago de Compostela], Departamento de Xeografía. Universidade de Santiago de Compostela, 2006.

[54] M.G. Pereira, Climate Variability and its Impacts on Wildfire Activity, Doutoramento (PhD) em Ciências Geofísicas, Faculdade de Ciências, Universidade de Lisboa, 2005.

[55] Z. Naveh, The evolutionary significance of fire in the mediterranean region, Vegetatio 29 (1975) 199−208.

[56] S.J. Pyne, Fire in America: A Cultural History of Wildland and Rural Fire, University of Washington Press, 2017.

[57] M.G. Pereira, B.D. Malamud, R.M. Trigo, P.I. Alves, The history and characteristics of the 1980−2005 Portuguese rural fire database, Nat. Hazards Earth Syst. Sci. 11 (2011) 3343−3358.

[58] P. Mateus, P.M. Fernandes, Forest fires in Portugal: dynamics, causes and policies, World Forests (2014) 97−115.

[59] J.E. Nisa, Incêndios florestais: prevenção e investigação criminal, REVISTAS Do S.M.M.P 51 (1992).

[60] F.L.G. Varnhagen, Manual de instruções práticas sobre a sementeira, cultura e corte dos pinheiros e da conservação da madeira dos mesmos; indicando-se os métodos mais próprios para o clima de Portugal, 1836.

[61] J. Parente, M.G. Pereira, M. Tonini, Space-time clustering analysis of wildfires: the influence of dataset characteristics, fire prevention policy decisions, weather and climate, Sci. Total Environ. 559 (2016) 151−165.

[62] F. Moreira, O. Viedma, M. Arianoutsou, T. Curt, N. Koutsias, E. Rigolot, A. Barbati, P. Corona, P. Vaz, G. Xanthopoulos, F. Mouillot, E. Bilgili, Landscape − wildfire interactions in southern Europe: implications for landscape management, J. Environ. Manag. 92 (2011) 2389−2402.

[63] F. Moreira, A. Águas, A. Ferreira, F.X. Catry, F.C. Rego, J.S. Silva, M. Bugalho, FIREREG - Factors Affecting the Post-fire Natural Regeneration Variability in *Pinus pinaster* and Eucalyptus Globulus in Portugal: Implications for Biodiversity and Post-fire Management, Centro de Ecologia Aplicada - Profa Baeta Neves, 2012.

[64] J.M. Omernik, Ecoregions of the conterminous United States, Ann. Assoc. Am. Geogr. 77 (1) (1987) 118−125.

[65] W. Cronon, Changes in the Land, Indians, Colonists, and the Ecology of New England, Hill and Wang, New York, USA, 1983.

[66] D. Foster, F. Swanson, J. Aber, I. Burke, N. Brokaw, D. Tilman, A. Knapp, The importance of land-use legacies to ecology and conservation, Bioscience 53 (1) (2003) 77−88.

[67] D.J. Bain, G.S. Brush, Placing the pieces: reconstructing the original property mosaic in a warrant and patent watershed, Landsc. Ecol. 19 (8) (2005) 843−856.

[68] M.R. Coughlan, D.R. Nelson, Influences of native American land use on the colonial Euro-American settlement of the South Carolina piedmont, PLoS One 13 (3) (2018) e0195036.

[69] L.A. Schulte, D.J. Mladenoff, The original US public land survey records: their use and limitations in reconstructing presettlement vegetation, J. For. 99 (10) (2001) 5−10.

[70] W.E. Shands, R.G. Healy, The Lands Nobody Wanted. The Lands Nobody Wanted, 1977.

[71] S.J. Pyne, Fire in America: A Cultural History of Wildland and Rural Fire, Princeton University Press, Princeton, New Jersey, USA, 1982.

[72] D.E. Conrad, J.H. Cravens, The Land We Cared for–: A History of the Forest Service's Eastern Region: USDA-Forest Service, Region 9, 1997.

[73] P. Gude, R. Rasker, J. van den Noort, Potential for future development on fire-prone lands, J. For. 106 (4) (2008) 198–205.

[74] V.C. Radeloff, R.B. Hammer, S.I. Stewart, J.S. Fried, S.S. Holcomb, J.F. McKeefry, The wildland–urban interface in the United States, Ecol. Appl. 15 (3) (2005) 799–805.

[75] V. Spyratos, P.S. Bourgeron, M. Ghil, Development at the wildland–urban interface and the mitigation of forest-fire risk, Proc. Natl. Acad. Sci. 104 (36) (2007) 14272–14276.

[76] A.D. Syphard, J.E. Keeley, A.B. Massada, T.J. Brennan, V.C. Radeloff, Housing arrangement and location determine the likelihood of housing loss due to wildfire, PLoS One 7 (3) (2012) e33954.

[77] A.D. Syphard, A.B. Massada, V. Butsic, J.E. Keeley, Land use planning and wildfire: development policies influence future probability of housing loss, PLoS One 8 (8) (2013) e71708.

[78] C.C. Klein, Community Wildfire Planning and Design: A Review and Evaluation of Current Policies and Practices in the Western United States, 2017.

[79] https://www.nfpa.org/Public-Education/By-topic/Wildfire/Firewise-USA.

[80] A.D. Bright, R.T. Burtz, Firewise activities of full-time versus seasonal residents in the wildland-urban interface, J. For. 104 (6) (2006) 307–315, https://doi.org/10.1093/jof/104.6.307.

[81] J.E. Keeley, Ecological Foundations for Fire Management in North American Forest and Shrubland Ecosystems (General Technical Report, PNW-GTR-779), USDA Forest Service, Pacific Northwest Research Station, 2009.

[82] PEA, Pre-Construction Environmental Assessments, National Environmental Policy Act., 1969.

[83] O.C. Stewart, Forgotten Fires: Native Americans and the Transient Wilderness, University of Oklahoma Press, Norman, 2002.

[84] G.W. Williams, Aboriginal use of fire: are there any "natural" plant communities, in: C. Kay, R.T. Simmons (Eds.), Wilderness and Political Ecology: Aboriginal Land Management-Myths and Reality, University of Utah Press, Logan, Ut, 2002.

[85] M.R. Coughlan, Unauthorized firesetting as socioecological disturbance: a spatiotemporal analysis of incendiary wildfires in Georgia, USA, 1987–2010, Fire Ecology 9 (3) (2013).

[86] M.C. Stambaugh, J.M. Marschall, E.R. Abadir, B.C. Jones, P.H. Brose, D.C. Dey, R.P. Guyette, Wave of fire: an anthropogenic signal in historical fire regimes across central Pennsylvania, USA, Ecosphere 9 (5) (2018) e02222, https://doi.org/10.1002/ecs2.2222.

[87] K.C. Ryan, E.E. Knapp, J.M. Varner, Prescribed fire in North American forests and woodlands: history, current practice, and challenges, Front. Ecol. Environ. 11 (s1) (2013) e15–e24, https://doi.org/10.1890/120329.

[88] S.,L. Stephens, Forest fire causes and extent on United States forest service lands, Int. J. Wildland Fire 14 (3) (2005) 213–222.

[89] H.L. Stoddard, Use of fire in pine forests and game lands, in: Paper Presented at the First Annual Tall Timbers Fire Ecology Conference, Tallahassee, 1962.

[90] H.L. Stoddard, Use of controlled fire in Southeastern upland game management, J. For. 33 (1935) 346–351.

[91] A.G. Way, Burned to be wild: herbert Stoddard and the roots of ecological conservation in the southern longleaf pine forest, Environ. Hist. 11 (3) (2006) 500–526.

[92] M.R. Coughlan, Wildland arson as clandestine resource management: a space—time permutation analysis and classification of informal fire management regimes in Georgia, USA, Environ. Manag. 57 (5) (2016) 1077—1087.

[93] S.L. Stephens, L.W. Ruth, Federal forest-fire policy in the United States, Ecol. Appl. 15 (2) (2005) 532—542.

[94] A. Leopold, Paiute forestry vs. forest fire prevention, Southwestern Magazine 2 (1920) 12—13.

[95] T.R. Vale, The Pre-European landscape of the United States: pristine or humanized? in: T.R. Vale (Ed.), Fire, Native Peoples, and the Natural Landscape Island Press, Washington D.C., 2002, pp. 1—39.

[96] H.H. Biswell, The big trees and fire, Natl. Parks Conserv. Mag. 35 (1961) 11—14.

[97] J.R. Sweeney, H.H. Biswell, Quantitative studies of the removal of litter and duff by fire under controlled conditions, Ecology 42 (3) (1961) 572—575, https://doi.org/10.2307/1932244.

[98] R.W. Mutch, Fighting fire with prescribed fire: a return to ecosystem health, J. For. 92 (11) (1994) 31—33.

[99] C. Sun, State statutory reforms and retention of prescribed fire liability laws on US forest land, For. Policy Econ. 9 (4) (2006) 392—402.

[100] C. Sun, B. Tolver, Assessing administrative laws for forestry prescribed burning in the southern United States: a management-based regulation approach, Int. For. Rev. 14 (3) (2012) 337—348.

[101] L. Dale, Wildfire policy and fire use on public lands in the United States, Soc. Nat. Resour. 19 (3) (2006) 275—284, https://doi.org/10.1080/08941920500460898.

[102] J. Cohen, The wildland-urban interface fire problem: a consequence of the fire exclusion paradigm, Forest History Today. 20—26 (2008) 20—26.

[103] J.B. Kauffman, Death rides the forest: perceptions of fire, land use, and ecological restoration of western forests, Conserv. Biol. 18 (4) (2004) 878—882.

Safety enhancement in extreme wildfire events

Fantina Tedim [1,2], Vittorio Leone [3], Sarah McCaffrey [4], Tara K. McGee [5], Michael Coughlan [6], Fernando J.M. Correia [1], Catarina G. Magalhães [1]

[1]Faculty of Arts and Humanities, University of Porto, Porto, Portugal; [2]Charles Darwin University, Darwin, NWT, Australia; [3]Faculty of Agriculture, University of Basilicata (retired), Potenza, Italy; [4]Rocky Mountain Research Station, USDA Forest Service, Fort Collins, CO, United States; [5]Department of Earth and Atmospheric Sciences, University of Alberta, Edmonton, AB, Canada; [6]Institute for a Sustainable Environment, University of Oregon, Eugene, OR, United States

5.1 Wildfire disasters: Trends and patterns

Wildfires can cause enormous, often long-term, impacts on communities, killing or seriously injuring people, adversely affecting mental health of survivors (e.g., posttraumatic stress, depression, and anxiety), and destroying homes, livelihoods, industries, agricultural, and forest assets. In the two first decades of the 21st century, wildfires have become increasingly deadly and destructive. Events that spread under catastrophic weather conditions, exceeding all previous records, have exacted the highest death toll. Although media stories suggest that wildfire fatalities are on the rise, a definitive assessment of this is difficult due to inconsistencies, as well as failure to collect, in how fatality data are developed.

In Europe, an initial version of a wildfire fatalities database was developed by Molina-Terrén et al. [1]. Gathering data from Portugal, Spain, Greece, and the region of Sardinia, in Italy, the database combines official information and other sources. For Portugal, the official wildfire database does not include fatalities. Existing fatality data since 1986 were collected by researchers [2−4] through fieldwork at the accident sites, interviews, media, and official or unofficial reports [1].

In the US, there is no national database that tracks civilian wildfire fatalities. The United States Fire Administration does track fire deaths, but without always separating wildfire from structural fires. In addition, the records are based on information provided by local authorities. Firefighter fatalities are tracked a bit more closely through the National Interagency Fire Center.

In Australia, the Life Loss database was developed by collating different available data on wildfire fatalities over the past 110 years (1901−2011); although civilian deaths were the key focus, efforts were also made to partially analyze firefighter fatalities [5].

In Canada, the number of wildfire fatalities can be found in Canada's Disaster Database (https://www.publicsafety.gc.ca/cnt/rsrcs/cndn-dsstr-dtbs/index-en.aspx), which

Extreme Wildfire Events and Disasters. https://doi.org/10.1016/B978-0-12-815721-3.00005-9

includes natural, technological, and conflict events since 1900. Here a constraint on understanding wildfire deaths is that only events conforming to at least one of the following five criteria are reported (10 or more people killed, 100 or more people affected/injured/evacuated or homeless, an appeal for national/international assistance, historical significance, and significant damage/interruption of normal processes such that the community affected cannot recover on its own). Considering that usually the number of fatalities per fire is low and does not fit the established criteria, it is likely underestimated.

A similar issue occurs with fatality information from the International Disaster Database (EM-DAT), which was launched in 1988 by the Center for Research on the Epidemiology of Disasters [6]. It contains essential core data on the occurrence and impact of natural and technological disasters across the world, from 1900 to the present day, conforming to at least one of the following criteria: 10 or more people dead; 100 or more people affected; the declaration of a state of emergency; and a call for international assistance. A weakness of the EM-DAT database when recording wildfire fatalities is that the figures are not necessarily related with a single event but with a fire season. This is evident in the case of Portugal; in 2003 the database included a wildfire event with 14 fatalities, but in fact there were 21 fatalities in 18 different fires. The same happened with the 2017 fires; EM-DAT reports two disasters with 109 fatalities, while on the official reports requested by Portuguese Government and Parliament [7−10], 112 fatalities in nine different wildfire events are mentioned.

Even though EM-DAT is not an accurate source of the number of wildfire fatalities, it still provides useful insight at the global level because the most relevant wildfire disasters are included. From 1911 to 1964 just 10 wildfire disasters with 1508 fatalities were recorded (Fig. 5.1). After 1964, 422 events and at least one per year were reported. In this period, 2754 victims and 8215 injured people were registered. This rise can be explained by a better data recording in the database, but also it is likely a result of greater human exposure due to increased population, as no significant trend in the number of fatalities at global [6,11], regional [1], or country [5,12] level was identified. In fact, there is a high interannual variability in the number of fatalities, which confirms the complexity of ecological and human circumstances that surround fatalities. The countries with the highest number of victims (including civilians and operational staff) from 1900 to 2018 are US (1435), Australia (501), Greece (232), and Portugal (224).

The direct fatalities from wildfires are consistently lower in the EM-DAT database than the figures from the Global Fire Monitoring Center (GFMC) annual reports (http://gfmc.online/media/bulletin_news.html). For example, for 2010, 2014, and 2017 the GFMC numbers for global wildfire fatalities were 205, 217, and 93, respectively, while in EM-DAT for the same years the values were 166, 16, and 39. These differences likely reflect the fact that fatalities for an individual wildfire event often occur at numbers below the thresholds established by EM-DAT, while the GFMC reports all the fatalities. According to the GFMC, an annual average of 192 direct victims caused by wildfires (including civilians, firefighters, aerial crews, and military) is reported between 2009 and 2017 at worldwide level, with a total of 1729 direct fatalities [13]. However, the consequences of wildfires on mortality can be larger if we consider

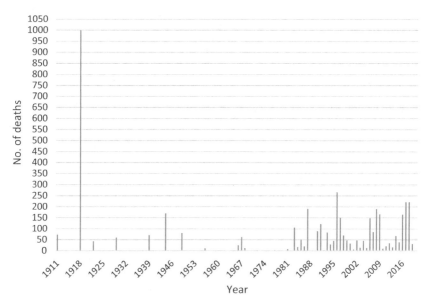

Figure 5.1 Between 1900 and 2018 at world level 432 wildfire disasters were recorded, with 4262 fatalities and 8115 injured people. Most of the fatalities occurred after the 1980s. *Source*: Data from EMDAT (2019).

the impact of smoke. For example, in Russia in 2010, about 56,000 more people died in July−August [14], due to the combination of smoke and heat wave.

According to the GFMC, between 2009 and 2017, civilians represent the highest percentage of fatalities (74%), but there are different patterns at the country level. For instance, in Australia and Greece civilians have been the most affected group [5,15,16], whereas firefighters, aerial crews, and military were the most affected group in Spain and Portugal [1,17].

There is a long list of tragic incidents involving the deaths of operational staff and mainly firefighters all over the world. Some of the most recent ones are presented in Table 5.1, which is not exhaustive. Many of the firefighters deaths are related with entrapment and burnover that can affect even well-trained and equipped operational staff, as it happened in 2013, in the Yarnell Hill Fire, in Arizona (US), where 19 firefighters were overrun by a sudden change of fire behavior, and killed, although they had deployed individual fire shelters [21]. Concerning wildfires that mostly affect civilians (Table 5.1), recent years have seen a growing number of events with significant number of fatalities: for instance the Kilmore East and Murrindindi fires, in 2009, occurred in the worst fire danger conditions ever recorded in the state of Victoria, Australia, with 160 fatalities, fireline intensities up to 88,000 kWm^{-1} and massive spotting [29,30]; the 2017 Pedrógão Grande fire in Portugal reached an intensity up to 60,000 kWm^{-1}, with 66 deaths [7−10]; the 2018 Neos Voutzas and Mati fire in Greece where 100 people perished [28]; and the 2018 Camp Fire in California, which caused 86 fatalities with the destruction of 18,804 structures, within the 62,053 ha burned. It is the deadliest event in California fire history [31].

Table 5.1 Wildfire single events with >20 fatalities of civilians and >10 operational staff since 1965.

| Date | Location, country | Fatalities | | References |
		Civilians	Operational staff (Firefighters, soldiers, aircrew)	
1966	Sintra, Portugal		25	[18]
	The Loop fire Disaster Forest fire, Los Angeles, California, US		11	[19]
1985	Armamar, Portugal		14	[18]
1994	South Canyon, Colorado, US		14	[20]
2013	Yarnell Hill, Arizona, US		19	[21]
2019	China		24	[22]
1983	Cockatoo, Australia	21		[23]
1991	Tunnel, Oakland Hills, US	25		[24]
2007	Artemida, Greece	21	3	[3]
2009	Kilmore East and Murrundindi, Victoria, Australia	160		[23]
2010	Mount Carmel, Israel	41	3	[25,26]
2017	Tubbs, California, US	22		[24]
2017	Pedrógão Grande, Portugal	65	1	[7]
2018	Camp fire, California, US	86		[27]
2018	Neos Voutzas and Mati, Greece	100		[28]

Although we cannot conclude that the number of wildfire disasters and fatalities has an increasing trend, some evidence can be underlined. First, the deadliest events across the world involved civilians and occurred in this century, arguably reflecting more people living in fire-prone landscapes, as well as the increasing occurrence of extreme wildfire events (EWEs) [32]. Second, fatalities occurred in a few major events and were associated with very severe weather conditions. Third, in the most recent tragic events, the highest number of victims are civilians. In Portugal the highest death toll ever recorded occurred in Pedrógão Grande Fire (2017), but just one out of 66 victims was a firefighter. In Black Saturday fires (Australia, 2009) 173 people lost their lives, but just one of the victims was a firefighter. The 2018 Camp Fire in California killed 86 civilians with no fatalities among the firefighters [24]. Finally, a larger number of fatalities occur when wildfires burn into more urban settings, resulting in house-to-house ignition, and/or have such a rapid onset that only last-minute evacuation is possible, exposing people to fire along evacuation routes. The 2017 Tubbs Fire and the 2018 Camp Fire in Northern California are examples of both dynamics at play.

Particularly for civilian fatalities, there is a need for more consistent attention, information gathering, and an accurate understanding of the critical underlying conditions and circumstances to reduce the number of deaths.

5.2 Causes and circumstances leading to fatalities

5.2.1 Causes of death

The main causes of death of firefighters and other operational staff are burns, health problems including heart attacks, physical trauma, respiratory problems, exhaustion, aerial accidents, and terrestrial accidents. In the US the number of firefighters' fatalities slightly increased and the causes of death have greatly shifted [33,34]. Between 2007 and 2016, 170 firefighters died during wildfire operations. Four major causes of death (heart attacks, 24%; vehicle accidents, 20%; aircraft accidents, 18%; and entrapments, 17%) were responsible for 79% of the total number of fatalities [34]. The leading causes of fatal aircraft crashes were engine, structure, or component failure (24%); pilot loss of control (24%); failure to maintain clearance from terrain, water, or objects (20%); and hazardous weather (15%) [35]. The increased use of aircraft operating in high-risk conditions and low-altitude flight (e.g., turbulence of convective plume, hot air, embers, ashes, and obstacles such as high voltage cables, cableways, cable cars and antennas) is accompanied by a growing number of aircraft crashes. Aircraft-crew fatalities are significant in Europe, particularly in Spain, where 72 out of 96 reported fatalities occurred. This can be explained to some extent by the heavy use of aerial-firefighting resources in Spain [1].

During a fire, survival may depend on every possible source of cover or protection from heat, for example, depressions in the ground, large rocks, or logs. If a root cellar or cave is used, it is important to vacate into the open as soon as possible, because of potential problems associated with accumulation of smoke and carbon monoxide [36]. Suffocation is a potential threat, although scarcely reported in long-time statistics. For instance, in the US wildland fire fatalities by year [37] for 1910–2017, suffocation and asphyxiation represent 0.2% and 0.3%, respectively.

In Europe, the activity being performed by victims at the time of the fatal incident most frequently was firefighting (42%), followed by evacuation (13%), agricultural activities (10%), being aboard an aircraft (10%), and protecting property (8%) [1]. In the deadliest events, civilians were mostly killed by the exposure to heat and smoke during unplanned evacuations, as it occurred in Artemida 2007 (Greece), Pedrógão Grande 2017 (Portugal), and Neos Voutzas and Mati 2018 (Greece).

In Australia most of the fatalities occurred during late evacuations [15,38]; however, the 2009 Black Saturday fires in Victoria (Australia) represent a shift of trend, as the majority of the victims were sheltering inside a building at the time of their death [39].

5.2.2 Mechanisms conducting to fatalities

Heat and smoke are the main killers and can cause: (1) acute health problems (i.e., dehydration, heat stress, heat exhaustion, heat stroke, heart attack, stroke); (2) inhalation of toxic gases that impair biological functions; (3) inhalation of hot gases, resulting in tissue swelling to the point of obstructing air exchange to the lungs; and (4) thermal injury to skin either through convective or radiative heating [40]. Magnitude of heating and time of exposure determine the amount of burn injury whose threshold of survivability can be approximately retained to be $7-10 \text{ kW m}^{-2}$, but injuries occur with heating levels greater than $6-7 \text{ kW m}^{-2}$ and durations longer than 60 s [40–43].

Wildfire smoke provokes detrimental respiratory and cardiovascular health effects, exhibits lung toxicity, and is associated with mortality and morbidity. Exposure to smoke can provoke short-term and long-term effects that can impact not just local communities but even others located several hundreds of kilometers away from the fire, as convective plumes can be injected into the troposphere and travel over long distances [44–49]. An average of 339,000 deaths per year, attributed to wildfire smoke exposure, is estimated at the global level. This order of magnitude is higher than the direct fatalities that occur at the wildfire location. Wildfire smoke can significantly decrease the ability to sense changes in fire behavior [20]; it reduces visibility, affecting the sense of direction and can cause road accidents. For example, road accidents originated by thick smoke from uncontrolled native *fynbos* (fine bush) burning were registered in South Africa, with up to nine fatalities per year [50]. Wildfire smoke is strongly influenced by fuel characteristics and the physical and chemical processes during combustion (i.e., flaming or smoldering conditions) and is characterized by a varying toxicity of emissions from different types of vegetation [44–49,51].

5.2.3 Circumstances of fatalities

5.2.3.1 Entrapment and burnover

The exposure to heat and smoke can be a result of different circumstances, such as entrapment and burnover, following sudden changes of fire behavior or manifestations of extreme fire behavior. The likelihood of entrapment and burnover increases with extreme fire weather conditions and strong winds [52].

Many deaths and injuries result from firefighters being surprised and overtaken by a fast-moving fire, because of underestimation of fire rate of spread, and/or overestimation of distance, and/or assessments about likely fire danger [53]. Factors other than simple visual perception need to be taken into account and include social factors such as (1) the perceived competence of the leader; (2) group cohesion; and (3) group norms about risk and danger [53]. Viegas et al. [3] analyzing the deadliest events occurred in Europe concluded that they were related to a sudden change of wildfire behavior. It modifies moment by moment, but a sudden change of rate of spread, direction of fire, and massive spotting activity is usually influenced by wind. In many situations fatalities occurred after a rapid wind change. On Ash Wednesday Fire (Australia, 1983), 46 out of 47 deaths, were caused by injuries suffered immediately after the wind changed [54]. The same happened in 2017 in Portugal, in Pedrógão

Grande Fire, where 34 out of 64 people died on a stretch of a national road and the others in 19 different places in a 10 km^2 area. Most of the victims (59 out of 64) died in 15 min when they were fleeing during the fire storm caused by a downburst. In that time the rate of spread of fire was about 15 kmh^{-1} and the fire intensity about 60,000 kWm^{-1} [7].

In addition to unexpected fire behavior, or under appreciation or misjudgment of fire behavior potential, entrapments and burnover can also result from physical inca-pability or debility generated before or during the fire; obstacles (e.g., fences, fallen tree, barbed wire), as it, respectively, happened in Greece (2018, Neos Voutzas and Mati fire), in Israel (2010, Mount Carmel fire), and in Italy (1983, Sardinia, Curraggia fire); lack of escape routes (1994, Mpumalanga Province, South Africa); vehicle fail-ures or traffic accidents (2017, Pedrógão Grande fire, Portugal); poor leadership, wrong and imprudent orders/instructions to firefighters (1993, Linguaglossa, Sicily, Italy); vital safety measures not being applied and lack of training in basic firefighting procedures (1994, Mpumalanga Provice, South Africa); unexpected massive spotting, where the firebrand shower can produce simultaneous ignitions on houses and in vege-tation [9,25,26,28,50,55,56].

5.2.3.2 Last-minute evacuation

Last-minute evacuation is a common behavior in many parts of the world during a fire and contributes to the majority of deaths [7−10,57,58]. Last-minute evacuation can be dangerous and difficult with heat, flames, smoke, darkness, ember storms, strong winds, fallen trees, traffic, and the urgency of the situation increasing the likelihood of accidents [59,60].

Therefore, when possible, early warning is a crucial point for citizens' safety as it can provide the time needed to prepare, ensuring all family members are informed, identify the best course of action for an individual situation, and evacuate well ahead of the fire. In Fort McMurray and Slave Lake fires 2016, Alberta, Canada, residents received the official evacuation order after the fires had already entered the towns and started burning houses, which is why they ended up evacuating late [61].

5.2.3.3 Buildings loss

Houses loss statistics are dominated by EWEs, usually occurring in severe weather days with high atmospheric instability. Homes can ignite as a result of high intensity flames as well of firebrands igniting lower-intensity surface fires adjacent to and/or spreading to contact the home, or accumulating on the home, and/or an adjacent burning structure. In other cases, firebrands ignite spot fires in vegetation that could subsequently spread to homes. Embers, radiant heat, and flame contact are the most common source of buildings' ignition [62−64]. Spotting can broadcast firebrands at a distance of kilometers ahead of the flaming front, dropping them on people and build-ings [65,66]. In the fires outbreak in October 15th, 2017, in Portugal, spotting started new fires at a distance of 6 to 21 km from the fire front (see Chapter 2); a value of 33 km is reported for 2009 Black Saturday fires [29]. Embers can attack a building before a fire front arrives, during or even after its passage [67]. In Guejito Fire, California, in 2007, the arrival of the wildland fire front caused the majority of the damage; the embers that arrived before the main front contributed to less than 5% of the total destroyed structures [68]. House-to-house fire spread is not significant in Europe but can occur in other countries such as Australia and the US. In Grass Valley Fire, 2007, California, 97% of homes ignited from fire spreading through surface fuels that contacted homes, from firebrands, and/or from thermal exposures directly related to burning residences [55].

Houses loss are correlated with (1) the design (e.g., number of re-entrant corners or roof valleys where embers may accumulate and ignite the building), and construction (e.g., type of wall material) of the buildings and surrounding elements [62,63,67,69−72]; (2) the density/composition of adjacent vegetation. The presence of defensible space and the distance to forest [73,74] can affect the survival of the houses. During the Black Saturday fires in 2009, about 25% of destroyed buildings in Kinglake and Marysville were located within the bushland boundary, and 60% and 90% within 10 and 100 m of bushland, respectively [75]. Blanchi and Leonard [67] show that also houses far enough from the forest can be directly impacted by flames from the fire front, in particular conditions of wind and large amount of burning debris; (3) the fire intensity and the ember exposure, which have a significant role in the survivability and destruction of structures [11,30,76,77]. In addition, exposure was found to play a significant role in structure survivability with respect to the effectiveness of defensive actions. Defensive actions were over two times more effective in saving structures in low exposure areas compared with high-exposure area [77]; (4) the strong winds that influence fire behavior can have a significant role in the vulnerability of the houses. Winds can cause either superficial or major damage to buildings before, during, or after the passage of fire front (e.g., damage to the building façades, breakage of windows, and lifting of roof materials such as tiles or sheeting). Wind can facilitate embers attacks, starting spotting even at a distance from fire front [65,66] or igniting buildings, when they enter through openings such as ventilators, eaves, and windows [67,74,78,79]. Wind can also force flames through small gaps in the building envelope and increase the intensity of embers and combustible debris that enters the building [80]. About 13% of houses surveyed after Black Saturday fires were affected by wind [81]; (5) the fire self-defense measures and fighting techniques. In the Guejito fire, 70% of the destroyed homes were not defended while 60% of defended structures on fire perimeter were saved. Over 50% of the structures were ignited within 3 hours after the main front of the fire hit the community [68].

5.2.3.4 People vulnerability, attitudes, and behaviors

The limited research focused on the social impacts of wildfires means that there is little empirical evidence about how behaviors contribute to poor outcomes. What is known emerges primarily from Australian research which has shown that human vulnerability, i.e., the intrinsic characteristics, attitudes, and behaviors of people or a community that make them susceptible to suffer damaging effects of a wildfire, can result from a range of circumstances. For example, the Black Saturday fires occurred on a very hot

Saturday, when people were exhausted by the heat, inside their air-conditioned houses. As a consequence, many people simply were unaware of the approaching fire [82,83]. Vulnerability is associated with physical and mental conditions related to age, disability and/or handicap, lack of mobility, lack of stamina, lack of knowledge or decision-making capacity under intense stress [83], and lack of preparedness. People inadequately prepared are more likely to be pressed into dangerously late evacuations, untenable defense strategies, and passive sheltering that contribute for high levels of fatalities and injuries [82].

The Black Saturday Fires fatalities resulted from a lack of risk awareness and a low level of planning and preparedness [83]. Awareness of living in an area at risk does not necessarily translate into understanding the characteristics of the hazard, the threats it can represent to individuals, households and communities, necessary precautions to be undertaken and how each action can increase safety and promote the protection of people and buildings, even in the case of EWEs.

The most common measures people undertake to protect their houses are often the easiest and lowest-cost actions (e.g., clearing leaves, buy buckets and mops) [82] which are not effective enough when dealing with EWEs. In Australia, O'Neill and Handmer [84] found plans within the "Stay or Go" policy that contain fatal flaws, for instance, the suggestion to shelter in interior rooms such as bathrooms, where people cannot observe what is happening outside or in the rest of the house and have no way of escaping if the house catches fire.

5.3   The safety protocols

5.3.1   For operational staff

Firefighting is a hazardous job that requires specialized equipment and clothing to stay protected from dangerous environments. The jacket and trousers worn by a firefighter have to be constructed of materials that will withstand intense heat. It is very dangerous to wear personal garments, mainly if made of cotton. A firefighter's helmet made from materials that favor protection from hard impacts protects the wearer from head injury.

Since the 1960s in the US, firefighters have an emergency safety equipment called *fire shelter,* a portable, folding one-person tent, made of heat reflecting aluminum foil, silica cloth, and fiberglass. Its use is rather debated, as it could allow firefighters to take biggest chances than they should. As a last resort, when inside a safety zone, a firefighter should dig into the ground, deploy his shelter, hunker down, hoping to be safe until fire has passed [85].

Data suggest that wildland firefighters die at a higher rate than those involved in structural fire response [86]. Thus to reduce firefighters' fatalities, a consensus emerged in the US wildland fire community to develop safety concepts and procedures. First to be proposed was the *10 Standard Fire Fighting Orders* in 1957, standard orders, organized in a sequential way, to be applied to all fire situations as firm rules of engagement [56]. Shortly after the *10 Standard Fire Fighting Orders* were incorporated into firefighters' training, the *18 Watch Out* situations were developed [56], producing the *10 & 18*, universal principles that guide fire suppression practices. The *18 Watch Out* call attention to various conditions that, if not mitigated, can have potentially serious or fatal consequences [33]. They were complemented by the LCES protocol (the acronym stays for Lookouts, Communications, Escape Routes, and Safety Zones; Gleason [87]) a series of recommendations to adhere to a procedure where safety from the hazard is assured. The LACES safety protocol is a slight modification to LCES, internationally recognized and adopted in a number of countries to improve and manage safety at wildfire incidents. LACES is an acronym for Lookouts, Awareness (or Anchor point, as proposed by Alexander & Thorburn [88]), Communications, Escape routes, and Safety zones. The statements in LACES are rules that must be adhered [33]. Its correct implementation helps to ensure that suppression personnel are appropriately supervised, informed and warned of risks and potential hazards. LACES can be adjusted to different circumstances, scale, and complexity of wildfire events. It is clear, concise, and easily understood and can also be easily applied through the *Incident Command System* [89]. The last two points of LACES highlight the importance of *Safety zones* and *Escape routes*.

Safety zones provide freedom from danger, risk, or injury [34,37,90−92]; they are a preplanned area of sufficient size and suitable location, expected to protect fire personnel from an approaching wildfire and avoid entrapment and burnover, without using fire shelter. *Safety zones* are separated from fuels by a *safe separation distance* (SSD) derived from flame height [93]. Given the difficulty and imprecision to quantify this value, Page and Butler [41] defined a *safety zone separation distance*, the distance from flames that would result in a low probability (1 or 5%) of any injury to entrapped firefighters. It can be assumed to be 8*Vegetation height *Δ w,s. The multiplier Δ w,s (wind and slope factor to take into account the convective heat from wind or slope) is between 1 and 10 [43,90,91,94].

A *safety zone* must be large enough to hold firefighting personnel and equipment, providing the aforementioned SSD. Safety zones can be already burned areas, natural (rock areas, water, meadows) or constructed areas (clear-cuts, roads, helicopter landing zones) [91], large enough to accommodate both personnel and equipment (e.g., engines). About 5 m^2 for each crew member and 28 m^2 for each engine are necessary.

Escape routes are the preplanned and understood routes or paths that firefighters must travel to reach a safety zone or other low-risk area in the event of a change in fire behavior [40]. To run away to avoid being overtaken by an advancing flame front and continue running can result a deadly option [95,96]. Fires burning upslope spread at a rate greater than what a firefighter is typically capable of traveling at [97], and certainly greater than what a firefighter can sustain for any significant period of time.

During the day, the average rate of travel for firefighters over rough but flat terrain would average about 80 m/min, with faster rates as high as 128 m/min possible given stable footing [97]. As the slope steepens, a firefighter's rate of travel decreases proportionally: for a gentle slope (i.e., 10%−20%), it is approximately 55 m/min, for slopes of 20%−40%, it is approximately 37 m/min. For slopes greater than 40%, travel rates would diminish to less than 18 m/min [97]. At night, rates are affected by reduced heat stress and poor vision. The aforementioned values are sustainable for short time and before fire overruns people trying to escape, and they are caught by smoke and hot air. Fires on steep terrain are thus capable of making exceedingly fast upslope runs and surpass the distances firefighters can travel [97].

The *margin of safety* [92] is the difference between time taken for the fire to reach the safety zone (T1) and the total time taken for firefighters to reach the same area (T2). A positive safety margin implies that the firefighter can reach the safety zone before being overtaken by the fire [98]. Although the *margin of safety* concept is useful in fire suppression activities, no practical guides or tools assist firefighters with such assessments [98].

Another concept of safety is the *Dead Man Zone* [99], the area around a wildfire that is likely to burn within 5 minutes given the current wind conditions or an anticipated change in wind direction. The distance from the fire front can range from under 100 m to well over 1 km. Firefighters should stay out of the *Dead man zone*. Crews operating in this area need to be able to see or be warned immediately about any change in fire behavior, have a refuge available close by, and go to it straight away if safety is threatened [100].

In many incidents the fire behavior could have been predicted, given the environmental conditions, but the current fire danger, the fuel conditions, and predicted weather were not taken fully into account in the majority of burnover [42]. Scholz [56] reports that in seven fatality events, 84 separate hazardous conditions or events were identified in the fatality reports, and in each of the events a single overlooked *Watch Out* appeared to be the major contributing factor. Knowledge and awareness are thus strategic elements for a reduction of incidents and must enter as components of a program to successfully reduce fatalities and injuries that include: (1) Individual responsibility: The amount of fatalities by heart attack and the number of fatalities during tests on the ground in US [34] largely denounce the necessity that each firefighter follows a physical conditioning program and pass a fitness test according standard rule, staying fit, maintaining bodyweight, and having regular checkups; (2) Training and qualifications: Firefighters need to understand the hazards resulting from environmental conditions, from weather and its influence on fire behavior, from fuels and their changing flammability. There is a need to produce and make compulsory courses for the different positions in firefighting. A good training is to participate to prescribed burning sessions [101], where the influence of the different pyrometrics can be easily understood and applied at a real scale. To train leaders, the use of simulation tools such as sand tables and computers should be encouraged [102]; (3) Strategy and tactics: Safety guidelines as the *10 & 18* and *LACES*, can help to understand conditions of risk and avoid it. In 1976–1999 a total of 875 entrapments has been recorded in US [34], with an average of 36.4 per year, and a total of 2330 near misses. Although equipment and technology have added better tools for more effective firefighting, too many victims of fire entrapments are still registered [34]; (4) Personal protective clothing: Mainly in some Mediterranean countries, where any intervention of suppression is done in the same time by different and heterogeneous categories of firefighters, some of them being occasional and noncareer people, personal protective clothing and equipment (considered rather burdensome) are frequently absent or partially used. Also, individual shelters are absent from the European scenario of extinction.

5.3.2   Wildfire safety policies for civilians

5.3.2.1   In Europe

In Portugal, the most fire-prone country in Europe, the central authorities completely ignore the social context where fires occur and their causes. Wildfire risk communication, through passive means, is orientated in two directions: (1) messages directed to fuel management. The approach is clearly a top-down *"one-size-fits-all"* type, based on coercive and punitive messages; (2) generalist mobilization campaigns, using passive means, that pretend to change people behaviors without clear messages with solutions.

No attention is given to individual preparedness and safety, which is of paramount importance as preparedness is very low [103]. Preparedness is not a major issue also in other European countries.

Wildfire policies mandate that when direct attack to flames becomes impossible, priority is to defend people and assets. Evacuation is the last option to avoid injuries and fatalities, when there is no possibility to protect all people. Evacuation can be the individual option of people recognizing they do not have physical and/or mental capabilities/conditions to face a fire and evacuate using their own means; in other cases, people accept the recommendation of evacuation and comply with it, with the help of the authorities that direct them to temporary safe locations. In some cases, evacuation is decided by authorities that oblige people to comply with it. However, there is a strong recalcitrance of people and usually only sick/handicapped people, children, and old people evacuate. This behavior is due to the effort of people to defend their house, which many times is not secured and is the result of a life of work, and contains the memories that people want to preserve.

5.3.2.2   In North America and Australia

In US, the wildfire policy is orientated to risk management (insurance), property protection (defensible space and fire-resistant materials in the houses), personal protection (evacuation), and early warnings to support the people actions. Timely evacuation, which range from voluntary to mandatory according to the states, is considered the safest option [104,105] for residents threatened by a wildfire. However, alternatives to evacuation have been proposed [105–107].

In Canada evacuation is the public safety proposal to face serious wildfire threats [108].

In Australia, Krusel and Petris [109] identified poor decision-making and/or a lack of understanding as two major causes of death on Ash Wednesday. In line with the principles guiding Victoria's emergency services, it was agreed that the training and education of the public would reduce future incidence of fire-related losses [110]. The official position to community wildfire safety was established in 2005 and the approach encouraged people to prepare, stay, and defend their properties and avoid last-minute evacuation—*prepare, stay, and defend or leave early policy* [82,111,112]. After Black Saturday fires this policy drew heavy criticism and the Australasian Fire and Emergency Services Authorities Council stated that early self-evacuation is the safest option for residents in the case of a fire. Currently, no Australian organization recommends to shelter in place, although recognizing that evacuation is not always possible and suggesting residents to have a contingency plan for taking last-resort survival shelter [113].

5.4 BESAFE: A safety framework for citizens

Coexistence with wildfire has been proposed as an alternative paradigm to overcome the weaknesses of fire suppression policies [114–117]. This coexistence requires a particular attention to preparedness that reduces the likelihood of injury and death to family members, increases opportunities for people to remain living in their home and neighborhood, assists their family and neighbors, and decreases exposure and vulnerability. This is of paramount importance as in many wildfires, residents cope without direct assistance from fire services [82]. Enhancing people's ability to cope, adapt, and recover reduces response and recovery costs, accelerates social and economic recovery in affected areas, and also contributes to lessen the rising costs of hazard events [118]. Preparedness needs to be developed to increase knowledge and capabilities and enhance the likelihood to survive, even in the most powerful EWEs.

Citizens safety behavior requires a collective effort where residents, public powers, organizations not just share responsibility but work together for the common goal of reducing the hazard, decreasing vulnerability, avoiding loss of lives, mitigating impacts, facilitating the recovery, and reducing economic and social costs.

In building safety there are some constraints such as: (1) The uncertainty related with the occurrence and characteristics of wildfires, which require different strategies to be addressed, taking into account the powerful EWEs. Handmer and Dovers [119] identified three types of emergencies (routine, nonroutine, and complex). The routine or nonroutine situations are relatively easy to manage, whereas complex emergencies pose huge challenges to management and survival. This is evident when people are expecting a *normal* fire and need to face an EWE without knowing what to do; (2) it is impossible to maintain the highest level of promptness of citizens for all the time of critical periods because there are so many competing everyday life activities that take priority over preventing or mitigating wildfire risk [75,120,121].

It is particularly challenging to ensure safety conditions mainly in the case of infrequent events, such as EWEs, and people must be aware of the operational constraints and limits of emergency services. As no single or isolated measure (e.g., fuel management, that is currently considered the panacea) provides an absolute solution to increase safety, reduce wildfire risk, and facilitate to cope with and recover from fire, it is necessary to create synergies of actions among different actors. Considering the different and competing activities which people (civilians and organizations) need to deal with in their everyday life, a concise and comprehensive framework is necessary to orientate the different steps civilians and operational staff must follow to enhance preparedness and safety. We labeled this framework BESAFE.

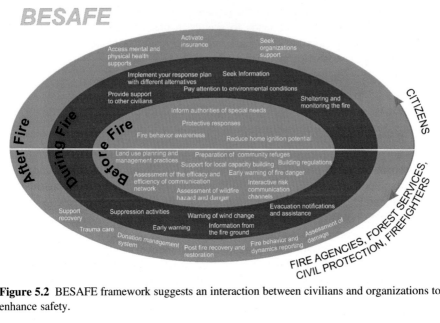

Figure 5.2 BESAFE framework suggests an interaction between civilians and organizations to enhance safety.

BESAFE (Fig. 5.2) has two components: (1) Civilians and (2) operational staff and decision-makers; it is organized in three phases: before, during, and after the fire. For each phase a nonexhaustive list of modules covering scientific/technical information (e.g., building ignition that explains how a building ignites and it is destroyed) and action-based protective activities (e.g., reduce home ignition potential) is proposed for both civilians and organizations. BESAFE is a tool putting together knowledge, making understandable the contribution of each action to enhance safety.

BESAFE suggests that civilians and organizations more than sharing responsibility work together based on an interactive communication, collaboration, respect of roles, acknowledgment of capacities, and participatory design of data and solutions for the local wildfire problems. Concerning the civilians, BESAFE looks for helping people to understand the wildfire problem and look for the most adequate measures to their own situation, instead of providing a check list of activities that people can hardly understand and implement.

5.4.1 Before the fire

Organizations should comply with the following tasks: (1) **Land use planning and management practices** (e.g., regulations about settlements and defensible spaces, fuel reduction, routes of access and egress) required to avoid EWEs and disasters [12,122–124]; (2) **assessment of wildfire hazard and danger** to produce accurate and updated information about the level of fire danger, fire behavior potential in a certain area, defensibility conditions by firefighters that must be communicated to civilians and inform all the planning activities of fire agencies. Many people have no previous fire experience and many others have the experience of small fires and just feel to be prepared to respond to small, *normal* fires [125], but they are not aware of the power of EWEs; (3) **assessment of the efficacy and efficiency of communication network** to ensure that alerts and early warning reach all concerned people. It is crucial to insure the coverage of the mobile phones network or of alternative channels of communication; (4) **early warning of fire danger** to communicate fire danger level, although during summer time there are so many days with the highest level of fire danger that many people believe impractical to evacuate on all days with such level of fire danger (e.g., Catastrophic Fire Danger) [60,82,126]; (5) **interactive risk communication channels** to enhance preparedness and safety measures in case of evacuation, buildings defense, and sheltering practices; (6) **preparation of community refuges and fire shelters** to expand the range of options to residents [82]; (7) **building regulations** to reduce the rate of tenability loss and eliminate weaknesses in the buildings where fires can develop [74]; and (8) **support for local capacity building** as communities have different needs, and capabilities. Organizations should help people to prepare fire response plans.

Civilians should comply with the following tasks: (1) **Fire behavior awareness** to know what to expect the fire will do and look for adequate preparedness measures. Civilians must understand the mechanisms through which fire can attack people and structures and how they can increase tenability and protection of properties and structures as well as to enhance personal safety. In addition, it is important to understand and avoid risk behaviors as well as to pay attention to conditions that lead to high hazard; (2) **reduce home ignition potential** as a key to effectively reducing home destruction, whose conditions and responsibility exclusively lie with homeowners rather than public land managers [127]. It is necessary to understand how houses and other structures are impacted by fires to help residents make more informed decisions about sheltering [74]; (3) **protective responses** effectiveness requires the knowledge about the different type of fires to prioritize the protective measures (i.e., defend property, sheltering, or evacuate) and prepare alternatives in the case of previous failures of the adopted strategies; (4) **inform authorities of special needs** concerning vulnerable people (e.g., people with disabilities and/or handicap).

5.4.2 During the fire

Organizations should comply with the following tasks: (1) **Suppression activities** managing the human resources and the technical means in the best way; (2) **early warning** of the existence of a fire and the potential fire behavior characteristics; (3) **information from the fire ground** about where fire is, its rate and direction of spread, and the likelihood of future behavior although this latter continuously changes; (4) **warning of wind change:** usually this information is known by firefighters but not by citizens. To be aware of this change and of its timing, it is important for people to take protective measures and to decide evacuation; (5) **evacuation notifications and assistance** to help people to become safer.

The civilians should: (1) **Seek information** but pay attention to environmental conditions, for not relying only on official instructions that for any reason could never arrive; (2) **implement a response plan with different alternatives**. Evacuation as early as possible is one of the protective responses if that is the plan, or prepare for shelter if the plan is to not evacuate. A prompt evacuation does not guarantee that fleeing residents will be safe from harm, and it is not simply a matter of convincing residents to leave, or to leave sooner. The means to allow safe egress from the area must be available, the road and transport infrastructures should cope with a large volume of vehicles and/or pedestrians all leaving at the same time [113]. For example, during the 2016 Fort McMurray fire the major egress route (Canada Highway 63) became jammed with traffic due to a mass evacuation, with some vehicles running out of fuel, and those evacuees being immobilized and potentially exposed to hazardous conditions as the fire advanced [74,106,128,129]; (3) **shelter and monitor the fire** when people take property defense and protective measures. The likelihood of survival is higher if active sheltering is adopted, which requires a continuous monitoring of the conditions inside and outside the shelter, even though fire front seems very far away [74]; and (4) **provide support to other civilians** that need help and psychological aid.

5.4.3 After the fire

Organizations must comply with several activities: (1) **Trauma care** to civilians but also to operational staff; (2) **assessment of damage** and disruption: It is crucial to have a protocol to record damage and costs; (3) **fire behavior and dynamics reporting**, to understand the lessons that need to be learned; (4) **postfire recovery and restoration** of ecosystems and forested areas must be based on the ecological knowledge of species, rather than on emotional options that often dictate irrational choices, such as early salvage logging in order to avoid dangerous fuel accumulation, but substantially delaying recovery, removing the elements of recovery, or accentuating the damage [130,131]. Allow natural recovery and recognize the temporal scales involved with ecosystem evolution: Human intervention should not be permitted unless and until it is determined that natural recovery processes are not occurring. Ecologically speaking, fires do not require a rapid human response, thus postfire treatments must consider resistance and resilience to fire. Although destructive in many cases, fire is an inherent part of the disturbance and recovery patterns to which native species have adapted [130]. With respect to the need for management treatments after fires, there is generally no need for urgency, nor is there a universal, ecologically based need to act at all. By acting quickly, we run the risk of creating new problems before we solve the old ones ([130]; p.5). Salvage logging should be prohibited in sensitive areas; (5) **support recovery** activities of households and communities; (6) **donation management system** to maintain transparence and efficacy in subsidies attribution.

Civilians should: (1) **Seek organizations to claim help**; (2) **access mental and physical health supports** as wildfire events can be very stressful and long-lasting; and (3) **activate insurance** to rebuild a home and/or to keep business going one.

5.5 Conclusion

The recent deadly wildfire events in Portugal (2017), Greece and California (2018), mainly involving civilians revealed in a very dramatic way the need to prepare society to cope with EWEs, otherwise the death toll will continue to rise.

Plants and ecosystems can rely on efficacious natural mechanisms to recover from damages caused by fire, whereas people do not have the same capacity, thus building preparedness should be a priority to increase citizens safety, mitigate losses and damage, and reduce the cost of wildfire management. It is of paramount importance because it is impracticable to control all the conditions that contribute to the occurrence of EWEs.

Considering that EWEs create very complex emergencies, which require strong knowledge and capabilities to know what to do to mitigate losses and damage, fire agencies and civilians need to have a tool that systematizes all the main components that it necessary to take in consideration to enhance safety and mitigate damage.

BESAFE framework responds to this need, showing the need of civilians and fire agencies and other organizations to work together to mitigate the consequences of fires including EWEs.

References

[1] D.M. Molina-Terrén, G. Xanthopoulos, M. Diakakis, L. Ribeiro, D. Caballero, G.M. Delogu, D.X. Viegas, C.A. Silva, A. Cardil, Analysis of forest fire fatalities in Southern Europe: Spain, Portugal, Greece and Sardinia (Italy), Int. J. Wildl. Fire. 28 (2019) 85, https://doi.org/10.1071/WF18004.

[2] D.X. Viegas, C. Rossa, D. Caballero, L.P.C. Pita, P. Palheiro, Analysis of accidents in 2005 fires in Portugal and Spain, for, Ecol. Manage. 234 (2006) S141, https://doi.org/10.1016/j.foreco.2006.08.188.

[3] D.X. Viegas, A. Simeoni, G. Xanthopoulos, C. Rossa, L. Ribeiro, L. Pita, D. Stipanicev, Recent Forest Fire Related Accidents in Europe, 2009, https://doi.org/10.2788/50781.

[4] D.X. Viegas, M. Ribeiro, M.A. Almeida, R. Oliveira, M. Teresa, P. Viegas, J.R. Raposo, V. Reva, A.R. Figueiredo, S. Lopes, Os grandes incêndios florestais e os acidentes mortais ocorridos em 2013, 2013. https://www.bombeiros.pt./wp-content/uploads/2013/12/Relatório_IF2013_parte1.pdf.

[5] R. Blanchi, J. Leonard, K. Haynes, K. Opie, M. James, F.D. de Oliveira, Bushfire fatalities and house loss in Australia: exploring the spatial, temporal and localised context, in: Adv. for. Fire Res, Imprensa da Universidade de Coimbra, Coimbra, 2014, pp. 685−695, https://doi.org/10.14195/978-989-26-0884-6_77.

[6] D. Guha-Sapir, P. Hoyois, P. Wallemacq, R. Below, P. Wallemacq, Annual Disaster Statistical Review 2016 the Numbers and Trends, Brussels, 2017, http://www.cred.be/sites/default/files/ADSR_2016.pdf.

[7] CTI, Análise e apuramento dos factos relativos aos incêndios que ocorreram em Pedrógão Grande, in: C.de Pêra, A. Alvaiázere, F.dos Vinhos, A. Góis, P. Pampilhosa da Serra (Eds.), Oleiros e Sertã entre 17 e 24 de junho de 2017, 2017. www.parlamento.pt./Documents/2017/Outubro/RelatórioCTI_VF .pdf.

[8] CTI, Avaliação dos Incêndios ocorridos entre 14 e 16 de outubro de 2017 em Portugal Continental, Relatório Final, Lisboa, 2018. https://www.parlamento.pt./Documents/2018/Marco/RelatorioCTI190318N.pdf.

[9] X. Viegas, M.F. Almeida, M. Ribeiro, E. De Investigação Domingos, J. Raposo, M.T. Viegas, R. Oliveira, D. Alves, C. Pinto, H. Jorge, A. Rodrigues, D. Lucas, S. Lopes, F. Silva, O complexo de incêndios de Pedrógão Grande e concelhos limítrofes, iniciado a 17 de junho de 2017, Coimbra, 2017, https://tretas.org/20171016 RelatorioIncendioPedrogaoGrande?action=AttachFile&do=get&target=20171016_ RelatorioIncendioPedrogao.pdf.

[10] X. Viegas, M.F. Almeida, M. Ribeiro, E. De Investigação Domingos, M.A. Almeida, J. Raposo, M.T. Viegas, R. Oliveira, D. Alves, C. Pinto, A. Rodrigues, C. Ribeiro, S. Lopes, H. Jorge, C.X. Viegas, Análise dos incêndios florestais ocorridos a 15 de outubro de 2017, Coimbra, 2019, https://www.portugal.gov.pt./download-ficheiros/ficheiro.aspx?v=c2da3d7e-dcdb-41cb-b6ae-f72123a1c47d.

[11] S.H. Doerr, C. Santín, Global trends in wildfire and its impacts: perceptions versus realities in a changing world, Philos. Trans. R. Soc. B Biol. Sci. 371 (2016), https://doi.org/10.1098/rstb.2015.0345, 20150345.

[12] R.P. Crompton, K.J. McAneney, K. Chen, R.A. Pielke, K. Haynes, Influence of location, population, and climate on building damage and fatalities due to Australian bushfire: 1925–2009, Weather. Clim. Soc. 2 (2010) 300–310, https://doi.org/10.1175/2010WCAS1063.1.

[13] Global Fire Monitoring Center, Global Wildland Fire Network Bulletin 9th Annual Global Wildland Fire Fatalities and Damages Report-2016, 2017. http://gfmc.online/wp-content/uploads/GFMC-Bulletin-01-2017.pdf.

[14] Global Fire Monitoring Center, Wildfire Fatalities January – December 2010: Overview by Countries, 2011. http://www.fire.uni-freiburg.de/media/bulletin_news.htm. http://www.fire.uni-freiburg.de/media/2010/01/news_20100104_au6.htm.

[15] K. Haynes, J. Handmer, J. McAneney, A. Tibbits, L. Coates, Australian bushfire fatalities 1900–2008: exploring trends in relation to the 'Prepare, stay and defend or leave early' policy, Environ. Sci. Policy. 13 (2010) 185–194, https://doi.org/10.1016/J.ENVSCI.2010.03.002.

[16] M. Diakakis, G. Xanthopoulos, L. Gregos, Analysis of forest fire fatalities in Greece: 1977–2013, Int. J. Wildl. Fire. 25 (2016) 797–809.

[17] A. Cardil, D.M. Molina, Factors causing victims of wildland fires in Spain (1980–2010), Hum. Ecol. Risk Assess. An Int. J. 21 (2015) 67–80, https://doi.org/10.1080/10807039.2013.871995.

[18] F. Ferreira-leite, L. Lourenço, A. Bento-Gonçalves, Large forest fires in mainland Portugal, brief characterization, Méditerranée (2013) 53–65, https://doi.org/10.4000/mediterranee.6863.

[19] NFPA, Top 10 Deadliest Wildland Firefighter Fatality Incidents, 2019. https://www.nfpa.org/News-and-Research/Data-research-and-tools/Emergency-Responders/Firefighter-fatalities-in-the-United-States/Top-10-Deadliest-Wildland-Firefighter-Fatality-Incidents.

[20] B.W. Butler, J.D. Cohen, Firefighter safety zones: a theoretical model based on radiative heating, Int. J. Wildl. Fire. 8 (1998) 73–77.

[21] J. Karels, The Following Information Is Preliminary and Subject to Change, Yarnell, AZ, 2013. https://www.wildfirelessons.net/HigherLogic/System/DownloadDocumentFile.ashx?DocumentFileKey=1745d109-6104-4f45-93f0-6ded0dd98a96&forceDialog=0.

[22] EMDAT, The Emergency Events Database, 2019. https://www.emdat.be/emdat_db/.

[23] S. Harris, W. Anderson, M. Kilinc, L. Fogarty, Establishing a Link between the Power of Fire and Community Loss: The First Step towards Developing a Bushfire Severity Scale Fire and Adaptive Management Report No. 89, 2011. www.dse.vic.gov.au.

[24] CALFIRE, Camp Fire General Information, 2019. http://cdfdata.fire.ca.gov/incidents/incidents_details_info?incident_id=2277.

[25] H. Kutiel, D. Malkinson, Weather conditions during the Mount Carmel 2010 forest fire and other fires in Israel, in: Abstr. Proc. Clim. Chang. for. Fires Mediterr. Basin Manag. Risk Reduct., January 24−26 , NirEtzion Israel, 2012.

[26] T. Shavit, S. Shahrabani, U. Benzion, M. Rosenboim, The effect of a forest fire disaster on emotions and perceptions of risk: a field study after the Carmel fire, J. Environ. Psychol. 36 (2013) 129−135, https://doi.org/10.1016/J.JENVP.2013.07.018.

[27] U. Irfan, California's Wildfires Are Hardly "Natural" — Humans Made Them Worse at Every Step, Vox, 2018. https://www.vox.com/2018/8/7/17661096/california-wildfires-2018-camp-woolsey-climate-change.

[28] G. Xanthopoulos, M. Athanasiou, Observations on wildfire spotting occurrence and characteristics in Greece, Wildfire Mag 28 (2) (2019) 18−21, https://doi.org/10.14195/978-989-26-16-506_65.

[29] M.G. Cruz, A.L. Sullivan, J.S. Gould, N.C. Sims, A.J. Bannister, J.J. Hollis, R.J. Hurley, Anatomy of a catastrophic wildfire: the Black Saturday Kilmore East fire in Victoria, Australia, For. Ecol. Manage. 284 (2012) 269−285, https://doi.org/10.1016/J.FORECO.2012.02.035.

[30] S. Harris, W. Anderson, M. Kilinc, L. Fogarty, The relationship between fire behaviour measures and community loss: an exploratory analysis for developing a bushfire severity scale, Nat. Hazards. 63 (2012) 391−415, https://doi.org/10.1007/s11069-012-0156-y.

[31] CALFIRE, Top 20 Deadliest California Wildfires, 2018. https://calfire.ca.gov/communications/downloads/fact_sheets/Top20_Deadliest.pdf.

[32] F. Tedim, V. Leone, M. Amraoui, C. Bouillon, M. Coughlan, G. Delogu, P. Fernandes, C. Ferreira, S. McCaffrey, T.K. McGee, J. Parente, D. Paton, M. Pereira, L. Ribeiro, D. Viegas, G. Xanthopoulos, F. Tedim, V. Leone, M. Amraoui, C. Bouillon, M.R. Coughlan, G.M. Delogu, P.M. Fernandes, C. Ferreira, S. McCaffrey, T.K. McGee, J. Parente, D. Paton, M.G. Pereira, L.M. Ribeiro, D.X. Viegas, G. Xanthopoulos, Defining extreme wildfire events: difficulties, challenges, and impacts, Fire 1 (2018) 9, https://doi.org/10.3390/fire1010009.

[33] R. Mangan, Wildland Fire Fatalities in the United States: 1990-1998, Missoula, 1999, https://www.fs.fed.us/t-d/pubs/pdfpubs/pdf99512808/pdf99512808pt.01.pdf.

[34] NWCG, NWCG Report on Wildland Firefighter Fatalities in the United States: 2007−2016, USA, 2017, https://www.nwcg.gov/sites/default/files/publications/pms841.pdf.

[35] C.R. Butler, M.B. O'Connor, J.M. Lincoln, Aviation-related wildland firefighter fatalities–United States, 2000−2013, MMWR. Morb. Mortal. Wkly. Rep. 64 (2015) 793−796. http://www.ncbi.nlm.nih.gov/pubmed/26225477.

[36] M.E. Alexander, Ou are about to be entrapped or burned over by a wildfire: what are your survival options? in: D.X. Viegas (Ed.), V Int. Conf. for. Fire Res. Elsevier, 2006 https://doi.org/10.1016/j.foreco.2006.08.020.

[37] NIFC, Wildland Fire Fatalities by Year, n.d. https://www.nifc.gov/safety/safety_documents/Fatalities-by-Year.pdf (accessed May 13, 2019).

[38] R. McLeod, Inquiry into the Operational Response to the January 2003 Bushfires in the ACT, Aust. Cap. Territ, Canberra, ACT, 2003.

[39] J. Whittaker, R. Blanchi, K. Haynes, J. Leonard, K. Opie, Experiences of sheltering during the Black Saturday bushfires: implications for policy and research, Int. J. Disaster Risk Reduct. 23 (2017) 119–127, https://doi.org/10.1016/J.IJDRR.2017.05.002.

[40] B.W. Butler, Wildland firefighter safety zones: a review of past science and summary of future needs, Int. J. Wildl. Fire. 23 (2014) 295, https://doi.org/10.1071/WF13021.

[41] W.G. Page, B.W. Butler, An empirically based approach to defining wildland firefighter safety and survival zone separation distances, Int. J. Wildl. Fire. 26 (2017) 655, https://doi.org/10.1071/WF16213.

[42] S. Munson, Wildland Firefighter Entrapments 1976 to 1999, 0051-2853-MTDC, Missoula, Montana, 2000, https://www.fs.fed.us/t-d/pubs/htmlpubs/htm00512853/.

[43] USFA (Unites States Fire Administration), Wildland Firefighting Safe Separation Distances, 2018. https://www.usfa.fema.gov/current_events/051018.html.

[44] F. Reisen, S.M. Duran, M. Flannigan, C. Elliott, K. Rideout, Wildfire smoke and public health risk, Int. J. Wildl. Fire. 24 (2015) 1029, https://doi.org/10.1071/WF15034.

[45] Y.H. Kim, S.H. Warren, Q.T. Krantz, C. King, R. Jaskot, W.T. Preston, B.J. George, M.D. Hays, M.S. Landis, M. Higuchi, D.M. DeMarini, M.I. Gilmour, Mutagenicity and lung toxicity of smoldering vs. Flaming emissions from various biomass fuels: implications for health effects from wildland fires, Environ. Health Perspect. 126 (2018) 017011, https://doi.org/10.1289/EHP2200.

[46] D.M.J.S. Bowman, F.H. Johnston, Wildfire smoke, fire management, and human health, Ecohealth 2 (2005) 76–80, https://doi.org/10.1007/s10393-004-0149-8.

[47] F.H. Johnston, S.B. Henderson, Y. Chen, J.T. Randerson, M. Marlier, R.S. DeFries, P. Kinney, D.M.J.S. Bowman, M. Brauer, Estimated global mortality attributable to smoke from landscape fires, Environ. Health Perspect. 120 (2012) 695–701, https://doi.org/10.1289/ehp.1104422.

[48] R.W. Picard, A society of models for video and image libraries, IBM Syst. J. 35 (1996) 292–312, https://doi.org/10.1147/sj.353.0292.

[49] T.E. Reinhardt, R.D. Ottmar, Baseline measurements of smoke exposure among wildland firefighters, J. Occup. Environ. Hyg. 1 (2004) 593–606, https://doi.org/10.1080/15459620490490101.

[50] C. De Ronde, Wildland Fire-Related Fatalities in South Africa—a 1994 Case Study and Looking Back at the Year 2001, For. Fire Res. Wildl. Fire Saf. '(Ed DX Viegas) CD-ROM, Elsevier BV Amsterdam, Netherlands, 2002.

[51] C.A. Alves, A. Vicente, C. Monteiro, C. Gonçalves, M. Evtyugina, C. Pio, Emission of trace gases and organic components in smoke particles from a wildfire in a mixed-evergreen forest in Portugal, Sci. Total Environ. 409 (2011) 1466–1475, https://doi.org/10.1016/J.SCITOTENV.2010.12.025.

[52] S. Lahaye, T. Curt, T. Fréjaville, J. Sharples, L. Paradis, C. Hély, What are the drivers of dangerous fires in Mediterranean France? Int. J. Wildl. Fire. 27 (2018) 155, https://doi.org/10.1071/WF17087.

[53] J. McLennan, Underestimating speed and overestimating distance of a wildfire - avoiding entrapment and burn-over, BFCRC Legacy, Fire Aust. Mag. (2009). http://www.bushfirecrc.com/news/news-item/underestimating-speed-and-overestimating-distance-wildfire-avoiding-entrapment-and-bu.

[54] D. Packham, C. Pierrehumbert, Bushfires in Australia: a problem of the weather, OMM Bol. (1990).

[55] J. Cohen, R. Stratton, Home Destruction Examination: Grass Valley Fire, Lake Arrowhead, California, Missoula, Montana, 2008. https://www.fs.fed.us/rm/pubs_other/rmrs_2008_cohen_j001.pdf.

[56] B. Scholz, The 10 standard firefighting orders and 18 watch out situations: we don't bend them, we don't break them...We don't know them, Fire Manag. Today Manag. Wildl. Fire. 70 (2010) 29–31. http://www.fs.fed.us/fire/fmt/index.html.

[57] T.J. Cova, F.A. Drews, L.K. Siebeneck, A. Musters, Protective actions in wildfires: evacuate or shelter-in-place? Nat. Hazards Rev. 10 (2009) 151–162, https://doi.org/10.1061/(ASCE)1527-6988(2009)10:4(151).

[58] S.L. Stephens, M.A. Adams, J. Handmer, F.R. Kearns, B. Leicester, J. Leonard, M.A. Moritz, Urban–wildland fires: how California and other regions of the US can learn from Australia, Environ. Res. Lett. 4 (2009) 014010, https://doi.org/10.1088/1748-9326/4/1/014010.

[59] J. Handmer, A. Tibbits, Is staying at home the safest option during bushfires? Historical evidence for an Australian approach, Environ. Hazards. 6 (2005) 81–91, https://doi.org/10.1016/j.hazards.2005.10.006.

[60] A. TIBBITS, J. WHITTAKER, Stay and defend or leave early: policy problems and experiences during the 2003 Victorian bushfires, Environ. Hazards 7 (2007) 283–290, https://doi.org/10.1016/j.envhaz.2007.08.001.

[61] T.K. McGee, Preparedness and experiences of evacuees from the 2016 Fort McMurray horse river wildfire, Fire 2 (2019), https://doi.org/10.3390/FIRE2010013. Page 13. 2 (2019) 13.

[62] J.D. Cohen, Preventing disaster: home ignitability in the wildland-urban interface, J. For. 98 (2000) 15–21, https://doi.org/10.1093/jof/98.3.15.

[63] R. Blanchi, J. Leonard, Property safety: judging structural safety, in: J. Handmer, K. Haynes (Eds.), Community Bushfire Safety', 2008, pp. 77–85.

[64] J. Leonard, N. McArthur, A history of research into building performance in Australian bushfires, in: B. Lord (Ed.), Proc. Aust. Bushfire Conf., Charles Sturt University, Albury, Albury, Australia, 1999, p. 8. http://www.csu.edu.au/special/bushfire99/papers/leonard/.

[65] E. Koo, P.J. Pagni, D.R. Weise, J.P. Woycheese, Firebrands and spotting ignition in large-scale fires, Int. J. Wildl. Fire. 19 (2010) 818, https://doi.org/10.1071/WF07119.

[66] J. Martin, T. Hillen, J. Martin, T. Hillen, The spotting distribution of wildfires, Appl. Sci. 6 (2016) 177, https://doi.org/10.3390/app6060177.

[67] J. Leonard, R. Blanchi, Investigation of Bushfire Attack Mechanisms Involved in House Loss in the ACT Bushfire 2003, 2005.

[68] A. Maranghides, W.E. Mell, A Case Study of a Community Affected by the Witch and Guejito Fires, Citeseer, 2009.

[69] J. Leonard, P. Bowditch, Findings of studies of houses damaged by bushfire in Australia, in: 3rd Int. Wildl. Fire Conf., 2003, pp. 3–6.

[70] R. Blanchi, J.E. Leonard, R.H. Leicester, Lessons learnt from post-bushfire surveys at the urban interface in Australia, For. Ecol. Manage. 234 (2006) S139, https://doi.org/10.1016/j.foreco.2006.08.184.

[71] W.E. Mell, S.L. Manzello, A. Maranghides, D. Butry, R.G. Rehm, The wildland - urban interface fire problem - current approaches and research needs, Int. J. Wildl. Fire. 19 (2010) 238, https://doi.org/10.1071/WF07131.

[72] N.A. McArthur, P. Lutton, Ignition of exterior building details in bushfires: an experimental study, Fire Mater 15 (1991) 59–64, https://doi.org/10.1002/fam.810150204.

[73] K. Chen, J. McAneney, Quantifying bushfire penetration into urban areas in Australia, Geophys. Res. Lett. 31 (2004), https://doi.org/10.1029/2004GL020244.

[74] R. Blanchi, J. Whittaker, K. Haynes, J. Leonard, K. Opie, Surviving bushfire: the role of shelters and sheltering practices during the Black Saturday bushfires, Environ. Sci. Policy. 81 (2018) 86–94, https://doi.org/10.1016/J.ENVSCI.2017.12.013.

[75] B. Teague, R. McLeod, S. Pascoe, Final report, 2009 Victorian bushfires royal commission, Parliam. Victoria, Melb. Victoria, Aust (2010).

[76] A.A. Wilson, I.S. Ferguson, Predicting the probability of house survival during bushfires, J. Environ. Manag. 23 (3) (1986) 259−270, in: https://publications.csiro.au/rpr/pub?list=BRO&pid=procite:df4f609f-bac6-4a6e-95ed-e4721923e7d0.

[77] A. Maranghides, D. McNamara, W. Mell, J. Trook, B. Toman, A Case Study of a Community Affected by the Witch and Guejito Fires: Report# 2: Evaluating the Effects of Hazard Mitigation Actions on Structure Ignitions, Natl. Inst. Stand. Technol. US Dep. Commer. US For. Serv., Gaithersburg, MD, 2013.

[78] R. Blanchi, C. Lucas, J. Leonard, K. Finkele, Meteorological conditions and wildfire-related houseloss in Australia, Int. J. Wildl. Fire. 19 (2010) 914, https://doi.org/10.1071/WF08175.

[79] R.A. Bradstock, A.M. Gill, Living with fire and biodiversity at the urban edge: in search of a sustainable solution to the human protection problem in southern Australia, J. Mediterr. Ecol. 2 (2001) 179−195, in: https://publications.csiro.au/rpr/pub?list=BRO&pid=procite:a19bd5a1-d0b8-4eaa-958d-5b65add3d555.

[80] R. Blanchi, J. Leonard, K. Haynes, K. Opie, M. James, M. Kilinc, D.F. de Oliveira, R. van den Honert, Life and House Loss Database Description and Analysis, CSIRO, Bushfire CRC report to the Atturney-General's Department, 2012. http://www.bushfirecrc.com/sites/default/files/managed/resource/life_house_loss_report_final_0.pdf.

[81] J. Leonard, R. Blanchi, F. Lipkin, G. Newnham, A. Siggins, K. Opie, D. Culvenor, B. Cechet, N. Corby, C. Thomas, N. Habili, M. Jakab, R. Coghlan, G. Lorenzin, D. Campbell, M. Barwick, G. Australia, Bushfire CRC Building and Land-Use Planning Research after the 7 Th February 2009 Victorian Bushfires Preliminary Findings Kilmore East-Murrindindi-Churchill-Bunyip-Maiden Gully, Victoria, Australia, 2009, http://www.bushfirecrc.com/sites/default/files/managed/resource/chapter-3-building-and-land-web.pdf.

[82] J. Whittaker, K. Haynes, J. Handmer, J. McLennan, Community safety during the 2009 Australian "Black Saturday" bushfires: an analysis of household preparedness and response, Int. J. Wildl. Fire. 22 (2013) 841, https://doi.org/10.1071/WF12010.

[83] J. Handmer, S. O'Neill, D. Killalea, Review of the Fatalities in the 7 February 2009 Bushfires: Final Report, Cent. Risk Community Saf. RMIT Univ. Bushfire CRC Melbourne, Victoria, 2010.

[84] S.J. O'Neill, J. Handmer, Responding to bushfire risk: the need for transformative adaptation, Environ. Res. Lett. 7 (2012) 014018, https://doi.org/10.1088/1748-9326/7/1/014018.

[85] E. Chung, Why emergency fire shelters aren't used in Canada, CBC News Technol. Sci. (2016). https://www.cbc.ca/news/technology/why-emergency-fire-shelters-aren-t-used-in-canada-1.1319366.

[86] T. Harbour, A report on a national meeting of wildland fire leaders meeting − united toreduce line-of-duty deaths and injuries of wildland firefighters − International Association of Wildland Fire, Int. Assoc. Wildl. Fire. (2018). https://www.iawfonline.org/article/a-report-on-a-national-meeting-of-wildland-fire-leaders-meeting-united-toreduce-line-of-duty-deaths-and-injuries-of-wildland-firefighters/.

[87] P. Gleason, Lookouts, communication, escape routes and safety zones "LCES", wildl, Fire Leadersh. Dev (1991). https://www.fireleadership.gov/toolbox/documents/lces_gleason.html.

[88] M.E. Alexander, W.R. Thorburn, LACES: adding an 'A'for anchor point (s) to the LCES wildland firefighter safety system, in: B. Leblon, M.E. Alexander (Eds.), Current

International Perspectives on Wildland Fires, Mankind and the Environment, 2015, pp. 121–144.

[89] NOG (National operational Guidance), The LACES Safety Protocol, 2019. https://www. ukfrs.com/guidance/search/laces-safety-protocol.

[90] B. Gabbert, A fresh look at the tragic dude Fire, Wildfire Today — Wildfire News Opin (2013). https://wildfiretoday.com/2013/06/30/a-fresh-look-at-the-tragic-dude-fire/.

[91] M.J. Campbell, P.E. Dennison, B.W. Butler, Safe separation distance score: a new metric for evaluating wildland firefighter safety zones using lidar, Int. J. Geogr. Inf. Sci. 31 (2017) 1448–1466, https://doi.org/10.1080/13658816.2016.1270453.

[92] M. Beighley, Beyond the safety zone: creating a margin of safety, Fire Manag. Notes. (1995).

[93] P.E. Dennison, G.K. Fryer, T.J. Cova, Identification of firefighter safety zones using lidar, Environ. Model. Softw. 59 (2014) 91–97, https://doi.org/10.1016/ J.ENVSOFT.2014.05.017.

[94] USDA FS, Firefighter Safety, 2018. https://www.firelab.org/project/firefighter-safety.

[95] C. Chandler, P. Cheney, P. Thomas, L. Trabaud, D. Williams, Fire in Forestry. Volume 1. Forest Fire Behavior and Effects. Volume 2. Forest Fire Management and Organization, John Wiley & Sons, Inc., 1983.

[96] D.X. Viegas, L.P. Pita, L. Ribeiro, P. Palheiro, Eruptive fire behaviour in past fatal accidents, in: B. Butler, M. Alexander (Eds.), Eighth Int. Wildl. Fire, 2005, p. 8. Missoula, www.adai.pt./docs/Papers_CEIF/2005_Eruptive Fire Behaviour in Past Fatal Accidents.pdf.

[97] M.E. Alexander, G.J. Baxter, G.R. Dakin, Travel rates of Alberta wildland firefighters using escape routes, in: Eighth Int. Wildl. Fire Saf. Summit., 2005, p. 11.

[98] M.E. Alexander, S.W. Taylor, W.G. Page, Wildland firefighter safety and fire behavior prediction on the fireline, in: Proc. 13th Int. Wildl. Fire Saf. Summit 4th Hum. Dimens. Wildl. Fire Conf, International Association of Wildland Fire, Missoula, Montana, 2015. http://www.wildfirelessons.net/yarnellhill.

[99] P. Cheney, J. Gould, L. McCaw, The dead-man zone—a neglected area of firefighter safety, Aust. For. 64 (2001) 45–50, https://doi.org/10.1080/00049158.2001.10676160.

[100] A. Stark, The Dead Man Zone at the 2009 Belimbla Fire, vol. 31, BUSH FIREbulletin - J. NSW Rural FIRE Serv, 2009, 8,9, https://www.rfs.nsw.gov.au/__data/assets/pdf_file/ 0006/4002/Bush-Fire-Bulletin-2009-Vol-31-No-3.pdf.

[101] G. Bovio, D. Ascoli, Fuoco prescritto: stato dell'arte della normativa italiana, L'Italia For. e Mont. 67 (2012) 347–358.

[102] B. Gabbert, How do we reduce the number of firefighter fatalities? Wildfire Today (2016). https://wildfiretoday.com/2016/01/18/how-do-we-reduce-the-number-of-firefighter-fatal- ities/.

[103] D. Paton, F. Tedim, Enhancing forest fires preparedness in Portugal: integrating com- munity engagement and risk management, Planet@ Risk 1 (2013).

[104] FEMA, How to Prepare for a Wildfire, USA, 2017, https://www.fema.gov/media-library- data/1409003859391-0e8ad1ed42c129f11fbc23d008d1ee85/how_to_prepare_wildfire_ 033014_508.pdf.

[105] T.B. Paveglio, M.S. Carroll, P.J. Jakes, Adoption and perceptions of shelter-in-place in California's rancho Santa Fe fire protection district, Int. J. Wildl. Fire. 19 (2010) 677, https://doi.org/10.1071/WF09034.

[106] S. McCaffrey, A. Rhodes, M. Stidham, Wildfire evacuation and its alternatives: per- spectives from four United States' communities, Int. J. Wildl. Fire. 24 (2015) 170, https:// doi.org/10.1071/WF13050.

[107] T. Paveglio, M.S. Carroll, P.J. Jakes, Alternatives to evacuation—protecting public safety during wildland fire, J. For. 106 (2008) 65—70, https://doi.org/10.1093/jof/106.2.65.

[108] J.L. Beverly, P. Bothwell, Wildfire evacuations in Canada 1980—2007, Nat. Hazards 59 (2011) 571—596, https://doi.org/10.1007/s11069-011-9777-9.

[109] N. Krusel, S.N. Petris, A study of civilian deaths in the 1983 Ash Wednesday bushfires, Victoria, Australia, Ctry. Fire Auth. Melb. (1992).

[110] B. Reynolds, A History of the Prepare, Stay and Defend or Leave Early Policy in Victoria, School of Management College of Business RMIT University, 2017. https://researchbank.rmit.edu.au/view/rmit:162075.

[111] J. Handmer, J. Abrahams, R. Betts, M. Dawson, Towards a consistent approach to disaster loss assessment across Australia, Aust. J. Emerg. Manag. 20 (2005) 10.

[112] Australasian Fire Authorities Council, Position Paper on Bushfires and Community Safety, East Melbourne, 2005, http://www.afac.cm.au.

[113] J. McLennan, B. Ryan, C. Bearman, K. Toh, Should we leave now? Behavioral factors in evacuation under wildfire threat, Fire Technol 55 (2019) 487—516, https://doi.org/10.1007/s10694-018-0753-8.

[114] M.A. Moritz, E. Batllori, R.A. Bradstock, A.M. Gill, J. Handmer, P.F. Hessburg, J. Leonard, S. McCaffrey, D.C. Odion, T. Schoennagel, Learning to coexist with wildfire, Nature 515 (2014) 58.

[115] C.I. Roos, A.C. Scott, C.M. Belcher, W.G. Chaloner, J. Aylen, R.B. Bird, M.R. Coughlan, B.R. Johnson, F.H. Johnston, J. McMorrow, T. Steelman, F. the, M.D. Group, Living on a flammable planet: interdisciplinary, cross-scalar and varied cultural lessons, prospects and challenges: Table 1, Philos. Trans. R. Soc. B Biol. Sci. 371 (2016) 20150469, https://doi.org/10.1098/rstb.2015.0469.

[116] International Union of Forest Research Organizations, Global Fire Challenges in a Warming World Summary Note of a Global Expert Workshop on Fire and Climate Change, IUFRO, Vienna, 2018. https://www.iufro.org/uploads/media/op32.pdf.

[117] R.L. Olson, D.N. Bengston, L.A. DeVaney, T.A.C. Thompson, Wildland Fire Management Futures: Insights from a Foresight Panel, Gen. Tech. Rep., 2015, pp. 1—44. NRS-152. Newt. Square, PA US Dep. Agric. For. Serv. North. Res. Station. 44 P. 152.

[118] D. Paton, Disaster risk reduction: psychological perspectives on preparedness, Aust. J. Psychol. (2018), https://doi.org/10.1111/ajpy.12237.

[119] J. Handmer, S. Dovers, The Handbook of Emergency and Disaster Policies and Institutions, Earthscan, London, 2007.

[120] C. Eriksen, N. Gill, Bushfire and everyday life: examining the awareness-action 'gap' in changing rural landscapes, Geoforum 41 (2010) 814—825, https://doi.org/10.1016/J.GEOFORUM.2010.05.004.

[121] K. Koksal, J. McLennan, D. Every, C. Bearman, Australian wildland-urban interface householders' wildfire safety preparations: 'Everyday life' project priorities and perceptions of wildfire risk, Int. J. Disaster Risk Reduct. 33 (2019) 142—154, https://doi.org/10.1016/J.IJDRR.2018.09.017.

[122] J. McAneney, K. Chen, A. Pitman, 100-years of Australian bushfire property losses: is the risk significant and is it increasing? J. Environ. Manage. 90 (2009) 2819—2822, https://doi.org/10.1016/J.JENVMAN.2009.03.013.

[123] J. Whittaker, D. Mercer, The Victorian bushfires of 2002—03 and the politics of blame: a discourse analysis, Aust. Geogr. 35 (2004) 259—287, https://doi.org/10.1080/0004918042000311313.

[124] R. Vélez, La sylviculture préventive des incendies en Espagne, Rev. For. Française. (1990) 320—331.

[125] J. Whittaker, M. Taylor, Community Preparedness and Responses to the 2017 NSW Bushfires: Research for the New South Wales Rural Fire Service, Melbourne, Aust. Bushfire Nat. Hazards CRC, 2018.

[126] P.F. Johnson, C.E. Johnson, C. Sutherland, Stay or go? Human behavior and decision making in bushfires and other emergencies, Fire Technol 48 (2012) 137−153, https://doi.org/10.1007/s10694-011-0213-1.

[127] D.E. Calkin, J.D. Cohen, M.A. Finney, M.P. Thompson, How risk management can prevent future wildfire disasters in the wildland-urban interface, Proc. Natl. Acad. Sci. U. S. A. 111 (2014) 746−751, https://doi.org/10.1073/pnas.1315088111.

[128] D. Thurton, Traffic-clogged highway during Fort McMurray wildfire spurs call for 2nd highway, CBC News (2016). https://www.cbc.ca/news/canada/edmonton/traffic-clogged-highway-during-fort-mcmurray-wildfire-spurs-call-for-2nd-highway-1.3821671.

[129] T.B. Paveglio, C. Kooistra, T. Hall, M. Pickering, Understanding the effect of large wildfires on residents' well-being: what factors influence wildfire impact? For. Sci. 62 (2015) 59−69, https://doi.org/10.5849/forsci.15-021.

[130] R.L. Beschta, C.A. Frissell, R. Gresswell, R. Hauer, J.R. Karr, G.W. Minshall, D.A. Perry, J.J. Rhodes, Wildfire and Salvage Logging: Recommendations for Ecologically Sound Post-fire Salvage Management and Other Post-fire Treatments on Federal Lands in the West, Oregon State Univ, Corvallis, OR, 1995.

[131] F. Moreira, O. Viedma, M. Arianoutsou, T. Curt, N. Koutsias, E. Rigolot, A. Barbati, P. Corona, P. Vaz, G. Xanthopoulos, F. Mouillot, E. Bilgili, Landscape − wildfire interactions in Southern Europe: implications for landscape management, J. Environ. Manage. 92 (2011) 2389−2402, https://doi.org/10.1016/J.JENVMAN.2011.06.028.

Firefighting approaches and extreme wildfires

Gavriil Xanthopoulos[1], Giuseppe Mariano Delogu[2], Vittorio Leone[3], Fernando J.M. Correia[4], Catarina G. Magalhães[4]
[1]Hellenic Agricultural Organization "Demeter", Institute of Mediterranean Forest Ecosystems, Athens, Greece; [2]Former Chief Corpo Forestale e di Vigilanza Ambientale (CFVA), Autonomous Region of Sardegna, Italy; [3]Faculty of Agriculture, University of Basilicata (retired), Potenza, Italy; [4]Faculty of Arts and Humanities, University of Porto, Porto, Portugal

6.1 Wildfire fighting approaches

6.1.1 Development history

Fires have always been a natural component of the earth's ground ecosystems. As a result, man has lived with fire and has used fire since prehistoric times. In addition to using fire for heating and cooking, man also used fire for a variety of other tasks, ranging from agriculture and foraging to hunting and even warfare. Fire was used to clear land (slash and burn) and to improve grazing quality and wildlife habitats for hunting [1]. In doing so, man learned how to start, control, and extinguish fire.

In Europe, for centuries, forest destruction was considered a neutral operation [2], so people were concerned about wildfires mainly when persons and assets were threatened [3]. It can be hypothesized that lack of fuel accumulation around settlements, due to agricultural cultivation, livestock grazing, and use of wood as timber and fuel, created unfavorable conditions for the growth of destructive wildfires. On the other hand, in the United States, where in the 19th century a relatively small population was spreading in the pristine forest lands of the country, fuels were much more substantial, increased further by fire exclusion policies in forests where native populations had been frequently using low-intensity fires, and the potential for a wildfire disaster was much higher. The most tragic manifestation of this potential was the 1871 Peshtigo fire with more than 1500 fatalities. After that, firefighting started becoming more serious. In 1886, in Yellowstone National Park, army soldiers became the first US wildland firefighters to be paid for their service [4]. Additional disasters that followed, most notable being the Great Fire of Idaho and Montana in 1910 [5], increased emphasis on wildfire suppression. When the US Forest Service was established, in 1905, wildfire suppression became its primary task. This trend finally led to the establishment, in 1935, of the U.S. Forest Service's "10:00 a.m." fire management policy, which stipulated that all wildfires were to be suppressed by 10 a.m. the morning after they were first spotted [6].

Extreme Wildfire Events and Disasters. https://doi.org/10.1016/B978-0-12-815721-3.00006-0

Before 1930s, forest protection agencies relied on pick-up firefighters who were hired on an "as-needed" basis. These men came from every walk of life. They were mostly young and had very little fire suppression training or experience. The first organized US Forest Service wildfire suppression crew was a "40-man" crew established in 1939, on an experimental basis, and located on the Siskiyou National Forest in southwestern Oregon [7]. The first reference to specialized training is made in regard to the first smokejumpers who were trained in Missoula, Montana and started operating experimentally in early 1940s, becoming operational in 1944 [8].

The huge wildland areas that had to be protected from wildfire in the US, combined with the scarcity of roads, especially in areas of sparse population, and the lack of capacity to bring water to the fireline led to the development of ad-hoc solutions in firefighting, tailored to the conditions. Forest firefighting became quite different from urban firefighting. In urban fire control, as a rule, nearly every fire poses a potential threat to life and property. Saving lives is the absolute priority of city firefighters, and they are prepared for that, wearing fire-resistant turnout coats, heavy helmets, and self-contained breathing apparatus as they often have to operate inside structures. In general they can count on roads and good availability of water. In wildlands, on the other hand, the firefighting paradigm evolved quite differently. Firefighters work in many different types of vegetation adjusting their methods accordingly. They wear lighter clothes than their urban counterparts, and hardhats, work with hand tools such as shovels and Pulaskis and use chainsaws to break vegetation continuity. More important, they learn how to put out the fire even when water is not available, throwing dirt on the flames or moving back, when the flames are too intense to attack them directly, and securing perimeters by setting fires to remove flammable vegetation between the flaming front and their control lines (suppression fire [or backfire] and burnout) [9].

Fighting fire in the vast landscapes of the US, often in roadless areas, entails a very high risk to firefighters, as they may be overtaken by fast-moving flames or may be trapped if they are not very careful and in full alert. The Mann Gulch fire of August 5, 1949 in Montana, which resulted in the death of 13 smokejumpers, made a huge sensation and brought many changes in wildfire fighting. Among these, it led, in 1957, to the development of the 10 Standard Fire Orders (SFOs) and the recognition of the 18 Watch Out Situations, in a United States Forest Service (USFS)-commissioned "Report of Task Force To Recommend Action to Reduce the Chances of Men Being Killed by Burning While Fighting Fire" (https://www.fs.usda.gov/Internet/FSE_DOCUMENTS/stelprdb5393525.pdf). The same report spurred the interest for research into and use of fire behavior knowledge in wildland firefighting.

In 1961, the US organized the first Interregional Fire Suppression Crews. They were a new type of highly trained crews, ready to move by airplane between regions, making it possible to apply the concept of quick mobilization and dispatching [7]. With these crews as the obvious example, specialized quality training started becoming a recognized need. This led to development and subsequent improvements in training content, separately for the various organizations. Then, in 1976, the National Wildfire Coordinating Group (NWCG) was established through a Memorandum of Understanding between the Department of Agriculture and the Department of the Interior. The purpose of the NWCG is to coordinate programs of the participating agencies so as to avoid wasteful duplication and to provide a means of constructively working together to achieve more effective execution of each agency's fire management program. The Group provides a formalized system to agree upon standards of training, equipment, aircraft, suppression priorities, and other operational areas. Twelve "working teams" and a number of subteams, comprised of member agency leaders and experts in various fields, were established in functional areas such as fire equipment, fire weather, incident operations, training, and incident business. These teams led the initial effort to achieve a broad national standardization in key areas of wildland fire management (http://www.nwcg.gov/history). After 1970, the Incident Command System (ICS) was developed as a standardized approach to the command, control, and coordination of emergency response providing a common hierarchy within which responders from multiple agencies can be effective.

In Europe, wild landscapes are not as extensive as in the US. Presence of dense population in rural areas, a mosaic of agricultural cultivations interrupting forest continuity, traditional use of fire in landscape management, removal of fuel wood from forests for use in heating and cooking, and scientifically based forest management by professional foresters are among the reasons why wildfires were relatively infrequent and benign until 1970s. In central and northern Europe wildfire was quite rare, while in the Mediterranean countries of Europe, where the vegetation is drier and the weather conditions are more conducive for fire in the summer, fires were more frequent but rarely became large and devastating thanks to the reduced dead biomass loads and to the existing fuel discontinuities. Fire suppression, as a rule, was not too demanding. The rural populations participated actively, trying to protect their own properties and production, using their traditional fire knowledge (T.F.K.), agricultural hand tools and equipment, in the general sense of traditional ecological knowledge [10]. They often resorted to the use of fire as a suppression tool when the flames were too long for direct fighting with hand tools [11].

The rural exodus that started in 1960s and continues to a large extent until today brought significant changes in wildfire suppression. Initially, fire trucks started being used for suppression. As the urban fire trucks were unable to move in the narrow forest roads, specialized fire trucks with off-road movement capability were introduced in the fire suppression services. Gradually the use of water for forest fire suppression became the norm. The use of hand tools and fire as a suppression tool were all but forgotten. The introduction of aerial firefighting, from the 1970s onward, with continuously increasing capacity and cost, further reduced the capability to fight fire using indirect attack, hand tools, and backfires. However, it gradually became evident that the firefighter's "toolbox," i.e., the options they have for controlling the different types of fires, had clearly been weakened. Only experienced locals, such as forest workers, shepherds, bee keepers, resin tappers, and similar had the knowledge, skills, and mentality to use these tools.

In 1993, in Greece, use of hand tools was reintroduced with the organization of the first groups of firefighters that were transported to the vicinity of starting fires by helicopters (helicrews). They were trained to fight with hand tools with the support of aerial drops by helicopters. They operated until 1997. Similarly, starting in 1994, Spain created its Reinforcement Brigade for Forest Fires (BRIF) with the intention to act as reinforcement of local ground resources. Since then it has grown into 10 bases all over Spain staffed by 600 firefighters [12]. The BRIF were originally patterned after the hotshot crews of the US Forest Service and received the same type of training by former members of US hotshot crews. According to the regional rules in all regions of Spain, a complex system of coordination was developed and original units of firefighters were created, such as the GRAF unit into the firemen corps of Catalonia after the Solsona wildfire in 1998.

The need to reintroduce fire as a prevention and suppression tool in fire management has been illustrated by "Fire Paradox," a large research project (2006—10), funded by the European Commission Research and Development program. The project aimed to promote the full integration of fire use in the prevention and suppression of wildfires. During the project, numerous fire use demonstrations helped to achieve reintroduction of fire use, even at small scale, in a few of the participating European countries [13].

6.1.2 How firefighting approaches are different and why

Although the physics of fire are universally the same, wildfires are quite different in various parts of the world and in various ecosystems, as a result of differences in the fuels, the topography, the weather, and the fire regime. Firefighting approaches are also different, because of not only differences in fire characteristics but also differences regarding the countries and the people involved in fire management. The reasons for these differences include

- firefighting history and tradition;
- size of protected areas especially when compared with the population size;
- level of participation of the public (volunteerism, empowerment of rural populations);
- training;
- available equipment;
- financial capacity of the country;
- sophistication level and professionalism;
- types of fuels;
- characteristics and importance of the wildfire problem;
- values at risk (e.g., extent of wildland—urban interface [WUI] areas).

6.1.2.1 Emphasis and capacity for indirect attack

Nowadays, firefighting is usually related to the use of water as the main extinction component. Many national and regional fire suppression forces are organized on the basis of different fire trucks with more or less water (from 400 to 10,000 L) and different mobility. In addition, aerial support is based on growing water capacity. This is very effective in the majority of wildfires, mainly in the initial attack phase. The upper fire size limit or a duration-based threshold for this phase is not universal, and what constitutes a "large fire" is somewhat arbitrary [14]. In the US, organizations working on fire management usually set the limit for successful initial attack at 4.0 ha (10 acres). For example, the California State fire control objective is to keep 95% of all initial attack wildland fires at 10 acres or less (County of Santa Barbara California Fire Department 2018). On the other hand, for statistical purposes, a fire is considered as large when it exceeds 121.4 ha (300 acres) in size [15].

When wildfire behavior exceeds the firefighter's capacity of extinction, with flames pushing firefighters back and a rate of spread faster than the fire trucks can follow, then a different approach is needed. It is a kind of proactive approach: not to follow the flames but anticipate them when and where fighting the fire is easier and safer.

Indirect attack is a typical tool for advanced firefighting organizations and—in part—it derives from old knowledge and skills of rural people, who did not have fire trucks and resorted to the use of fire as a firefighting tool. Rural people in Europe and native populations (as previously said) had built skills and confidence in using fire as a tool for indirect attack, their T.F.K. As wildfires with extreme behavior are becoming more common nowadays, many advanced fire suppression organizations realize the need for using the proactive approach for

- planning extinction operations in emergency;
- identifying appropriate locations where they can organize safely indirect attack;
- defining safe areas where they can build effective firelines for indirect attack;
- using preventive silviculture to make landscapes safer, by reducing the intensity and rate of spread of fires and by offering opportunities for better planning and organizing for extended attack.

6.1.2.2 Professional versus volunteer

There is a significant variety between countries regarding the work status of the personnel of wildfire suppression organizations. Tradition, characteristics of the fire problem, resources at risk, and financial considerations influence the reality that has developed in each country. Some countries depend mainly or completely on professional firefighters, whereas others mostly rely on volunteers.

As wildfires can be dangerous at any time and firefighting is a very demanding task, there are no differences in skills needed for professionals and volunteers in firefighting: Wildfire firefighting is a critical activity that requires many common components:

- physical and athletic aptitudes (normally checked into standard program);
- analysis capacity about wildfire behavior;
- very good knowledge of firefighting methods;
- knowledge of safety protocols (LACES - (Lookouts, Awareness, Communications, Escape Routes, Safety Zones) and the 10 standard firefighting orders);
- to know and use standard terminology in communication.

In most countries, there are both types of firefighters. It is critical how they are organized and trained and how they cooperate with each other. In some countries where there exists a complete devolution of competences from state to regions (such as in Italy), such cooperation suffers as a law on standardization is missing, and the same is true for guidelines for common operations.

6.1.2.3 Emphasis to ground versus aerial firefighting

Aerial firefighting appeared before the Second World War, especially in the US, and gradually increased in the 1950 and 1960s, using to a significant extent surplus military aircraft. From 1970s specially designed amphibian water bombers, most notably the Canadair CL-215, became a very important firefighting tool, especially in the countries of Mediterranean Europe. In 1970 and 1980s, helicopters started being widely used in aerial firefighting operations, but their role increased steeply as heavy helicopters found their way in the state fleets of aerial firefighting resources or became available for private contracting, especially in 1990s. Many of these helicopters originated from the military surplus of the USSR.

The increase of the role of aerial resources has had some marked influences regarding the firefighting reality. It has improved the effectiveness of initial attack as the aerial resources arrive to fires quickly, even at inaccessible places. It has also allowed firefighters to perform direct attack on fires with more intense behavior than in the past. The percentage of fires that are stopped in their initial stages has increased in most countries. On the negative side, it has contributed greatly to the steep increase of the cost of firefighting in the last decades, something that has affected the funds available for fire prevention, especially through active management of forest landscapes. Smaller burned area in relatively easy fire seasons combined with lesser emphasis on fire prevention result in biomass built-up and higher fuel continuity, which, under adverse weather conditions, become one of the main contributors for the appearance of fires with extreme behavior. As firefighters have become generally used to rely on direct attack having the support of aerial resources and have little experience with indirect attack, the situation often gets completely out of control. Aerial resources often prove unable to fight extreme fires due to smoke, extreme wind, and turbulence and when intensity exceeds a certain threshold that, according to most authors, is roughly around $3000\,kWm^{-1}$, but with specific exceptions bringing it to $8300\,kWm^{-1}$ [16,17]. Furthermore, when many simultaneous fires escape initial attack, the number of aerial resources often proves inadequate, as they are often not used in an integrated way with ground crews but are expected to function as their substitute, inadequate or ineffective fighting from the air contributes further to the high growth and huge destructive potential of extreme fires.

6.1.2.4   Type of organization (land management vs. urban/civil protection)

Another significant difference regarding how wildfire firefighting is approached has to do with the type of organization that carries the responsibility for this task. At one extreme, this responsibility is carried by a land management agency, as a rule the Forest Service. At the other end, a firefighting agency that is responsible for urban firefighting also undertakes the responsibility for wildfire suppression. There are also intermediate organization schemes, where a system allows many organizations to work together under specific cooperation rules. The National Incident Management System (NIMS) of the United States is a good example [18] in spite of having certain drawbacks [19]. In addition, in some cases, the responsibility is assigned to different organizations within the same country, usually at regional level. Such differences exist between the autonomous regions of Spain and of Italy.

It should be noted that for a number of reasons, which include the weakening of land management organizations, as rural populations abandon the countryside, there is a general tendency to move the responsibility to urban firefighting and civil protection agencies.

One example is the California Department of Forestry. In 1970s it was a land management agency, albeit one with serious fire responsibilities. In 1974, under the pressures of postwar development, it became the California Department of Forestry and Fire Protection. In 2007 it collapsed that mission into Cal Fire, "*which operates like an urban fire service in the woo*ds" [9]. It is now the largest full service all-risk fire department in the Western United States and the second largest municipal fire department in the United States, behind only the New York Fire Department. In another example, in Greece, the responsibility for forest firefighting was transferred overnight, in May 1998, from the Forest Service to the (Urban) Fire Service [20]. The result of such changes is usually an increased cost for firefighting (more ground and aerial resources, more technology, more use of direct attack) and a neglect for fire prevention, leading to overall negative results in the long term. Considerations about which type of agency should be responsible for forest fires are discussed by Xanthopoulos [21].

6.1.3   Examples of currently existing organizational and firefighting approaches

6.1.3.1   Greece

In Greece, the responsibility of fire suppression was transferred by the Government from the Forest Service to the Fire Service through Law 2612/1998. After that time, the Fire Service, which is a professional agency with semimilitary structure, received increased funding that led to increased ground personnel and fire trucks and aerial resources. Initial obvious weaknesses due to lack of knowledge and experience were gradually remedied, but the operation became very costly especially because of the heavy reliance on strong national and contracted aerial resources. Furthermore, this reliance became a major weakness when aerial resources were not adequate: In the summer of 2007, on August 24, under very adverse conditions (drought, very high temperature, extremely low relative humidity, strong wind), a series of fires that escaped initial attack and spread with extreme fire behavior quickly surpassed the capacity of aerial resources. The situation got out of hand for 4 days, resulting in huge burned areas, heavy damages to properties and infrastructures and, most important, numerous fatalities. About 80 people died due to the wildfires in that summer [22]. Eleven years later, on July 23, 2018, a wildfire in North East Attica, 20 km from the center of Athens, fanned by a gale force wind that made most aerial resources unable to operate effectively, burned a whole WUI area called Mati causing 102 fatalities [23].

The Forest Service has a minor role in prevention, mainly regarding forest road and firebreak maintenance, and limited fuel management. Its performance is inadequate because of reduced and aging personnel and severely inadequate funding. More funding for fire prevention, especially in WUI areas, is distributed by the General Secretariat for Civil Protection to the municipal authorities around the country. However, owing to lack of expertise of the personnel of the municipalities, political criteria, and poor control over spending, these funds are not used optimally.

The armed forces offer some contribution to fire prevention, mainly contributing in patrolling during high fire danger days, and also participate in fire suppression. Soldiers are often sent to help with mopping-up and guarding large fires for potential restarts. The Hellenic Air Force is deeply involved in firefighting as it operates the national fleet of Canadair amphibian water bombers and smaller PZL M18 Dromader single-engine planes. The Army Air Force also makes available a varying number of helicopters. These resources are available year-round, whereas private helicopters are contracted only for the peak of the fire season.

Another resource that participates in the fire management scheme is the volunteers. There are three types: those working in the fire stations with the Fire Service, volunteer groups formed by the municipalities, and independent volunteer groups formed through private initiatives. Except for the Fire Service volunteers who are added to regular fire truck crews, utilization of the others by the Fire Service is poor to this day.

As a final note, after the 2018 disaster in North Eastern Attica, the fire management system of the country is under reconsideration, aiming to improve effectiveness and efficiency [24].

6.1.3.2 Italy

The first time public institutions officially talked about wildfires in Italy was on the occasion of a national Conference on "The buildup of forest resources and their defense from wildfire" held in Bergamo in 1967. Previously rules about fire management and wildfires were only present in some local laws. The first regional law about prevention and extinction of wildfires was approved by Lombardia Region (1972). It was followed by Tuscany and Umbria (1973) and finally by other regions.

In 1975 the framework Law "Integrative rules for woodland defense from wildfires," L. 47/1975 was approved. Its main goals were reordering of local laws, techniques, and means adopted to prevent wildfires, reconstitution of burned forest areas, and definition of tasks and responsibilities. A definition of "forest wildfire" was not provided.

In this framework, planning was a regional responsibility; structures and facilities provided in regional plans were borne by the State. The concept of "structures" and "facilities" included silvicultural options such as introduction of less flammable species in forests located in dry-hot climate, grazing in forests, firebreaks and water reservoirs, lookout towers, radio communication webs, aerial resources, etc. In the concept of "planning," the "recovery of burnt forests" was also included. The law introduced for the first time a prohibition to build in partially or totally burned forests, as well as a prohibition for any kind of change of designated land use for 10 years after a fire; penalties were of administrative nature. The provisions of L. 47/75 remained partially unrealized or deeply reduced due to lack of funds.

In 2000, a new framework law (L. 353/2000) was enacted, replacing L. 47/75. The most important change was introduction in the Penal Code of a new article concerning forest wildfire crime with a penalty of up to 10 years imprisonment for offenders. But also important is the definition of forest wildfire: *"Forest wildfire is a fire with susceptibility to spread into wooded areas, with trees or brushes, including eventual structures and anthropic infrastructures located inside those areas, or in cultivated or fallows or grazed soils neighboring these areas."* This definition introduces wildfires in the field of Civil Protection.

The resulting framework can be synthetized as follows:

Planning is a regional responsibility. The Ministry of Agriculture disappears, and the Civil Protection Dept. is the main agency involved in operations and coordination guidelines.

The core of the law is focused on "prevision, prevention, and extinction operations."

All the competences in connection with forest fires are transferred to the Regions, leaving only the water bombers fleet management at national level.

Silviculture and land management initiatives (improvement of preventive silviculture, grazing as a tool for reducing fuels and so on), are only a small part of Regional Planning: This tool is not present in all regions, and normally, actions in emergency clearly prevail over prevention and long-term land management.

The "nonstructural" activities of Civil Protection are mainly focused on achieving a rapid rescue after the incident, not in advance, paying little attention on systematic and consistent policies of preventive silviculture and the relevant management tools.

A great difference exists between Regions, where organizational solutions are different, mainly regarding the participation of volunteers: In the Piemonte Region, only a well-trained, organized, and structured group of volunteers (Corpo volontari AIB) operates in extinction and prevention. In other Regions, such as Sardinia, there are many and differently organized groups of volunteers, which complicates coordination regarding engagement procedures, terminology, communications, safety, and so on.

Last but not least, prohibition and penalties, the important issue of the repealed law 47/75 (recovery of burnt forests), now is a prohibition (Art. 10): Forested areas and ranges affected by fire must maintain their designated use for 15 years. No building initiative is allowed in such areas for 10 years after fire. In those areas, for 5 years after wildfires, plantations and environmental engineering activities supported with public funds are prohibited, while grazing and hunting are prohibited for 10 years. The application of these restrictions is only possible through a record of the georeferred perimeter of all the burned areas that the municipalities are obliged to set up and maintain (Wildfires cadastre).

In 2016, a law on "Simplification of Public Administration" (L. 177/2016) provided the dissolution of State Forestry Corps and its passage into the Carabinieri, a member of the Italian national police force organized as a military unit. The competences about wildfire coordination of the disbanded State Forestry Corps were transferred to the National Fire Corps, without, however, changing general skills about wildland wildfires for a Corps that mainly has competences in urban/industrial wildfires.

6.1.3.3 Portugal

It was after the big wildfires of 2003 and 2005 that the Portuguese State, realizing that it was facing a growing problem, took the initiative to reform and restructure the public entities responsible for combating rural fires.

Currently, the structure of the forest defense against wildfires is based on three institutions: the Institute of Nature Conservation and Forests (ICNF), the National Republican Guard (GNR), and the National Authority for Emergency and Civil Protection (ANEPC). The ICNF is the entity responsible for the Forest and Protected Areas, also having an important role in the prevention, awareness, and maintenance of infrastructures, as well as in the recording and analysis of the statistics of rural fires. The GNR holds the Protection and Assistance Intervention Group (GIPS), which is responsible for the initial attack on rural fires, and the Protection Service of Nature and the Environment (SEPNA) which is responsible for surveillance and detection. The Portuguese National Authority for Civil Protection (ANPC) is the entity responsible for coordinating the response of the operatives and for the whole management of the combat either by land or by air [25].

Despite these three pillars, whenever a rural fire occurs, the main operational combat force is formed by the Fire Brigade (CB) whose entities are the Humanitarian Associations of Voluntary Firefighters (AHBV). CB, in addition to urban and rural fires, offers various services to the community, such as transportation of patients not considered urgent, and emergency medical services in partnership with the National Institute of Medical Emergencies (INEM). Its responsibility to combat rural fires has been established for a long time (Decree Law no. 55/1981) and is considered as the main operational force of fire suppression in Portugal.

The year 2017 was another milestone regarding the forest and rural fire management approach in Portugal. It was characterized by huge fires that took place on days outside the formerly considered critical period of fires: The Pedrogão Grande fire on 17−19 June, 2017, and caused 64 fatalities, and the fires of October 15−16, 2017, in Central and Northern Portugal that killed more than 50 people. In response, the Government adopted new measures such as establishment of the Agency for Integrated Rural Fires (AGIF) and greater flexibility of land and air firefighting response in case of risk. The professionalization of the system was also strengthened, and a Single Prevention and Combat Directive was created. It is in this directive that is defined a Special Rural Fire Fighting Device (DECIR), defined by the ANPC, that articulates the entities involved in fire suppression, guaranteeing an operational response at all times. The device is organized and operated throughout the year, being reinforced according to the levels of operational commitment, depending on the probability of fire occurrence. The DECIR presents two operational commitment levels where one is permanent and corresponds to the time of year where there is less likelihood of fire, and the other is the level "reinforced" because it requires more means and greater readiness in operational response.

In Portugal, despite the fact that there has been a greater investment in prevention, especially since the last extreme wildfire events (2017), a policy focused on fighting wildfires continues.

6.1.4 United Kingdom

A variety of structures and informal management solutions emerged in response to local needs. Knowledge of wildfire accumulated within regional and national wildfire forums and academic networks. Only later did the need for central emergency planning and the response to climate change produce a national policy response.

The UK has no single agency or firefighting force with specific responsibility to manage wildfires. Instead, statutory responsibility rests with individual Fire and Rescue Services (FRS) under the Fire and Rescue Services Act 2004, or equivalent for the devolved administrations. FRS training and equipment focuses on fighting fires in urban buildings, handling emergencies such as chemical spills and rescue from road traffic accidents. Most services had little knowledge and understanding of rural wildfires 10 years ago. The penetration of wildfire culture into organizations is improving but still depends on intergenerational championing and knowledge exchange [26].

6.2 Effectiveness and efficiency considerations

Effectiveness and efficiency are two important concepts in the field of forest fire management. Regarding fire suppression, effectiveness refers to the ability to accomplish the objective, which is fire control with the minimum possible damages without compromising firefighter safety, while efficiency is the ability to achieve this objective with the least possible cost.

Commonly, national statistics analysis at the end of each fire season are used to assess how the system was effective. Common parameters used for this task include burned surface, number of wildfires detected and extinguished, and average burned area per fire. Further analysis involves time of arrival to each event, people and resources involved in operations, and so on. Data on the cost of firefighting are generally hard to find, at least in Europe. Thus efficiency analyses at organizational level are relatively uncommon. It is rare to find analyses about avoided damages, harmonization of tactics to reach a better result (and reduced costs), correct use of aerial and terrestrial resources, etc.

Regarding fire suppression organizations, there can be many trade-offs between effectiveness and efficiency, and decisions, in general, are not clear-cut. There is no clear, universally accepted methodology on how to decide the level, composition, and organization of firefighting resources in a country or region. Main considerations include the mix, number, and strength of ground forces (professional, seasonal, and volunteer firefighters; capacity of fire trucks; other tools; adoption of technological advancements; etc.), and the type and capacity of aerial resources (fixed wing airplanes, amphibian water bombers, helicopters). However, neither effectiveness nor efficiency is linked one-to-one to these main choices. Especially efficiency can be affected strongly by the way fire suppression is organized (centrally controlled vs. regionally or locally based) and the level of cooperation between the various agencies (e.g., professional organizations with volunteer groups, ground firefighting organizations with aerial resources operators (e.g., Air Force), civil protection organizations with land management agencies, etc.). Quality of presuppression planning, personnel training and motivation, other details affecting operational efficiency (e.g., quality of decision making at the dispatching center, quality of on-site coordination, avoidance of wasted

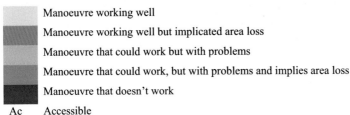

Fire behaviour	Direct attack with manual tools			Direct attack with water lines			Direct attack with aerial means			Technical fire and manual tools			Technical fire and water line			Technical fire and bulldozer			Technical fire: backfire		
	Ac	Aw	In	Ac	Aw	In	Ac	Aw	In	Ac	Aw	In	Ac	Aw	In	Ac	Aw	In	Ac	Aw	In
Low intensity																					
Medium intensity																					
High intensity																					
Torching																					
Crown fire																					

Where:

Manoeuvre working well

Manoeuvre working well but implicated area loss

Manoeuvre that could work but with problems

Manoeuvre that could work, but with problems and implies area loss

Manoeuvre that doesn't work

Ac   Accessible

Aw   Away, accessible with difficulties

In   Inaccessible

Figure 6.1 Relations between intensity of fire front, typology of attack maneuvers, and accessibility to maneuver site.
Castellnou and Miralles [27], modified.

flying hours of helicopters by organizing refueling base set-up near each large fire, etc.), and a host of other factors also play a significant role in the overall efficiency of each fire suppression organization.

On the other hand, a conceptual approach on correct evaluations about how to use effectively and efficiently all resources in fire suppression operations is defined in a scheme proposed by Castellnou and Miralles [27] (Fig. 6.1). Obviously, in this scheme there are possibilities for different maneuvers to overlap, so this model could change in different situations and for different operating crews. In this example, efficiency and effectiveness could be evaluated as a balance between fire intensity, kind of main tactics adopted or their combinations, and accessibility of the operations site. Other elements should also need to be considered such as available resources and crews in the area of operations, command capacity, experience, training, etc.

6.3   Firefighting approaches regarding extreme wildfires

Fire suppression organizations around the world are having to fight extreme wildfires that appear with higher frequency and with continuously worsening destruction potential [28]. Independently of the reasons behind the worsening trend, these fires are a reality that has been causing serious damage to properties, infrastructures and the environment, as well as increasing the number of fatalities (Australia 2009, Portugal 2017, Greece and California 2018). While science has pointed very clearly to the need to put, in the long term, emphasis on fire prevention and fuel management, the fire suppression organizations need to adopt and apply improvements that will make them to cope better now and in the future. In doing so, they have to address a number of important challenges that extreme wildfires pose:

- They need to improve their response regarding fires that can potentially become extreme. A good prediction capability of a fire to become extreme and appropriately tailored initial attack capacity is absolutely necessary.
- They need to increase the limits of fires (intensity, rate of spread, perimeter) that they can cope with successfully.
- They need to cope with "uncertainty" in firefighting planning. Uncertainty refers to a highly and rapidly changing scenario in forest and WUI wildfires and to the complex, often over-lapping information and circulation in the command chain, which make taking decisions difficult and prone to errors, if not impossible; assessing and selecting proper information is the basis for choosing a better strategy and for reducing uncertainty to an acceptable level.
- They need to address population concerns, set priorities, and act quickly to ensure safety (protection priorities, evacuation options, etc.).
- They need to perform all the above without compromising firefighter safety.

Defining an overall strategy to cope with an extreme fire, rather than resorting to tactics and maneuvers, is critical. In the context of firefighting, strategy refers to "the process of establishing the best plan to accomplish secure and efficient fire control when suppression forces arrive to the fire scene." From setting of priorities to deciding if, where, when, and how the resources should fight, developing a sound strategy is critical and challenging. Especially for such events, modern forecasting capabilities, from detailed meteorological previsions to fire behavior simulations in time and space, can help define "critical points" as well as "opportunity points," where firefighting forces will have an advantage (less intense fire behavior, increased safety, more effective fighting from the air and on the ground, etc.), maximizing the probability for successful fire control. Thus firefighting with a correct strategy becomes proactive rather than simply reactive and can lead to better and well-coordinated tactics that can be applied with safety.

To expand their capacity to fight extreme fires better, firefighting organizations need to have well-trained, experienced, and very capable coordinators to assign to each such event. Actually, it is best to have well-organized coordination teams ready to be dispatched immediately to a fire predicted to become extreme. Both the national and regional coordination centers and the on-site coordinator need to achieve the best possible coordination of ground and aerial resources. The on-site coordination team should assess continuously the available resources versus the challenges they will face from current and predicted fire behavior. More often than not reducing firefighting in expectation of better conditions (e.g., dusk) constitutes a rational option. Firefighters may have to focus on easier parts of the perimeter while at the front only offer protec-tion to homes and citizens in danger. Changing from direct to indirect attack is a commonly selected option: Extreme fires are a good reason for reintroducing the use of fire as a firefighting tool (backfire, burnout operations) where it has been forgotten or even forbidden.

Extreme fires present a further issue that firefighting organizations have to be prepared for: A region and even a country may not have adequate resources to fight a really large fire with extreme behavior, let alone more than one large simultaneous fires (e.g., Portugal, October 15th' 2017, Greece August 24th' 2007). In preparation for this, the organization in every country should prepare to draw resources from within the country (local authorities, volunteers, armed forces, etc.) as well as from abroad. In the European Union (EU), the Union Civil Protection Mechanism (UCPM) was established in 2001 to improve the EU response to natural and man-made disasters inside and outside Europe, offering mutual help. Seventeen years later, through the RescEU initiative, the EU is improving the UCPM further, developing European reserve of capacities. In addition, there can also be bilateral aid between two countries based on prior agreements or in response to humanitarian calls. A key issue for cooperation, at least on the ground, is the existence of a common organizational scheme. The ICS of the United States has been successfully used for sharing personnel and other resources between countries that work with it. The USA, Canada, Mexico, Australia, and New Zealand offer many examples of helping each other, using the ICS, in the last 2 decades.

A final point is that as extreme wildfires get even worse, knowledge about predicting them and fighting them effectively must be improved. In this respect, firefighting organizations in cooperation with researchers should strive to find out, through documenting and analyzing current and future extreme fires, which fire management approaches are better and under what conditions they can be most successful. In addition, they need to develop criteria for evaluation of the approaches used on each fire regarding their effectiveness and efficiency.

6.4   Conclusions

In conclusion, firefighting organizations and approaches are quite different from one country to the other as a result of many factors, including people and the environment. However, as extreme fires become a major issue internationally, with the number of wildfire caused fatalities trending upward, the necessity for improving the fire suppression system in all suffering countries becomes more obvious and urgent. Such improvements are much more likely to succeed if scientists and operational officers work together.

References

[1] F. Tedim, G. Xanthopoulos, V. Leone, Forest Fires in Europe: Facts and Challenges, Wildfire Hazards, Risks and Disasters, 2015, pp. 77—99, https://doi.org/10.1016/B978-0-12-410434-1.00005-1.
[2] B. Vecchio, Il bosco negli scrittori italiani del Settecento e dell'età napoleonica, PBE 235, Piccola Biblioteca Einaudi, Torino, 1974.
[3] A. Jacquot, Incendies en forêt. Evaluation des dommages, 1904.

[4] National Fire Fighter Wildland Corp, The History and Evolution of Wildland Firefighting, 2015. http://www.nationalfirefighter.com/blog/The-History-and-Evolution-of-Wildland-Firefighting.

[5] T. Egan, The Big Burn: Teddy Roosevelt and the Fire that Saved America, Houghton Mifflin Harcourt, 2009.

[6] S. Pyne, Fire in America: A Cultural History of Wildland and Rural Fire - Stephen J. Pyne, WA, University of Washington Press, Seattle, 1982. https://books.google.pt./books?hl=pt-PT&lr=&id=N3QkDwAAQBAJ&oi=fnd&pg=PP1&dq=Pyne,+S.J.+1982%3B+1997.+Fire+in+America:+a+cultural+history+of+wildland+and+rural+fire.+Seattle,+WA:+University+of+Washington+Press.+680+p.&ots=Oiphc4Bm3V&sig=re1igQq7S5XPdYSyTRd.

[7] M.E. Alexander, The interregional fire suppression crew, Fire Manag. USDA For. Serv. 35 (3) (1974) 14−19. http://www.cfs.nrcan.gc.ca/pubwarehouse/pdfs/33441.pdf.

[8] Aviation and Fire Management, History of Smokejumping, 1980, p. 33.

[9] S. Pyne, It's Time to Rethink How We Fight Forest Fires- Americans Learned Long Ago How to Keep Cities from Burning. And Then, it Seems, We Forgot, Pacific Stand, 2018. https://psmag.com/environment/only-a-rethink-can-prevent-forest-fires.

[10] P. Colorado, D. Collins, Western scientific colonialism and the re-emergence of native science, Pract. J. Polit. Econ. Psychol. Sociol. Cult. (1987) 50−65.

[11] M.R. Huffman, The many elements of traditional fire knowledge synthesis, classification, and aids to cross-cultural problem solving in fire-dependent systems around the world, Ecol. Soc. 18 (4) (2013) 3. https://doi.org/10.5751/ES-05843-180403.

[12] B. Gabbert, Wildland Firefighters in Spain Strike as Fires Burn Wildfire Today, Wildfire Today - Wildfire News Opin, 2015. https://wildfiretoday.com/2015/07/28/wildland-firefighters-in-spain-strike-as-fires-burn/.

[13] J.S. Silva, F. Rego, P. Fernandes, E. Rigolot, Towards Integrated Fire Management - Outcomes of the European Project Fire Paradox, Joensuu, Finland, 2010. https://www.repository.utl.pt./bitstream/10400.5/15236/1/REP-FIRE_Paradox-efi_rr23.pdf.

[14] M.P. Thompson, F. Rodríguez y Silva, D.E. Calkin, M.S. Hand, A review of challenges to determining and demonstrating efficiency of large fire management, Int. J. Wildland Fire 26 (2017) 562. https://doi.org/10.1071/WF16137.

[15] Glossary of Wildland Fire Terminology, Natl. Wildfire Coord. Gr., 2012. https://www.nwcg.gov/glossary/a-z.

[16] E. Stechishen, E. Little, M. Hobbs, W. Murray, Productivity of Skimmer Air Tankers, 1982. http://www.cfs.nrcan.gc.ca/bookstore_pdfs/12092.pdf.

[17] I.T. Loane, J.S. Gould, Aerial Suppression of Bushfires: Cost-Benefit Study for Victoria, National Bushfire Research Unit, CSIRO Division of Forest Research, Canberra, Australia, 1986. https://www.cabdirect.org/cabdirect/abstract/19860611619.

[18] J.F. Annelli, The National Incident Management System: A Multi-Agency Approach to Emergency Response in the United States of America, Riverdale, 2006. https://pubag.nal.usda.gov/download/36325/PDF.

[19] D.A. Buck, J.E. Trainor, B.E. Aguirre, A critical evaluation of the incident command system and NIMS, J. Homel. Secur. Emerg. Manag. 3 (2006). https://doi.org/10.2202/1547-7355.1252.

[20] G. Xanthopoulos, The 1998 forest fire season in Greece: a forest fire expert's account, Int. For. Fire News. 20 (1999) 57−60.

[21] G. Xanthopoulos, People and the Mass Media during the fire disaster days of 2007 in Greece, in: Proc. Int. Bushfire Res. Conf. Fire, Environ. Soc., 2008, pp. 494−506.

[22] G. Xanthopoulos, Olympic flames, Wildfire 16 (2007) 10−18.

[23] G. Xanthopoulos, M. Athanasiou, FIRE GLOBE: Attica Region, Greece (July 2018), in: Wildfire Mag, Internantional Association of Wildland Fire, 2019, pp. 18–21. https://digitalis.uc.pt./handle/10316.2/44582.

[24] J. Goldammer, G. Xanthopoulos, G. Eftichidis, G. Mallinis, I. Mitsopoulos, A.A. Dimitrakopoulos, Report of the Independent Committee Tasked to Analyze the Underlying Causes and Explore the Perspectives for the Future Management of Landscape Fires in Greece, 2019.

[25] M. Beighley, A.C. Hyde, Portugal Wildfire Management in a New Era Assessing Fire Risks, Resources and Reforms, 2018. https://www.isa.ulisboa.pt./files/cef/pub/articles/2018-04/2018_Portugal_Wildfire_Management_in_a_New_Era_Engish.pdf.

[26] R. Gazzard, J. McMorrow, J. Aylen, Wildfire policy and management in England: an evolving response from Fire and Rescue Services, forestry and cross-sector groups, Philos. Trans. R. Soc. Biol. Sci. 371 (2016) 20150341. https://doi.org/10.1098/rstb.2015.0341.

[27] M. Castellnou, M. Miralles, Evaluación de la capacidad de trabajo de las maniobras de extinction con fuego forestal del Cos de Bombers de la Generalitat, 2013.

[28] F. Tedim, V. Leone, M. Amraoui, C. Bouillon, M. Coughlan, G. Delogu, P. Fernandes, C. Ferreira, S. McCaffrey, T. McGee, J. Parente, D. Paton, M. Pereira, L. Ribeiro, D. Viegas, G. Xanthopoulos, F. Tedim, V. Leone, M. Amraoui, C. Bouillon, M.R. Coughlan, G.M. Delogu, P.M. Fernandes, C. Ferreira, S. McCaffrey, T.K. McGee, J. Parente, D. Paton, M.G. Pereira, L.M. Ribeiro, D.X. Viegas, G. Xanthopoulos, Defining extreme wildfire events: difficulties, challenges, and impacts, Fire 1 (2018) 9. https://doi.org/10.3390/fire1010009.

Part Three

Towards a New Approach to Cope with Extreme Wildfire Events and Disasters

The suppression model fragilities: The "firefighting trap"

7

Gavriil Xanthopoulos[1], Vittorio Leone[2], Giuseppe Mariano Delogu[3]
[1]Hellenic Agricultural Organization "Demeter", Institute of Mediterranean Forest Ecosystems, Athens, Greece; [2]Faculty of Agriculture, University of Basilicata (retired), Potenza, Italy; [3]Former Chief Corpo Forestale e di Vigilanza Ambientale (CFVA), Autonomous Region of Sardegna, Italy

7.1   The dominant fire management approach today: The wildfire suppression model

7.1.1   Wildfire suppression model: rationale

Chapter 6 presented a short account of the way in which wildfire fighting developed as the main fire management approach in the 20th century, initially in the U.S.A. and then in Europe, Australia, and elsewhere, gradually becoming dominant, leading to firefighting organizational growth and complexity, and finally to an ever-increasing cost. Today, the suppression model remains the universal response to wildfire occurrence [1]. It strategically deploys fire suppression resources (crews, vehicles, aerial means) to extinguish wildfires as quickly as possible to contain their spread and mitigate impacts on environment and assets. The rationale of such a model is an aggressive and prompt fire control, inspired by the principle that fire represents a universal threat to people, resources, and wildlands [2]. It is a typically reactive response, based on the paradigm of "war against fire" defined as *"an antiquated 'zero-tolerance' fire management paradigm that reigned for nearly a century and is only now starting to be replaced by more evidence-based strategies"* [3].

The current policy of war against fire, based on more firefighters, more airplanes, stricter rules, and stronger tactics [4–9], aimed to timely and strongly react to fire occurrence, can appear successful as it is likely to reduce damages in the short term [10], but it fails in addressing the roots of the increasing wildfire potential. The fire suppression paradigm has not been able to solve the wildfire problem anywhere in the world. On the contrary, at a global scale, the problem is expanding and previously immune countries, such as United Kingdom, Sweden, Germany, all Eastern Europe, etc. seem to be facing problems they were not familiar with in the recent past. The current tendency for such countries is to subscribe to the same suppression model that has failed to solve the problem elsewhere.

Understanding the fire suppression dominance model requires to carefully observe the mentality and terminology associated with it. The narrative uses military metaphors and language [11] to describe strategy, tactics, organization, and firefighting operations based on the Fight, Control, Exclude principle (command-and-control chain model

Extreme Wildfire Events and Disasters. https://doi.org/10.1016/B978-0-12-815721-3.00007-2

[12]). For instance, an explicit comparison of firefighting organization with an army is by Vélez [13], and the slogan *"hit hard, hit fast"* is evidently the motto *"Hit hard! Hit fast! Hit often!"* by Admiral William Frederick Halsey Jr. of the U.S. Navy in the Pacific war theater, during WWII. Put in another way, the aim is to control all wildfires through early detection and initial attack when they are still small in size [14,15]. The reaction must be strong and immediate (*"muscled attack"* dispatch policy [16]; *"un combate rapido e contundente"* i.e., aggressive, blunt, forceful, impressive, robust, vigorous fight [17,18]. The required resources are very large, and their effective management is a real challenge. The Incident Command System (https://www.ukfrs.com/foundation-knowledge/foundation-incident-command) developed in the U.S.A. and used in many countries is a response to this challenge, widely advocated for the management of the high number of firefighters intervening in large, intense, and complex wildfires.

7.1.2 Suppression model dominance: why is this model so widely accepted?

The broad acceptance of the fire suppression model can be surprising to anyone analyzing long-term fire statistics because of the realization that throughout the application of this paradigm in the last 100 years or so, the wildfire problem has been worsening. Therefore, the question immediately comes up: Why is this model so widely accepted? Obviously, there can be no unique and definitive explanation, but potential reasons include the following:

- Aggressive firefighting for immediate protection of threatened values (life, property, infrastructures, environment) is an obvious response to avoid or mitigate damages.
- The impressive nature of fire suppression operations (e.g., water bombing of fires), the heroism involved, and the sense of strength when the objectives are attained.
- There are many interests related to providing wildfire suppression services, from the agencies carrying the responsibility to the providers of services and technological products.
- The results of fire suppression are immediately obvious. This can be important to the public, as well as to politicians and other decision-makers. Having to work under the pressure of the next elections, they are more likely to choose options that will bring immediate results rather than invest in long-term strategies based on wise management of natural resources.
- Politicians and other decision-makers often lack specialized knowledge about the environment and socioeconomic issues as related to wildfires. Even more often, they are not motivated to devote the necessary time to learn about it. They want to "keep-it-simple." Wildfire suppression fills this need. It is an easy-to-explain concept that does not require deep knowledge about the environment and other underlying factors.

Given these reasons, it is quite clear that effective fire suppression is a realistic option and most likely will be such an option in the future.

7.2 Assessment of the fire suppression model

7.2.1 The failures of the fire suppression model

Fire exclusion policies can be efficacious on the short term, but they are not sustainable in the long run [19,20], as the combination of fuel accumulation, caused by a generalized fire exclusion policy, and severe weather, can overwhelm any suppression capacity, fostering larger and more severe fires [7,10,21].

The immediate and fast suppression, advocated by the 10 a.m. policy in the U.S.A. since 1935, started being questioned a few decades later. While in the context of the ecological knowledge of the time, fire exclusion was believed to promote ecological stability and also reduce damages and economic losses (Forest History Society), already by 1960 and 1970s, accumulating evidence revealed the negative changes that were taking place. This evidence varied from comparison of historic photos to current conditions [22,23], to ecological changes detection [24], to stand density and fuel load comparisons [25], to reports on the growing firefighting challenges, to the increasing number of very large fires and associated damages, and ultimately, to the disappointing trends of fire statistics regarding yearly number of fires, total burned area, and fire suppression costs (Fig. 7.1). The U.S. Forest Service currently faces fires much greater in size and ferocity than 20−30 years ago. The firefighting budget has grown to about 50% of the agency's entire budget, which limits funds available for land management activities such as land restoration and forest thinning that could aid in fire suppression.

In short, fire scientists and managers put in evidence that total suppression in the U.S.A. was producing forests with high fire hazard, and such forests were being burned by high-severity wildfire [26]. They showed that fire suppression was causing more harm than good, stressing the crucial role fire plays in forest ecosystems. A policy

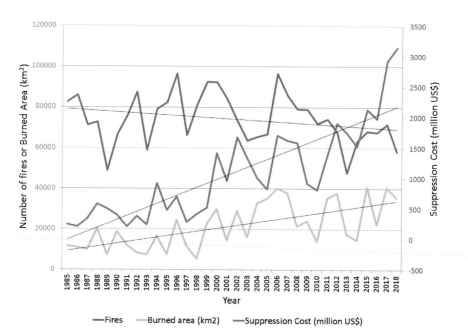

Figure 7.1 Number of fires, burned area, suppression cost, and linear trend lines in the U.S.A., for the 1985−2018 period.
Source of data: National Interagency Fire Center.

of total suppression certainly reduces the total burned area, at least initially, but has the unintended side effect of fuel accumulation in the form of more densely spaced trees and thicker undergrowth [27,28].

The fire suppression policy has been applied widely in many countries other than the U.S.A. including the countries of Mediterranean Europe, often assigning wildfire management to urban fire suppression organizations without considering first the specific conditions and the appropriateness of the model [29], usually with poor long-term results. For example, in Greece, the transfer of responsibility of wildfire fighting from the Forest Service to the Urban Fire Service in 1998 was followed by a steep increase in the cost of fire suppression, which more than doubled with the addition of more fire trucks and powerful aerial resources. At the same time the Forest Service weakened, time, effort, and funding for fire prevention diminished. A comparison of the average yearly burned area in the country in the years before and after the change shows that it has changed very little, from 44,805 ha in the 1977−97 period to 45,018 ha in the 1998−2017 period. However, in the last 2 decades, Greece has suffered two disastrous fire seasons that broke every previous record with 167,000 ha (in 2000) and 261,500 ha (in 2007) [30]. It also faced an increasing number of wildfire disasters, while wildfire-related fatalities that were missing before 1977 started becoming a regular event, exceeding 80 losses in 2007 and 100 losses in 2018 [31]. The situation, as it developed, has been assessed as a clear failure [30].

Although the total fire suppression policy has been strongly questioned, its legacy remains strong, not only in the U.S.A. but also in other countries around the world that chose to follow the example. This legacy includes support for the criminalization of fire and establishment of restrictive regulations on fire, making difficult its use (under the form of prescribed fire, backfire, suppression fires) as a management tool [32,33]. The restrictive approach is inspired by an urban-centric perspective, where the traditional use of fire [8], as a land management tool, is excluded. The potential contribution of fire in wildfire prevention is rejected, and the possibility of using fire for suppression by local communities is ignored [34]. This rooted antifire sentiment is based on the postulate of a fragile and threatened Mediterranean forest [35] where *".... forestry had demonized fire and the Mediterranean alliance between fire and herds, particularly goats"* [36].

7.2.2 Weaknesses of the suppression model

The evident reasons for the acceptance of the suppression model and the significant weaknesses outlined previously are best put in perspective through the "Strengths, Weaknesses, Opportunities, Threats" (SWOT) analysis of current and expected pros and cons regarding the suppression model, published by Tedim and Leone [37] (Table 7.1).

Examining Table 7.1, a marked unbalance between negative and positive points is evident, as $S < W$, $O < T$ (their number are 6 v. 24, 4 v. 10, respectively). Given the predominance of Weaknesses and Threats that largely prevail, a possible strategy is the WT one (mini-mini or defending strategy), exploiting the possible changes of the weaknesses [38]. This strategy is consistent with the shift of paradigm and

Table 7.1 SWOT analysis of the *war against fire* paradigm.

Strengths	Weaknesses
Hard and fast attack can contain the size of fires and limit damages and cascading effects Capacity and priority to protect people and assets, integrated by visible structural prevention infrastructures, fits the public expectation for intervention and gives people a feeling of safety even in high-risk areas A better deployment of resources, equipment, and technology can improve results It is consistent with the short "political life" of decision-makers; short-terms results are visible and politically expedient Gets efficacious support from dedicated legislative corpus. Allows clear-cut identification of those responsible for starting fires (generally called "arsonists"): This can create a sense of security for people and institutions	Incapacity to reduce fire occurrence Symptomatic approach, where wildfire is merely a civil protection emergency Scarce attention to land management measures of prevention and risk awareness Neglects the ecological role of fire in ecosystems Ignores the traditional fire knowledge (T.F.K.) or rural people and criminalizes the use of fire as a management tool Episodic use of prescribed burning, by institutional actors, without acknowledging the T.F.K. of rural people Reluctance for use of fire-for-fire-control methods ("suppression fires") Attack based on water use Wildfires easily overwhelming the extinction capacity of water use Excess of expectation on the efficacy of the aerial intervention Easily leads to misuse and overuse of aerial resources Suppression technology may create a false sense of safety, thus influencing the prefire mitigation or fire-saving measures by communities and individuals Unresponsive during multiple simultaneous emergencies, and in case of extreme wildfire events (EWEs) (Tedim et al. 2018) Difficult and costly intervention in wildland–urban interface (WUI) areas Reactive in short-term response to wildfire events, but diminished importance as they are distanced by time Approach to new challenges only based on the reinforcement of equipment and new technologies, with increasing costs Resources mobilization concentrates on response instead of reinforcing prevention and looking for the integration of the two State-centric and top-down legislation, with little if any decision-making powers at the local level Scarce consultation with local populations and lack of participatory approach A top-down bureaucratic structure Disaster legislation focused on the response, with scarce attention to reduce vulnerability and enhance resilience Command-and-control structures akin to civil defense and civil protection modes Focus more on building the institutional capacity of firefighters than of communities Communities and individuals abdicate from their responsibilities when public powers substitute them in coping with wildfires

Continued

Table 7.1 SWOT analysis of the *"war against fire"* paradigm.—cont'd

Strengths	Weaknesses
Opportunities	**Threats**
The political mandates appreciate and reinforce the suppression organization The legislative corpus sustains the activity and continuity of the suppression model Media and public opinion pressure support the suppression model The efforts of the organizational system to guarantee its own survival and resist to accept changes	Limited financial resources to sustain the increasing and staggering costs of suppression Increase of fire hazardous landscapes, due to the changes of connectivity of vegetal cover Expansion of the WUI Increased frequency and intensity of weather extreme conditions and longer fire seasons projected with climate change (CC) Decreasing availability of water resources Uncertainty of anthropogenic impacts on wildfire regimes Uncertainty on the contexts in which wildfire management will operate in future Increased vulnerabilities (e.g., economic, social) to wildfires Possible decrease of human resources in rural areas to support voluntary firefighters, due to depopulation and aging Escalating of EWEs occurrence

Adapted from the study by Tedim and Leone[37].

alternative sustainable ways to *"coexist with fire"* [19,39–41]. The following comments refer to the most relevant items evidenced by the SWOT analysis.

7.2.2.1 Operational limits

The main shortcoming of wildfire suppression policies lie in the incapacity of firefighting forces to control wildfires exceeding the commonly accepted fireline intensity limit of $10,000 \, \text{kW} \, \text{m}^{-1}$ (see chapter 1). Even at $4000 \, \text{kW} \, \text{m}^{-1}$, as most experts agree [42–45], control is already extremely difficult, and efforts at direct control are likely to fail. This means that firefighters cannot control those few fires, which although infrequent, often in the form of extreme wildfire events (EWEs), are responsible for the majority of damages and often for grand-scale disasters. Current statistics in most developed countries that face wildfire problems show that 1%–5% of all wildfires become large incidents, thus accounting for about 85% of total suppression expenditures and up to 95% of the total burned area [46]. While fire suppression organizations tend to boast about such statistics as proof of their success in initial attack of wildfires, the truth is that the overall destruction potential finally remains intact.

Assuming that wildfire intensity can reach peaks of $150,000 \, \text{kW} \, \text{m}^{-1}$, as observed in fires of 2009 in Victoria [47], much over the previous limit of $100,000 \, \text{kWm}^{-1}$ considered insurmountable [48,49], even the most advanced and well-organized suppression services only operate on about the lower 7% of the variability range of fireline intensity. A costly and complex organization, even having the availability of aerial re-sources, is thus only able to control fires of low to medium intensity that represent the majority of events but is unable to contain medium high to extreme values of intensity, whose number is expected to increase in the near future, becoming the *"new normal"* [50,51] and responsible for most of the impacts.

However, even fires with low to medium intensity can exceed control capacity, when a number of them occur at the same time, but dispersed in a wide territory, often obliging to split operational units, or to prioritize response, leaving some fires unattended. The pressure on firefighters, fostered in periods of maximum fire fre-quency to extinguish fires and quickly move on to new ones (the dual-duty problem [16]), may result in insufficient mop-up and consequently in repeated rekindles. Another case is the occurrence of wildfires in WUI or rural—urban interface (RUI) areas [52], where the priority of intervention largely favors defense of people and assets, so obliging to deploy human and technical resources in a very dedicated way.

After the threshold of 10,000 kWm^{-1}, firefighting actions are inefficacious against fires of levels 5 to 7 in the wildfire classification by Tedim et al. [21]. As intensity increases, fire behavior changes in a nonlinear way, control capacity gradually loses efficacy until it is overwhelmed by fire. Organization improvements that, at an ever-increasing cost, lead to better initial attack and successful suppression of less intense fires finally do not avert the major disasters caused by EWEs. Obviously then, where intensity values of such level are expected, the aim should not only be the control or containment of the flames front, when and if possible, but a coordinated series of pro-active actions directed to prevent and mitigate fire action and reduce its impacts. Because wildfires, and especially EWEs, are the resultant of a complex interplay be-tween natural and social conditions, processes and factors in all the wildfire chain (i.e., prevention and mitigation, ignition, spread, suppression, impacts, recovery and restoration) [21] a gradual shift from the suppression model to a different one, more characterized by prevention, is advocated by a multiplicity of authors [19,39—41].

7.2.2.2 Lack of integration with prevention

Firefighting centralized services are usually not integrated with prevention. Prevention under the total fire suppression model, if and where implemented, is directed toward structural measures (e.g., creation of water points, opening of roads, creation of fire-breaks and other fuel management network) and to raise public awareness by pro-grams, designed to inform target groups about the need for fuel management, protection around buildings and settlements and wise behaviors, through coercive measures. This type of intervention can rarely be effective because it ignores the social context where fires occur. The main form of sensitization is a passive diffusion of in-formation, but since 1980s, the scientific community demonstrated that this type of performance is ineffective [53—58] and that the most effective forms are based on the involvement of populations, which means interactive communication and collab-orative action between organizations and citizens [59].

7.2.2.3   Reduced attention to fire causes

Wildfire prevention cannot ignore why, when, and where fires are more likely to occur. Based on such knowledge, prevention strategies can be designed to directly target the root of the problem [60]. The lack of a strategic problem-solving approach results in short-term fixing of problems, or merely in suppression of their symptoms, rather than understanding and addressing the underlying factors that cause them. The long-term outcome is very poor as it does not take into account that wildfire risk is a socio-ecological "pathology" ([61] p. 276): *"a set of complex and problematic interactions among social and ecological systems across multiple spatial and temporal scales, that deviate from what is considered healthy or desirable,"* and that *vulnerability to environmental hazards depends on both biophysical and social factors that respectively determine where wildfire potential is elevated, and where and how people are affected by wildfire* [62].

Suppression of wildfires irrespective of their causes does not profit from prevention activity, aimed at reducing them. From one side, they are symptoms of social disruption or conflicts, from the other of a generalized negligence or lack of prudence. The two main groups of causes evidently demand different approaches of analysis and solution. Investigation of fire causes is not an easy task [63,64]. Efforts for better identification of ignition points by the Physical Evidences Method [65], and alternative efforts to assess fire causes such as the Delphi method [66,67] cannot result in high degree of certainty. As a matter of fact, a high percentage of the causes remain unknown and cannot be categorized [60]. In the European Union, the creation of a harmonized classification facilitates the understanding of the different realities at European level [68].

Correctly identifying, recording and analyzing fire causes can provide critical support to prevention efforts. For example, in many cases, a high number of simultaneous fires in some geographic areas are often a blatant expression of malicious fire setting, demanding specific targeted prevention activities. Assuming the problem as a social-ecological one helps to identify the critical cross-scale dynamics that are not currently part of much of the wildfire discussion [69].

Important changes in land use and land settlement (rural vs. urban settlements) in the last 30 years, in many rich areas of the planet, have caused a great amount of fuel accumulation or a typical change in fuel classes (from grasses to woody biomass for example) that create an unexpected new hazard near houses or residential villages. An obvious example are touristic areas such as those along the Mediterranean Sea coast, where the presence of a traveling "temporary" population in the summer totally changes the safety conditions. In this case, the concept of "prevention" is totally difficult to be applied by people not knowledgeable nor interested in the basics of fire risk. They can easily become the culprits of fire starts because of negligence, and they can also be exposed to danger in case of a fire event in their vicinity, because of their lack of knowledge on how to react regarding evacuation or escape, as tragically demonstrated by the Pedrogao Grande fire event in June 2017, in Portugal, and the eastern Attica fire in July 2018, in Greece.

7.2.2.4   Lack of communities' engagement

The top-down suppression approach often operates without consultation with local communities, which are excluded from any initiative of defense in the territory where they live. This approach once again vilifies the opportunity to exploit traditional fire knowledge (T.F.K.), the wisdom-based customary uses of fire [8], which can be of crucial importance in case of extreme fire events. As there is no possibility of direct attack because of intensity level, among the few possible and potentially effective interventions can be the use of suppression fires in the form of backfire or burning-out [32]. In addition, in the implementation of prescribed burning activities, the T.F.K. of operators can make the difference.

7.2.2.5   Lack of awareness

An effective risk communication program demands a participatory approach involving the community in the planning and solution process [70] and the need to tailor fire preparedness messages and actions to the audience [71]. The shift from an awareness approach to one focused on building resilience alters the top-down relationship, giving community an actively participant role [72]. It is necessarily a two-way, interactive, and long-term process, where agencies and communities are engaged in a dialog [73]. Moving beyond simple one-way messaging toward developing an outreach effort more likely to increase community preparedness requires an understanding of communities in relation to wildfire mitigation and management.

Awareness is only one component of behavior changes [55,74]. The goal of risk communication is to inform about fire hazard but also to offer guidance and opportunities (support, legal context) to people to take actions to prevent fire outbreaks, to reduce fire intensity, and to mitigate the risk. Preparedness results not just from providing information but also from ensuring that people can use information and resources to meet their needs. How individuals respond to information is a function of the content (pragmatic, tailored to needs, addressing constraints) and the overall credibility of the information sources.

7.2.2.6   Building resilience

To build resilient communities, risk communication should promote the engagement of communities and be integrated as a daily practice of development [37,70,75]. The challenge is how to maintain high preparedness over time. A well-prepared community also has increased capacity for recovery, which is likely to proceed more smoothly than in places where no prefire planning has taken place [76].

7.2.2.7   Short-term perspective

The lack of any provision for the long run, the inexistence of reference to the cost, and the unawareness of environmental constraints [77], the incapacity to master and be adapted to the dynamic characteristics of fire environment, emphasized by local (e.g., changes in forestry and agricultural production systems, WUI expansion) and global changes (e.g., climate change) [52], complete the list of weaknesses of the suppression model.

7.2.3 The firefighting trap

After decades of fire suppression policies, and very restrictive legal frameworks of traditional fire use, aimed to curb the increasing trend of fire occurrence, it is recognized [78–80] that the all-out suppression policy that dictates intervention on all fire events (i.e., the concept that fires are a suppression challenge, a Civil Protection task, rather than a symptom, underlying management or social problems [34]) seems more and more inadequate to cope with a complex phenomenon in which ecological, social, economic, historical, and cultural components compound [37], failing to solve the wildfire issue in the long term.

The approach merely directed to fire extinction, without due attention to prevention, can be fairly described by the metaphor of the firefighting trap [10,81,82] well known in the business management domain [10], where *"putting out fires"* allegorically expresses the concept to deal with problems (fires). Despite appearances, the metaphor is stranger to the wildfire domain: It is used by managers to describe a short-sighted cycle of problem-solving, dealing with "fires," or problems, as they arise, mainly by suppressing their symptoms, rather than understanding and addressing the factors that cause the problem. This approach increases the chance that the same problem will crop up again in the future.

The metaphor of fighting fires is widely used in literature, typically to indicate the allocation of resources to solve unanticipated "fires" and refers to the unplanned allocation of human and other resources to fix problems discovered late in a product's development cycle. It describes the impediments to successful process execution in new product development [82] and can be considered as a type of crisis management that persists as a steady-state phenomenon and is a principle source of low productivity [83]. Management researchers have identified for a long time now that "firefighting" constitutes a serious impediment to performance in many product development environments, but in spite of this widespread consensus on its evils, it paradoxically persists and can often be a self-reinforcing phenomenon. Owing to a combination of structural and psychological factors, changes are very difficult, especially in case of complex multistage environments [82].

An original transposition of the aforementioned concept to the domain of wildfire control is by Collins [84] who coincidentally explored the inspiration for the metaphor of a "quick-fix" management strategy when dealing with true firefighting [85], to describe the unintended consequences of decision-making focused on fixing rather than preventing problems.

In his thesis, analyzing fire control activity for Portugal [84], with the scope to draw up a model illustrating the relationships that contribute to forest-fire management, Collins identifies the same critical steps that correspond to the way a problem should be resolved in the domain of product development environments: *"Solving a problem takes time: an engineer must study the symptoms, confirm that the problem is real, conduct background research, diagnose its causes, search for a good solution, and implement the solution"* ([81], p 83).

All the aforementioned explain the apparently instinctive management response by political decision-makers to situations of worsening wildfire occurrence: Under the pressure by the public and the media, mainly exerted by people living in WUI areas, they are typically inclined to allocate significant resources in more suppression and advanced technology to control fires. They adopt seemingly rational decisions, with a logical immediate response of targeting symptoms, where energy and resources are mostly devoted to fire suppression, while less attention is paid to fire prevention, i.e., to the longer term actions that address the underlying causes [10,85].

There is no doubt about the appeal of this kind of intuitive decision that directly treats the symptoms of devastating fires and appeases the public [10] because it may certainly mitigate mild fire damages in the short term, but with the effect of reducing prevention resources. The result is to undermine fuel removal initiatives, such as pre-scribed fire, and fuel management in forests, aimed to the strategic reduction of fuel load, putatively leading to greater events, which act to further inflate suppression budgets. The obvious increase in fuel continuity and fuel load has expectable conse-quences in terms of higher intensity and rate of spread (ROS), in case of a fire event, and, of course, higher difficulty of control. This translates to more burned area, greater damages, and further pressure to manage fires with additional suppression forces ([10]; p. 6). As a result, an apparently sound management option can create a vicious self-reinforcing cycle, whereby the problem continues to get worse yet the so-lution remains the same ([84]; p. 34), as theorized by the firefighting trap metaphor.

On the other side, preventative fuel removal is a useful strategy for mitigating fire intensity, but its effectiveness may only be realizable at large scale, which demands appropriate budgeting and planning on a pluriannual basis.

In the perspective of new scenarios, characterized by an increasing number of EWEs, which can become the *"new normal"* [51], even the most aggressive suppression activ-ity, enhanced by the impetus of events on political decision-makers, may have success only on fires under the threshold of control capacity already mentioned. During extreme fire activity, in addition to the absolute incapacity to extinguish such fires, an insufficient mop-up can result, which increases the incidence of rekindles and thus number of fires to suppress ([84]; p.118). Once again, a vicious self-reinforcing cycle is the issue of deci-sions that merely focus on symptoms. The lesson to be learnt is that a preferential focus on fire suppression for fixing problems of wildfire occurrence provides immediate ben-efits but can become an inferior policy over time ([10]; p.6), because of its unintended negative consequence in terms of increasing total fire damage.

On the contrary, a balanced approach of suppression and prevention activities seems the obvious solution to minimize the total burned area, without partial emphasis [10] that only exacerbates the problem.

7.3 A proactive model as a possible alternative

Although simple and convincing, the gradual shift from the paradigm of fire suppres-sion to the alternative one of *"coexisting with fire,"* mainly based on prevention, as advocated by many authors [39–41,86], is not an easy bureaucratic initiative. Many facts and conditions represent strong barriers to this solution. First of all, the fire exclusion model is a social construct [87] strongly influenced by media, resulting in increasing expenditures on high-tech firefighting solutions: It emotionally and psychologically resonates with public opinion [10] but above all fits the public demand for total suppression, often driven by an irrational assessment of wildfire risk [37] and the negative perception of charred landscapes, emphasized by the mass media [88].

The advocated shift toward more prevention substantially changes the approach: suppression operates on the evident symptoms but not on the roots of the event, only aiming to reduce its impact, whereas prevention operates on the single components of fire occurrence, making their interplay difficult. Actually, the current suppression model is a simplistic reply to an extremely complex phenomenon, basically considering it almost as a mere combustion process, without proper attention to the presence of social components in all the phases of the wildfire chain (i.e., prevention and mitigation, ignition, spread, suppression, impacts, recovery and restoration), even in those traditionally considered as merely biophysical ones [21]. Prevention includes a wide array of actions and interventions, ranging from land-use regulations and law enforcement to forest management, fuel management, public education and awareness, building codes (engineering techniques and hazard-resistant construction), and infrastructures (e.g., lookout towers, water filling points, firebreaks, forest roads). Table 7.2 reports a comparison of the two paradigms [89], synthesizing the characteristic of the advocated proactive management.

Difficulties arise from a series of obstacles, some of them internal to the model of suppression, for instance, the negative reaction of the established firefighting organizations toward a different allocation of resources and budgets, which can turn into diminishing economic advantages. Investments in prevention (e.g., fuel management), although strongly advocated by research [6,9,39–41,90,91], are less attractive to the public and policymakers that, as a rule, have limited time in governance [10]. In prevention they lack immediate, visible, short-term benefits, while, on the contrary, fire suppression is an evident, short-term, tangible activity. Preventive initiatives are less attractive to policymakers as the result can never be attributed to them with certainty [10]. Consequently, as preventive actions are less visible than suppression equipment (e.g., fire trucks, air bombers, helicopters, UAVs), they receive less political attention and, subsequently, fewer resources [92].

The negative long-term effects of the fire suppression policy, the persistence of the paradigm, the negative reaction of the firefighting organizations, including a resistance to cooperate with other agencies, the complexity of the problem, have been recognized and documented in a recent study by an international committee that was appointed by the Government of Greece, in the aftermath of a wildfire in Eastern Attica with more than 100 fatalities in July 2018 [31]. The committee was tasked to investigate the reasons for the worsening trend regarding wildfires in Greece. In its report it clearly identified the existence of the firefighting trap. It also recognized the difficulty to escape from it, as this cannot be done through a quick and easy legislative fix. The report recommends establishment of a small, highly qualified and powerful coordinating organization to oversee a step-by-step process in which suppression and prevention will be

Table 7.2 Management paradigms and their main characteristics.

Current management paradigm *"War against fire"*	Proactive management paradigm *"Coexist with fire"*
Symptomatic approach (acting on results). Wildfires mainly as a Civil protection task	Etiologic approach (acting on problems' roots) wildfires mainly as a problem of resources management and of resolution of social conflicts, only marginally as a problem of Civil protection
Focus on forest	Focus on the territory
Local and sectorial perspective	Holistic perspective in the frame of coupled human−natural systems (CHNS)
Fire is always a threat	Wildfires are always a threat, whereas fire can be beneficial, as an ecological factor and land management tool
Quick and muscular attack, to keep fire as small as possible	Focus on prevention, search of complementarity and integration with prevention and mitigation, preparedness, emergency, recovery
Criminalization of the traditional fire use as a land management tool, of prescribed fire use. No acknowledgment of traditional fire knowledge.	Fostering collaboration with communities living in the territory, who have resources and knowledge for using fire in a wise and aware way
Sensitization mainly based on dissemination of info and best behavior rules	Sensitization based on participative involvement of communities, their empowerment for active participation and better resources exploitation
Emphasis on reinforcement of technology and improvement of extinction procedures	Directions for wildfire management entailing *"more resilience and a less combative relationship with nature"* [86]
Prevention focused on risk mitigation by infrastructures and evacuation of exposed people	Prevention based on the modification of causes, integrated in actions and activities of daily life also for promoting sustainable development
Scarce or null consultation of local communities	Collaboration among all actors living in the territory, composition of conflicting interests, search of synergies among actors
Concept of territory absent in policies of forest-fire fighting	Centrality of the concept of territory in risk reduction

balanced, both in terms of emphasis and funding, and a continuous monitoring and assessment process will assure that all actions will focus on both effectiveness and efficiency and on long-term results [30].

7.4 Conclusions

In conclusion, the dominance of the wildfire suppression model can be easily understood and explained. It is more than likely that strong fire suppression mechanisms will continue to be advocated in the future. However, the strong weaknesses of the model have been identified by scientists and have been demonstrated by statistics, and its failure to solve the wildfire management issue in the long term has become evident. As shown in this chapter, the situation of how countries and agencies respond to the worsening of the wildfire problems with ad-hoc, quick-fix, suppression-oriented measures has all the characteristics of the "firefighting trap" problem that is well known in the world of business management for a long time now. Accordingly, to escape from the trap is not easy, as there is strong resistance by fire suppression organizations, by people and politicians who cannot grasp the problem and are accustomed to simplistic solutions of the type "more wildfire problems, more firefighting." Any serious effort for change will need to be based on promoting understanding of the firefighting trap reality, and of ensuring long-term support for the necessary reorganization, priority setting, and redistribution of funds, followed by monitoring and assessment, by a highly specialized, open-minded and dedicated group of wildfire experts.

References

[1] T.J. Duff, K.G. Tolhurst, Operational wildfire suppression modelling: a review evaluating development, state of the art and future directions, Int. J. Wildland Fire 24 (2015) 735, https://doi.org/10.1071/WF15018.
[2] G.H. Aplet, Evolution of wilderness fire policy, Int. J. Wilderness. 12 (2006) 9−13. http://ijw.org/wp-content/uploads/2006/12/Apr-2006-IW-vol-12-no-1small.pdf#page=10.
[3] SciLine, Wildfire Trends in the United States, 2018. https://www.sciline.org/evidence-blog/wildfires.
[4] F. Moreira, O. Viedma, M. Arianoutsou, T. Curt, N. Koutsias, E. Rigolot, A. Barbati, P. Corona, P. Vaz, G. Xanthopoulos, F. Mouillot, E. Bilgili, Landscape - wildfire interactions in southern Europe: implications for landscape management, J. Environ. Manag. 92 (2011) 2389−2402, https://doi.org/10.1016/j.jenvman.2011.06.028.
[5] P.M. Fernandes, Creating fire-smart forests and landscapes, Forêt Méditerranéenne. XXXI (2010) 417−422, n° 4, http://documents.irevues.inist.fr/bitstream/handle/2042/39228/Fmnoabon.pdf?sequence=1.
[6] P.M. Fernandes, Fire-smart management of forest landscapes in the Mediterranean basin under global change, Landsc. Urban Plan. 110 (2013) 175−182, https://doi.org/10.1016/j.landurbplan.2012.10.014.

[7] P. Mateus, P.M. Fernandes, Forest Fires in Portugal: Dynamics, Causes and Policies, Springer, Cham, 2014, pp. 97−115, https://doi.org/10.1007/978-3-319-08455-8_4.

[8] M.R. Huffman, The many elements of traditional fire knowledge synthesis, classification, and aids to cross-cultural problem solving in fire-dependent systems around the world, Ecol. Soc. 18 (4) (2013) 3, https://doi.org/10.5751/ES-05843-180403.

[9] P. Corona, D. Ascoli, A. Barbati, G. Bovio, G. Colangelo, M. Elia, V. Garfi, F. Iovino, R. Lafortezza, V. Leone, R. Lovreglio, M. Marchetti, M. Marchi, G. Menguzzato, S. Nocentini, R. Picchio, L. Portoghesi, N. Puletti, G. Sanesi, F. Chianucci, Integrated Forest Management to Prevent Wildfi Res under Mediterranean Environments, vol. 38, 2015, pp. 24−45, https://doi.org/10.12899/ASR-946.

[10] R.D. Collins, R. de Neufville, J. Claro, T. Oliveira, A.P. Pacheco, Forest fire management to avoid unintended consequences: a case study of Portugal using system dynamics, J. Environ. Manag. 130 (2013) 1−9, https://doi.org/10.1016/J.JENVMAN.2013.08.033.

[11] D.M. Smith, Sustainability and Wildland Fire: The Origins of Forest Service Wildland Fire Research, US Department of Agriculture, Forest Service, Forest Health Protection, 2017.

[12] C.S. Holling, G.K. Meffe, Command and control and the pathology of natural resource management, Conserv. Biol. 10 (1996) 328−337, https://doi.org/10.1046/j.1523-1739.1996.10020328.x.

[13] R. Vélez, La Defensa Contra Incendios Forestales: Fundamentos Y Experiencias, second ed., McGraw-Hill nteramericana de España, Madrid, 2009.

[14] Canadian Forest Service, Canadian Wildland Fire Strategy: A Vision For An Innovative And Integrated Approach to Managing the Risks, Alberta, 2005, http://www.ccmf.org/pdf/Vision_E_web.pdf.

[15] Forest Practices Board, Fire Management Planning. Special Investigation, Victória, Canada, 2012, http://www.nss-dialogues.fr/IMG/pdf/5thIWFConference2011.pdf.

[16] A.P. Pacheco, J. Claro, T. Oliveira, Simulation analysis of the impact of ignitions, rekindles, and false alarms on forest fire suppression, Can. J. For. Res. 44 (2014) 45−55, https://doi.org/10.1139/cjfr-2013-0257.

[17] J. Guarque, Estrategias y tácticas, los puntos clave en la gestión de los incendios forestales en las WUI, in: Los Incend. for. En Interfaz Urbana Hacia Una Integr. Del Riesgo En La Planif. Del Territ., Barcelona, Spain, 2014. http://www.paucostafoundation.org/administrador/app/webroot/uploads/1415875280.pdf.

[18] G.M. Delogu, Dalla parte del fuoco, ovvero, Il paradosso di Bambi, Il maestrale, 2013. https://books.google.pt/books?id=KfX2ngEACAAJ.

[19] R.L. Olson, D.N. Bengston, L.A. DeVaney, T.A.C. Thompson, Wildland Fire Management Futures: Insights from a Foresight Panel, 2015, https://doi.org/10.2737/NRS-GTR-152.

[20] F. Tedim, V. Leone, G. Xanthopoulos, Wildfire risk management in Europe. the challenge of seeing the "forest" and not just the "trees, in: Proc. 13th Int. Wildl. Fire Saf. Summit 4th Hum. Dimens. Wildl. Fire, Manag. Fire, Underst. Ourselves Hum. Dimens. Saf. Wildl. Fire., 2015, pp. 213−238.

[21] F. Tedim, V. Leone, M. Amraoui, C. Bouillon, M. Coughlan, G. Delogu, P. Fernandes, C. Ferreira, S. McCaffrey, T. McGee, J. Parente, D. Paton, M. Pereira, L. Ribeiro, D. Viegas, G. Xanthopoulos, F. Tedim, V. Leone, M. Amraoui, C. Bouillon, M.R. Coughlan, G.M. Delogu, P.M. Fernandes, C. Ferreira, S. McCaffrey, T.K. McGee, J. Parente, D. Paton, M.G. Pereira, L.M. Ribeiro, D.X. Viegas, G. Xanthopoulos, Defining extreme wildfire events: difficulties, challenges, and impacts, Fire 1 (2018) 9, https://doi.org/10.3390/fire1010009.

[22] G.E. Gruell, Seventy Years of Vegetative Change in a Managed Ponderosa Pine Forest in Western Montana–implications for Resource Management, US Dept. of Agriculture, Forest Service, Intermountain Forest and Range, 1982.

[23] R.E. Keane, K.C. Ryan, T.T. Veblen, C.D. Allen, J. Logan, B. Hawkes, Cascading effects of fire exclusion in Rocky Mountain ecosystems: a literature review, Bark Beetles, Fuels, Fire Bibliogr (2002) 52.

[24] A.S. Leopold, Wildlife Management in the National Parks, US National Park Service, 1963.

[25] S.F. Arno, H.Y. Smith, M.A. Krebs, Old Growth Ponderosa Pine and Western Larch Stand Structures: Influences of Pre-1900 Fires and Fire Exclusion, Ogden, 1997. https://www.fs.fed.us/rm/pubs/rmrs_gtr292/int_rp495.pdf.

[26] S.L. Stephens, L.W. Ruth, Federal Forest-Fire Policy in the United States, Ecol. Appl. 15 (2005) 532–542, https://doi.org/10.1890/04-0545.

[27] D.E. Calkin, J.D. Cohen, M.A. Finney, M.P. Thompson, How risk management can prevent future wildfire disasters in the wildland-urban interface, Proc. Natl. Acad. Sci. USA 111 (2014) 746–751, https://doi.org/10.1073/pnas.1315088111.

[28] D.E. Calkin, M.P. Thompson, M.A. Finney, Negative consequences of positive feedbacks in US wildfire management, For. Ecosyst. 2 (2015) 9, https://doi.org/10.1186/s40663-015-0033-8.

[29] G. Xanthopoulos, People and the Mass Media during the fire disaster days of 2007 in Greece, in: Proc. Int. Bushfire Res. Conf. Fire, Environ. Soc., 2008, pp. 494–506.

[30] J. Goldammer, G. Xanthopoulos, G. Eftichidis, G. Mallinis, I. Mitsopoulos, A.A. Dimitrakopoulos, Report of the Independent Committee Tasked to Analyze the Underlying Causes and Explore the Perspectives for the Future Management of Landscape Fires in Greece, 2019.

[31] G. Xanthopoulos, M. Athanasiou, Attica region Greece (July 2018), Wildfire Mag. Int. Ass. Wildland Fire 28.2 (2019) 18–21. https://www.iawfonline.org/article/fire-globe-attica-region-greece-july-2018/.

[32] C. Montiel, D.T. Kraus, Best Practices of Fire Use: Prescribed Burning and Suppression: Fire Programmes in Selected Case-Study Regions in Europe, European Forest Institute, 2010.

[33] C. Montiel, P. Costa, M. Galán, Overview of suppression fire policies and practices in Europe, in: S. Silva, F. Rego, P. Fernandes, E. Rigolot (Eds.), Towar. Integr. Fire Manag. Eur. Proj. Fire Parad., . European, Joensuu, Finland, 2010, pp. 177–187.

[34] Food and Agriculture Organization of the United Nations, Community-based Fire Management: A Review, Food and Agriculture Organization of the United Nations, 2011.

[35] V. Clément, Les feux de forêt en Méditerranée : un faux procès contre Nature, Espace Géogr. 34 (2005) 289, https://doi.org/10.3917/eg.344.0289.

[36] S.J. Pyne, Vestal Fire: An Environmental History, Told through Fire, of Europe and Europe's Encounter with the World, University of Washington Press, Seattle, WA, U. S, 2000.

[37] F. Tedim, V. Leone, Enhancing resilience to wildfire disasters: from the "war against fire" to "coexist with fire Disaster resilience: an integrated approach, in: D. Paton, D. Johnston (Eds.), Resil. An Integr. Approach, Charles C Thomas, Publisher, 2017, pp. 362–383, pp. 362–383.

[38] V. Vulturescu, d. Ghiculescu, M. Țîțu, Computer aided swot analysis applied on evaluation of innovation potential in Central and South-East Europe, Manag. Sustain. Dev. 3 (2011) 11–19. http://www.cedc.ro/media/MSD/Papers/Volume 3 no 1 2011/MSD_Vulturescu_V_Ghiculescu_D_Titu_M.pdf.

[39] M.A. Moritz, E. Batllori, R.A. Bradstock, A.M. Gill, J. Handmer, P.F. Hessburg, J. Leonard, S. McCaffrey, D.C. Odion, T. Schoennagel, A.D. Syphard, Learning to coexist with wildfire, Nature 515 (2014) 58–66, https://doi.org/10.1038/nature13946.

[40] D. Paton, P.T. Buergelt, F. Tedim, S. McCaffrey, Wildfires: International Perspectives on Their Social—Ecological Implications, Wildfire Hazards, Risks and Disasters, 2015, pp. 1–14, https://doi.org/10.1016/B978-0-12-410434-1.00001-4.

[41] A.M.S. Smith, C.A. Kolden, T.B. Paveglio, M.A. Cochrane, D.M. Bowman, M.A. Moritz, A.D. Kliskey, L. Alessa, A.T. Hudak, C.M. Hoffman, J.A. Lutz, L.P. Queen, S.J. Goetz, P.E. Higuera, L. Boschetti, M. Flannigan, K.M. Yedinak, A.C. Watts, E.K. Strand, J.W. van Wagtendonk, J.W. Anderson, B.J. Stocks, J.T. Abatzoglou, The science of firescapes: achieving fire-resilient communities, Bioscience 66 (2016) 130–146, https://doi.org/10.1093/biosci/biv182.

[42] K. Hirsch, D. Martell, A review of initial attack fire crew productivity and effectiveness, Int. J. Wildland Fire 6 (1996) 199, https://doi.org/10.1071/WF9960199.

[43] P.M. Fernandes, H.S. Botelho, A review of prescribed burning effectiveness in fire hazard reduction, Int. J. Wildland Fire 12 (2003) 117, https://doi.org/10.1071/WF02042.

[44] B.M. Wotton, M.D. Flannigan, G.A. Marshall, Potential climate change impacts on fire intensity and key wildfire suppression thresholds in Canada, Environ. Res. Lett. 12 (2017) 095003, https://doi.org/10.1088/1748-9326/aa7e6e.

[45] M.E. Alexander, F. V Cole, Predicting and interpreting fire intensities in Alaskan black spruce forests using the Canadian system of fire danger rating, in: Soc. Am. for. Conv., 1995.

[46] J. Williams, D. Albright, A.A. Hoffmann, A. Eritsov, P.F. Moore, J.C. Mendes De Morais, M. Leonard, J. San Miguel-Ayanz, G. Xanthopoulos, P. Van Lierop, Findings and implications from a coarse-scale global assessment of recent selected mega-fires, in: 5th,International Wildl. Fire Conf., FAO, Sun City, South Africa, 2011.

[47] K. Tolhurst, Report on the Physical Nature of the Victorian Occurring on 7th February 2009 Fires, Victoria, 2009. http://royalcommission.vic.gov.au/getdoc/5905c7bb-48f1-4d1d-a819-bb2477c084c1/EXP.003.001.0017.pdf.

[48] A.M. Gill, P.H.R. Moore, Fire intensities in eucalypt forestsof south-eastern Australia, in: Int. Conf. for. FireResearch, Coimbra, 1990, pp. 1–12.

[49] A.M. Gill, S.L. Stephens, Scientific and social challenges for the management of fire-prone wildland—urban interfaces, Environ. Res. Lett. 4 (2009) 034014, https://doi.org/10.1088/1748-9326/4/3/034014.

[50] A. Beighley, M. Hyde, Gestão dos Incêndios Florestais em Portugal numa Nova Era Avaliação dos Riscos de Incêndio, Recursos e Reformas, 2018.

[51] D. Xavier Viegas, Are extreme forest fires the new normal? Conversat (2013). https://theconversation.com/are-extreme-forest-fires-the-new-normal-15824.

[52] F. Tedim, M. Garcin, C. Vinchon, S. Carvalho, N. Desramaut, J. Rohmer, Comprehensive vulnerability assessment of forest fires and coastal erosion: evidences from case-study analysis in Portugal, Assess. Vul. Nat. Hazards (2014) 149–177, https://doi.org/10.1016/B978-0-12-410528-7.00007-2.

[53] M.K. Lindell, D.J. Whitney, Correlates of household seismic hazard adjustment adoption, Risk Anal. 20 (2000) 13–26, https://doi.org/10.1111/0272-4332.00002.

[54] D. Paton, P. Burgelt, T.D. Prior, Living with bushfire risk: social and environmental influences on preparedness, Aust. J. Emerg. Manag. 23 (3) (2008) 41–48.

[55] D. Paton, J. McClure, P.T. Bürgelt, Natural hazard resilience: the role of individual and household preparedness, Disaster Resil. An Integr. Approach. 105 (2006) 27.

[56] T.A. Steelman, S. McCaffrey, Best practices in risk and crisis communication: implications for natural hazards management, Nat. Hazards 65 (2013) 683−705.

[57] K.J. Tierney, Disaster Preparedness and Response: Research Findings and Guidance from the Social Science Literature, 1993. http://udspace.udel.edu/handle/19716/579.

[58] R.E. Kasperson, P.J. Stallen, Communicating Risks to the Public: International Perspectives, Springer Science & Business Media, 1991, https://doi.org/10.1007/978-94-009-1952-5.

[59] D. Paton, J. McClure, Preparing for Disaster: Building Household and Community Capacity, Charles C Thomas Publisher, 2013.

[60] H. Hesseln, Wildland fire prevention: a review, Curr. For. Reports. 4 (2018) 178−190, https://doi.org/10.1007/s40725-018-0083-6.

[61] A.P. Fischer, T.A. Spies, T.A. Steelman, C. Moseley, B.R. Johnson, J.D. Bailey, A.A. Ager, P. Bourgeron, S. Charnley, B.M. Collins, Wildfire risk as a socioecological pathology, Front. Ecol. Environ. 14 (2016) 276−284.

[62] G. Wigtil, R.B. Hammer, J.D. Kline, M.H. Mockrin, S.I. Stewart, D. Roper, V.C. Radeloff, Places where wildfire potential and social vulnerability coincide in the coterminous United States, Int. J. Wildland Fire 25 (2016) 896, https://doi.org/10.1071/WF15109.

[63] V. Leone, N. Koutsias, J. Martínez, C. Vega-García, B. Allgöwer, R. Lovreglio, The Human Factor in Fire Danger Assessment, 2003, pp. 143−196, https://doi.org/10.1142/9789812791177_0006.

[64] V. Leone, R. Lovreglio, M.P. Martín, J. Martínez, L. Vilar, Human factors of fire occurrence in the mediterranean, in: Earth Obs. Wildl. Fires Mediterr. Ecosyst., Springer Berlin Heidelberg, Berlin, Heidelberg, 2009, pp. 149−170, https://doi.org/10.1007/978-3-642-01754-4_11.

[65] A. Marciano, R. Lovreglio, A. Patrone, A. Notarnicola, V. Leone, Tecniche di analisi delle motivazioni degli incendi, Appl. Nel Territ. Del Parco Naz. Del Cilento e Vallo Di Diano, Sherwood 162 (2010) 13−18.

[66] R. Lovreglio, M.J. Rodrigues, G. Silletti, V. Leone, Wildfire cause analysis through Delphi method: a study-case, Ital. J. For. Mt. Environ. 63 (2008) 427−447. http://ojs.aisf.it/index.php/ifm/article/view/201.

[67] O. Meddour-Sahar, R. Meddour, V. Leone, R. Lovreglio, A. Derridj, Analysis of Forest Fires Causes and Their Motivations in Northern Algeria: The Delphi Method, vol. 6, 2013, https://doi.org/10.3832/IFOR0098-006, 247, Http://Iforest.Sisef.Org/.

[68] A. Camia, T. Durrant, J. San-Miguel-Ayanz, Harmonized Classification Scheme of Fire Causes in the EU Adopted for the European Fire Database of EFFIS, Executive, Publications Office of the European Union, Luxembourg, 2013.

[69] T. Steelman, U.S. wildfire governance as social-ecological problem, Ecol. Soc. 21 (2016), https://doi.org/10.5751/ES-08681-210403 art3.

[70] D. Paton, F. Tedim, Wildfire and Community: Facilitating Preparedness and Resilience, Charles C Thomas Publisher, 2012.

[71] T.K. McGee, S. Russell, "It's just a natural way of life…" an investigation of wildfire preparedness in rural Australia, Environ. Hazards 5 (2003) 1−12, https://doi.org/10.1016/j.hazards.2003.04.001.

[72] P. O'Neill, Developing a Risk Communication Model to Encourage Community Safety from Natural Hazards, Citeseer, 2004.

[73] J.A. Bradbury, Risk communication in environmental restoration programs, Risk Anal. 14 (1994) 357−363, https://doi.org/10.1111/j.1539-6924.1994.tb00252.x.

[74] N. Dufty, A new approach to community flood education, Aust. J. Emerg. Manag. 23 (2008) 4−8. https://search.informit.com.au/documentSummary;dn=193124468114274;res=IELAPA.

[75] T.K. McGee, Public engagement in neighbourhood level wildfire mitigation and preparedness: case studies from Canada, the US and Australia, J. Environ. Manag. 92 (2011) 2524—2532, https://doi.org/10.1016/J.JENVMAN.2011.05.017.

[76] S. McCaffrey, Community wildfire preparedness: a global state-of-the-knowledge summary of social science research, Curr. For. Reports. 1 (2015) 81—90.

[77] G. Xanthopoulos, Who should be responsible for forest fires? Lessons from the Greek experience, in: Proc. Second Int. Symp. Fire Econ. Planning, Policy a Glob. View, 2004, pp. 19—22.

[78] S.F. Arno, S. Allison-Bunnell, Lessons from nature: will we learn? in: S.F. Arno (Ed.), Flames Our for. Disaster or Renewal? Island Press, Washington, DC, U.S., 2002, pp. 169—182.

[79] D. Carle, Burning Questions: America's Fight with Nature's Fire, Greenwood Publishing Group, 2002.

[80] P.N. Omi, Forest Fires : A Reference Handbook, ABC-CLIO, 2005.

[81] R. Bohn, Stop Fighting Fires, Harv. Bus. Rev., 2000. https://hbr.org/2000/07/stop-fighting-fires.

[82] N.P. Repenning*, Understanding fire fighting in new product development, J. Prod. Innov. Manag. 18 (2001) 285—300, https://doi.org/10.1111/1540-5885.1850285.

[83] L.A. Perlow, The time famine: toward a sociology of work time, Adm. Sci. Q. 44 (1999) 57, https://doi.org/10.2307/2667031.

[84] R.D. Collins, Forest Fire Management in Portugal : Developing System Insights through Models of Social and Physical Dynamics, 2012. https://dspace.mit.edu/handle/1721.1/72651.

[85] J. Chu, Study Finds More Spending on Fire Suppression May Lead to Bigger Fires |, MIT News, 2013. http://news.mit.edu/2013/forest-fire-management-1120.

[86] R.L. Olson, D.N. Bengston, A World on Fire, AAI Foresight Rep. 2; Spring/Summer. Free. WA AAI Foresight, 2015, pp. 1—16, 16 pp. http//www. aaiforesight. com/content/world-fire.

[87] S.J. Pyne, Problems, paradoxes, paradigms: triangulating fire research, Int. J. Wildland Fire 16 (2007) 271, https://doi.org/10.1071/WF06041.

[88] F. Tedim, V. Leone, G. Xanthopoulos, A wildfire risk management concept based on a social-ecological approach in the European Union: fire Smart Territory, Int. J. Disaster Risk Reduct 18 (2016) 138—153, https://doi.org/10.1016/J.IJDRR.2016.06.005.

[89] F. Tedim, The concept of "fire smart territory": contribution for a shift of the approach in wildfire management in Portugal, in: Geogr. Landsc. Risks B. Tribut. To Prof. António Pedrosa, Imprensa da Universidade de Coimbra, 2016, pp. 249—282, https://doi.org/10.14195/978-989-26-1233-1_12.

[90] K. Hirsch, V. Kafka, C. Tymstra, R. McAlpine, B. Hawkes, H. Stegehuis, S. Quintilio, S. Gauthier, K. Peck, Fire-smart forest management: a pragmatic approach to sustainable forest management in fire-dominated ecosystems, For. Chron. 77 (2001) 357—363, https://doi.org/10.5558/tfc77357-2.

[91] G. Xanthopoulos, D. Caballero, M. Galante, D. Alexandrian, E. Rigolot, R. Marzano, Forest fuels management in Europe, in: P.L. Andrews, B.W. Butler, Comps (Eds.), Fuels Manag. To Meas. Success Conf. Proceedings. 28—30 March 2006; Portland, OR. Proc. RMRS-P-41, CO US Dep. Agric. For. Serv. Rocky Mt., Fort Collins, 2006, 2006.

[92] FAO, Workshop on Forest Fires in the Mediterranean Region: Prevention and Regional Cooperation, Sabaudia, Italy, 2008. http://www.fao.org/3/k2891e/k2891e00.pdf.

Understanding wildfire mitigation and preparedness in the context of extreme wildfires and disasters

8

Social science contributions to understanding human response to wildfire

Sarah McCaffrey[1], Tara K. McGee[2], Michael Coughlan[3], Fantina Tedim[4,5]
[1]Rocky Mountain Research Station, USDA Forest Service, Fort Collins, CO, United States; [2]Department of Earth and Atmospheric Sciences, University of Alberta, Edmonton, AB, Canada; [3]Institute for a Sustainable Environment, University of Oregon, Eugene, OR, United States; [4]Faculty of Arts and Humanities, University of Porto, Porto, Portugal; [5]Charles Darwin University, Darwin, NWT, Australia

8.1 Introduction

Recent years have witnessed a growing number of stories about extreme wildfires that have had significant social impacts, from Australia to Portugal to California. Although this has heightened the call to find ways to better "coexist with fire," it must be recognized that wildfire—human interactions are as old as humanity itself. Humans around the world have ignited and used fire as a basic tool for millennia; people have and continue to use confined fires for a range of quotidian reasons including cooking, heating, and processing of materials (e.g., in the production of brick, ceramics, metals). Use of broadcast landscape fire for hunting, gathering, agriculture, and construction purposes is another age-old practice. Although more recently such broadscale burning has become frowned on in many places, particularly more industrialized countries, the current practice of prescribed fire derives from these traditions. The use of landscape fire was and continues to be indispensable to our evolution as a species and to the development of our many and diverse social and economic systems.

Even today only a small portion of wildfire-human interactions occur in contexts where fire presents a serious hazard to life and property. The use of and dependence on fire and fire-causing technologies creates an inherent and ubiquitous hazard for people inhabiting fire-prone environments. Although in many places significant resources have been directed toward prevention and suppression of wildfires, losses from wildfires persist and are increasing in some areas. A key reason for the increased attention to wildfire is the growing human exposure and negative impacts on human health, livelihoods, and well-being. The causes of the increased exposure vary geographically and

Extreme Wildfire Events and Disasters. https://doi.org/10.1016/B978-0-12-815721-3.00008-4

across socioeconomic gradients with the level of risk determined by a range of large-scale social factors including population growth, changing settlement patterns, and shifts in natural resource management practices. How these factors may contribute to extreme fire damage can vary considerably. In some areas, people moving from urban areas into more fire-prone rural landscapes are seen as increasing wildfire risk as landscape fragmentation and new settlements are seen to complicate fire and land management decisions in ways that increase wildfire hazards. In other places, such as Mediterranean Europe, individuals moving from rural to urban areas for economic opportunities can contribute to increased wildfire risk as unplanned afforestation resulting from agricultural land abandonment can increase fuel loading in rural communities where traditional land-use practices involved the use of fire. In other places the fire risk is effectively brought to established communities. For example, a key source of increased risk in parts of Portugal results from the initial introduction of eucalyptus in the 20th century and the more recent expansion of eucalyptus plantations for the pulp industry.

Similarly, the reasons individuals live in fire-prone areas are highly variable. In many cases economic considerations, such as livelihood opportunities and housing affordability, are a key factor influencing residency choices. In other cases, people may be attracted to fire-prone landscapes by natural amenities such as scenery, recreational opportunities, and solitude. Furthermore, many people such as First Nations in Canada and rural Portuguese have lived in fire-prone areas settled by their ancestors, long before wildfire risk became a significant challenge.

Ultimately, there is no single explanation for what has led to increased fire risk in a given location, and it is important to carefully assess the accuracy of the beliefs around how social dynamics contribute to extreme fire risk in a specific location. A challenge with understanding social issues around wildfire preparedness and mitigation at a global level is that local context is critical as both the level of wildfire risk and potential social outcomes can be contingent on specific local dynamics such as local culture, land management and building practices, and institutional histories [1]. Therefore this chapter will not focus on regional specifics but provide a broad overview of a range of factors and dynamics to consider in assessing specific local conditions.

8.2 Social science theoretical insights into preparedness and mitigation

Several fields of study provide useful insights into understanding the human—wildfire relationship: (1) natural hazards, (2) diffusion of innovations, and (3) risk and crisis communication. The first field of research provides a framework for how societies and individuals perceive and respond to the wildfire hazard, the second provides further insight into factors that may influence adoption of fire mitigation measures, and the third helps identify key dynamics to consider in effective outreach efforts.

All three fields address decision-making in the face of uncertainty. The uncertainty around when and where an event may occur, and if it does occur just how negative the outcomes will be is a key factor that informs and shapes human response to a hazard.

"Risk arises not just from how some future can be described, but from the uncertainty, actual or perceived, surrounding that description. Indeed, it is only because we need to act under conditions of uncertainty that the concept of risk is of any interest whatsoever. Living with natural processes that are periodically hazardous means that people have choices to make .... " [2]. It is this uncertainty that is an underlying focus of many scientific efforts: Studies about natural hazards and risk and crisis communication both focus on understanding how individuals interpret and respond to the uncertainty created by a potential hazard event, while diffusion of innovations has been described as "an uncertainty-reduction process" [3].

8.2.1 Natural hazards

By definition, a natural hazard results from human nature interactions: A hazard is simply a normal biophysical process that only becomes seen as a hazard when it begins to have a significant negative effect on something humans value, whether that is homes, water quality, or an endangered species. Water flowing in a stream is a beneficial process—providing everything from drinking water to recreational fishing and boating opportunities—until it begins to overflow the stream bed and damage crops or homes; at that point the natural process has become a natural hazard. Natural hazards research works to understand the range of factors that influence adoption of measures to decrease, or mitigate, potential damage and why certain responses to a hazard are favored over others.

The natural hazards field grew out of Gilbert White's work on early US flood control policy: specifically why, despite all the levees and dams built under the 1936 Flood Control Act, US flood damage continued to rise. At the time, the rational actor model of human behavior prevailed, and it was assumed that as long as individuals understood the risk, they would choose to make the most cost-effective or economically optimal choice. In the case of floods, they would recognize areas of higher flood danger, value that land less, and choose to live elsewhere [4]. Instead, White's work, and a plethora of subsequent natural hazards studies, demonstrated that human response to hazards was not based purely on hazard-related economic calculations but also was influenced by a range of factors such as sufficient resources to undertake protective actions, beliefs and attitudes toward the problem, and available mitigation options.

Over time, the field expanded its scope beyond a focus on individual decision-making to examine how larger scale factors, including mescoscale (mid-level) and macroscale variables, influenced the hazard itself as well as how humans responded to it. Of note is that important macrolevel and mesolevel variables often are not directly related to the hazard: "both institutional and cultural phenomena may buffer or focus damage, without being tied to specific vulnerabilities or agents of damage" [5]. This idea that a hazard may be exacerbated from external actions not directly related to the hazard is an important point for understanding the wildfire risk around the world. As indicated earlier, in some places the fire hazard has increased because of decreased agricultural burning, often as a result of larger scale economic drivers leading to rural depopulation while in other locations the hazard has increased because of the establishment of eucalyptus plantations by both industrial and smaller scale landholders.

8.2.2   Societal stages of response to natural hazards

Through a series of international case studies, natural hazards research developed a framework in the 1970s that identified four different societal stages for coping with a natural hazard: loss absorption, acceptance, reduction, and change [4,6]. Understanding which stage a society or an individual or community might be at can help decision-makers more readily identify disconnects between the current societal stage and the actions being taken and identify appropriate next steps.

The first stage, loss absorption, takes place when a hazard's effect is small enough to impose relatively few costs to society and adaptations are unconsciously made to absorb them. Carrying a raincoat if it looks like it might rain is a simple example of such an action. Once a hazard's effect begins to exceed a society's natural absorptive capacity, the effected group begins to see the biophysical process as something that is potentially hazardous and to make adjustments. At first these are fairly passive; the potential for loss is recognized but little is done to alter the hazard as bearing the cost is preferable to the effort and uncertainty of making any significant changes. Instead, the focus is on minimizing impacts by finding ways to help those most directly affected by an event to absorb the loss. After an event, these measures include governmental and charity disaster relief. Before an event, a primary mechanism is insurance which effectively spreads the risk of individual loss across a larger population. Here it is worth noting that although some may see insurance as a potential means of changing behavior, it is not the main intended function of insurance.

Once the societal costs of the hazard become too large to easily absorb, more active measures begin to be taken, alongside existing redistributional mechanisms, to actively reduce or mitigate potential damage. Initially the focus of these mitigation efforts tends to be on physical actions, generally engineering or technological fixes, to modify the environment (sometimes referred to as structural mitigation) to prevent or diminish the effect of the hazard by shifting its location, its timing, or the process that creates it. At this point, the hazard is generally seen as correctable with technology, often via larger scale engineering fixes (e.g., dams and levees). Such technical attempts to modify the environment are appealing because they can generally be accomplished directly through government action, avoiding the need for individual or community involvement [7].

However, for many hazards, including wildfire, such physical environmental modifications fail to effectively reduce the negative outcomes over time. Resulting from natural biophysical processes, it ultimately is not possible to completely eliminate a natural hazard. And for some hazards the physical modifications only serve to raise the hazard threshold—there may be fewer hazardous events overall, but when they do occur it will be because they overwhelm the structural safeguards which often then contributes to more extensive harmful consequences. The failure of the levees around New Orleans after Hurricane Katrina is a good example of this dynamic. Similarly, a singular focus on fire prevention and complete fire suppression as the main means of mitigating fire risk is a clear example of large-scale government attempts to modify the environment to eliminate or minimize the fire hazard. However, it is not possible to prevent all wildfires and, in many ecosystems, suppressing has led to fuel buildup that, overtime, can contribute to an increased rather than decreased long-term fire risk as higher fuel loads contribute to more extreme fire behavior that can overwhelm response capacity.

As the limitations of physical mitigation measures are recognized, mitigation actions begin to turn toward efforts to modify human behavior (sometimes referred to as nonstructural mitigation) through both voluntary and regulatory measures, as well as more indirect efforts to shift cultural norms and rules. *Voluntary measures can involve a range of outreach and financial or technical incentives measures such as one-on-one homeowner consultations or assistance with vegetative debris disposal.* Regulatory actions include tools such as building codes, local ordinances, and zoning. Building codes to help increase ability of buildings to withstand a hazard via both construction standards (e.g., nail spacing—particularly relevant for hurricanes and earthquakes) and material requirements (such as fire-resistant roofs) are one of the more widely used regulatory measures as, for a number of hazards, they are quite effective at mitigating risk, can be adapted to meet local norms/needs, and tend to be more feasible to implement. Local ordinances also can help regulate activities that may contribute to a hazard such as vegetation clearance requirements. Policy can also be written to provide economic incentives or sanctions to encourage desired behavior. Although use of regulatory measures may appeal as a "simple fix," enacting regulatory measures is generally a time-intensive and unpredictable process with effectiveness dependent on cultural acceptance of such mandates and the ability to enforce them. Social norms and rules are less tangible and harder to address directly but over time can have perhaps a longer and more significant influence on both hazard creation and mitigation. In the case of the wildfire hazard, studies have shown that belief that neighbors have positive views of fire mitigation activities is positively associated with other individuals adopting those actions [7a].

Besides the potential for larger scale physical mitigation measures to simply raise the hazard threshold, a less discussed concern with the tendency to focus on physical and technical fixes has been that historically this focus has often, intentionally and unintentionally, led to elimination of existing mitigative behaviors. In many places, traditional uses of fire to improve range conditions or minimize fire risk has been discouraged and is often labeled as arson [8]. In the United States, as the emphasis on suppression increased in the 1930s, education programs were developed to discourage and demonize local use of fire which had often been conducted in part to minimize the fire hazard. This has meant that, ironically, more recent outreach efforts have had to be targeted toward increasing local comfort with reintroduction of fire as a management tool. Therefore, although the tendency to focus on larger scale physical/technical mitigation measures is unsurprising given the greater ability for more centralized and governmental control of such endeavors, countries where fire is only beginning to become a significant hazard may want to resist the tendency to focus solely on technical solutions and move immediately to an approach that also values promoting existing as well as new adaptive behavioral responses. It also is important to note that while the structural (environmental modification) and nonstructural (behavior change) categories provide a neat division in mitigation approaches, in practice, the two are not distinct. Changing building characteristics in practice could be seen as an environmental or structural mitigation measure, however, ensuring it occurs at a meaningful scale may also require non structural measures to changing human behavior *(e.g., financial incentives, building codes)*.

The final of the four coping stages occurs only when the negative impacts from the hazard have become so extreme that, despite mitigation efforts, complete change—of land-use or living methods— is required. This stage is quite rare as most cultures and societies, particularly highly developed ones, are resistant to such large-scale change as the overall societal costs are too high. Arguing that individuals should simply not live in fire-prone areas ignores *both the broad geographic extent of such areas as well as* wide range of personal, economic, and social reasons why communities have developed in those areas.

8.2.3   Wildfire preparedness/mitigation measures

Preparedness, mitigation, and prevention are often used interchangeably in the wildfire response world. Although interrelated, they refer to specific dynamics in response to a hazard, and it is important to distinguish between the terms and clarify their meaning. Emergency response is generally divided into four distinct phases: mitigation, preparedness, response, and recovery. Preparedness generally refers to activities undertaken to be ready to respond to an actual event. With wildfires, this includes ensuring availability of equipment, such as engines and airplanes, and firefighting personnel with appropriate training. It also includes planning and coordination of response activities, such as evacuation. Mitigation focuses, as discussed previously, on actions to reduce vulnerability and potential impacts of an event. With wildfire, the primary focus has been on actions to reduce vegetative fuel, at multiple scales, and to increase fire resistance of structures and infrastructure. It can also include actions to decrease, or prevent, ignitions. Here it is important to note that while prevention is frequently conflated with mitigation, often used to describe a range of activities beyond preventing unwanted ignitions, in reality it is the inverse with wildfires where prevention effectively is a specific type of mitigation. While preventing, an event does reduce potential impact, many actions—such as vegetation management—that are often described as prevention in reality cannot prevent an event but do reduce (mitigate) its potential negative impacts. This is not a minor distinction as lack of clarity or confusion over the goal of an action can decrease the chances it will be adopted.

Scale is an important consideration in understanding natural hazard preparedness and mitigation. At larger macro spatial scales, the likelihood of a damaging wildfire in a given year is often high, and regional and national levels of government have significant incentives to develop and implement mechanisms to respond to the hazard. As the scale of focus decreases, the exposure also tends to decrease; from a pure probability perspective the odds of a wildfire occurring in a specific rather than general area are lower. In general, the smaller the scale the lower the incentive to expend significant attention or resources to mitigate the local risk to a specific hazard. As a result, higher levels of government are more likely to devote specific resources to assessing, planning for, and responding to a hazard, while local governments and individual residents are less likely to have concerns about a specific hazard high on their radar in comparison with the other competing priorities of everyday life. In her integrative framework for studying natural hazards, Palm [9] identifies three key scales to consider: macro, meso, and micro. These levels are useful to think about how wildfire preparedness and mitigation measures might be vary at different levels.

8.2.4 Societal (macroscale)

At the societal or national scale, a broad array of factors may come into play to influence preparedness and mitigation efforts for a particular hazard. This can be via funding priorities, strategic direction, establishment of national programs, and provision of resources and incentives targeted toward improving outcomes for the specific hazard. However, at this level, the influence of such efforts is complex and does not inherently lead to better outcomes. For instance, national policies may serve only to shift the risk to other areas or hazards (e.g., individuals not allowed to live in fire-prone areas may instead live in flood plains) or other timeframes (e.g., as indicated earlier in the United States a national policy of fire suppression minimized the immediate fire risk but in many places over time has increased the long-term risk). In addition, many large-scale socioeconomic factors not directly related to the hazard can influence preparedness and mitigation. Economic opportunities or constraints may draw individuals to hazardous environments and "institutional and cultural phenomena may buffer or focus damage, without being tied to specific vulnerabilities or agents to damage" [5]. As a result, any effort to minimize risk from a particular hazard needs to consider the larger social context in which the hazard is situated.

8.2.5 Intervening or mid-level factors (mesoscale)

Middle-level factors, often in the form of local programs and governmental organizations, can be key conduits between larger scale resources and on the ground preparedness and mitigation efforts. This level can be critical as it often acts as a convener and interpreter that determines how interactions between larger scale factors and individual decisions are negotiated. Whether or how a national or provincial/state policy is implemented can depend on how more local actors choose to interpret the policy which in turn can constrain or enable an individual's mitigation choices [9]. Convening entities also can play a key role in securing resources, gaining public and organizational support for mitigation activities, and coordinating activities between groups.

At the mesoscale, key mitigation actions can focus on both structural and nonstructural elements. Structural mitigation efforts to modify the environment can include land management activities to reduce fuels and increase landscape resiliency. Nonstructural elements tend to focus on two ways to change behavior. The first is through regulatory means such as zoning, building codes, and land-use regulations. Land-use regulations can be designed to encourage activities that decrease fire risk, such as limiting unmanaged eucalyptus plantations or how new development takes place. Zoning bylaws could require that wildland fire be addressed in general plans, and subdivision development plans could require use of fire-resistant building materials, adequate access, firebreaks, etc. A key challenge here is that, other than building codes, there is limited empirical evidence on the type of development that most effectively decreases fire risk and how this might differ depending on local context. It is likely that the best development practices to mitigate fire risk will vary depending on local fire regime, topography, or building practices. A second key more behaviorally focused mesoscale preparedness activity involves organizational support for the programs and individuals who can provide the information, resources, and coordination needed to build capacity to undertake mitigation activities such as defensible space or evacuation planning.

8.2.6   Individual/household (micro) scale

On an individual or household level, fire mitigation generally can be described as activities that are undertaken to increase the likelihood of both human and structural survival during a wildfire. There is good empirical evidence that actions to (1) modify the environment around a building to decrease fire intensity and minimize ignition sources (primarily embers) and (2) increase a structure's ignition resistance can greatly increase the chance a building will survive with or without active protection [10,11]. Environmental modification primarily involves activities to break up the continuity of vegetation and decrease the available fuel adjacent to the building. This includes pruning low-lying tree branches, thinning vegetation, and removing dead matter and excess groundcover. Generally, the area surrounding a structure is divided into a series of expanding zones, with the degree of vegetation modification needed decreasing as distance from the structure increases. Actual characteristics of each zone, distance and degree of needed vegetation modification, for a specific structure will vary depending on factors such as topography and type of vegetation. The most basic recommendation is to modify vegetation for a minimum of 30 feet around a structure. Given the habit of vegetation to grow, fuel management is an on-going effort, requiring some level of periodic maintenance to be effective.

Specific activities to increase resistance of structures to fire vary by regional building styles; areas where the default standard are stone buildings likely have less work to undertake than regions where the preference is for wooden structures raised a few feet from the ground on stilts. A key focus of any of these efforts is on actions that can protect the structure from ember attack, the dominant cause of structural ignition during wildfires. These actions can include use of fire-resistant roofing (the single most effective action) and siding materials; screening of vents, eaves, and other openings; *cleaning out and/or* enclosing overhanging space (such as under decks) where heat convection can draw embers into the structure; and availability of screens, heavy curtains, or plywood to cover windows (the most vulnerable part of a building to radiant heat) in the event of a fire.

During a fire, the key protective actions revolve primarily around evacuation decisions. Fire is a particularly challenging hazard when it comes to evacuation as the conditions can be more variable than other hazards in several ways. First, rapid changes in fire direction and speed due to weather, particularly wind changes, mean that predicting where and when a fire might impact a population can be difficult. It also means that individuals may have anywhere from only minutes of warning that a fire is imminent or several days to prepare. Second, these rapid changes mean that evacuation may not always be the safest option, such as when evacuation routes have been cut off or there has been little warning, particularly if the property has been well prepared and residents are knowledgeable and mentally and physically prepared to stay on their property safely. Third, with proper preparation and in non-extreme conditions, evidence indicates that staying and protecting the home increases the odds it will survive. Fourth, in extreme conditions, plans and preparations that might be sufficient property protection for most fires are less likely to be effective. This means that the best course of action for the same individual and property can be quite variable, depending on where the fire originates, level of mitigation and preparedness, and environmental conditions at the time. This variability suggests that no single approach is likely to ensure safety in all situations.

8.3   Factors that influence individual protective action decisions, with reference to specific fire research findings

Understanding individual perceptions of and response to a hazard is a central focus of natural hazards research. Perhaps more than many other natural hazards, effective fire mitigation is influenced by individual action—it is not just a case of enacting effective building codes (which are critical for tornado and earthquake mitigation) but of changing both behavior and opinions on more personal matters of home construction and siting, vegetation and esthetic preferences, and acceptance of large-scale vegetation management practices in the surrounding landscape. Although historically wildfire has not been a significant focus of natural hazards research [12], since the late 1990s a growing number of studies have examined various aspects of individual response to wildfire, particularly whether and why homeowners choose to undertake mitigation on their property, individual perspectives about fuels treatments, and evacuation decision-making [1]. Findings from this work by and large parallel those found for other hazards on key factors found to influence hazard response and preparedness. These are discussed below with specific examples drawn from the wildfire literature. It is important to note that most of the social wildfire research has come from the United States and more recently Australia and Canada. Although specific dynamics are likely to vary by location, comparing results across these studies suggests that the general dynamics in terms of which variables are most influential appear to be reasonably consistent across countries [1].

8.3.1   Risk interpretation

While hazard and risk are often used interchangeably, they do not inherently refer to the same process. While the definition of hazard is fairly variable, the most common perspective focuses on hazard as potential: that the term refers to the conditions that create the *potential* for loss or damage. Comparatively, risk is most often defined as the probability of an event with harmful consequences. While potential and probability seem similar, potential focuses on conditions that may contribute to various outcomes, while probability focuses on the likelihood of a specific outcome: Flipping a coin twice has the *potential* for it to land on heads both times but the actual *probability* of this occurring is only 25%. With wildfires, a hazard characteristic is the amount and configuration of fuels, both vegetation and buildings, while wildfire risk takes into account these conditions along with the likelihood of ignition (e.g., ignition sources, weather): If ignition is unlikely (e.g., high humidity), even highly hazardous conditions may have a low fire-risk.

Early risk-related research assumed that people's actions would be directly related to the calculated probability of the event and the magnitude of its consequences—the factors that risk analysis focuses on. However, research demonstrated that how individuals perceive and respond to risk is a complex dynamic that reflects a range of factors such as personal ability to influence, voluntariness of exposure, risk attitudes, economic considerations, and benefits of exposure. While these processes are often lumped together and described as risk perception, it is perhaps more accurate to think of the dynamic as risk interpretation; the term "perception" can suggest that it is possible for people to perceive the same risk, whereas the key item at issue is really how the same risk may be interpreted differently [2]. How risk is understood and responded to is not straightforward or consistent between different individuals: It is not a given that each individual, or organization, will be considering the same spatial and temporal factors when calculating probability, nor that they will be considering the same set of harmful consequences (e.g., house loss vs. specific environmental damage) in their assessment.

The result of this complexity is that there is ample evidence that recognition of a risk does not in and of itself lead to increased preparedness [12]. Numerous studies in the United States indicate that residents living in fire-prone areas are already well aware of the fire risk, that many are undertaking mitigation activities, and that a range of considerations in how the risk is interpreted can influence lack of action [13]. Studies show that the spatial scale and voluntariness of the risk exposure, as well as the benefits of the exposure (e.g., being near nature, economic opportunities), also can influence how an individual responds to wildfire risk [1]. Studies also suggest that although wildfire information can increase an individual's assessment of fire probability, the probability assessment is not associated with increased preparedness, whereas consideration of likely consequences is associated with increased preparedness [14,15]. Ultimately, recognizing a risk is a necessary but not sufficient condition for individuals to adopt mitigation measures as other factors also influence the decision process.

8.3.2 Experience

Although experience with a hazard is often thought to be an important influence in increased preparedness, the evidence for this is quite mixed. Studies have found that experience can both increase (generally via increased salience) and decrease (fatalism or lightning does not strike twice) risk perception and mitigation efforts, or it can have little to no effect [12,16]. The strongest relationship between experience and adoption of mitigation measures is *frequency* of experience: The more frequent the experience, the more likely one is to have a realistic assessment of the likely occurrence and potential impact of a hazard [17,18]. Even here, it is not inherent that frequent experience will lead to mitigation; some studies have found that repeated experience with a hazard (e.g., seasonal flooding) may lead to a "disaster subculture" where people become so used to the hazard that it simply becomes part of life and mitigation is not even considered [18a]. Similar to the dynamics around risk, individuals can choose to interpret a specific experience differently. For example, research indicates that a near-miss experience can be interpreted as indicating a successful outcome (no major losses) or as a close call (disaster narrowly averted): The former interpretation leads individuals to ignore or discount the need for more protective behaviors (it worked!), while the latter interpretation makes individuals more likely to consider taking additional protective measures [19]. After the 2009 Australia Black Saturday fires, three-fourths of surveyed individuals who indicated that they evacuated late (many of whom reported encountering significant challenges including poor visibility and fallen trees) also indicated that, as they ultimately were unharmed, they would undertake the same action in the future [20].

8.3.3   Efficacy (response and self)

Even when risk interpretation and experience do lead individuals to explore ways to mitigate their exposure, other considerations shape the process of choosing and implementing mitigation adjustments. Access to information is an important initial item; before an individual can consider mitigation, they need to know what mitigation options are available. Once the range of potential adjustments has been identified, individuals then engage in two types of evaluation related to efficacy: response efficacy and self-efficacy. Response efficacy relates to the perceived effectiveness of the action in mitigating the risk: A belief that an action will be effective has a positive association with adoption of the practice. Self-efficacy relates to the ability to actually implement the activity; lack of necessary resources such as time, money, or physical ability is generally associated with lower implementation rates. For example, wildfire studies indicate that lack of time rather than knowledge is likely a key constraint for part-time residents undertaking mitigation activities on their property and that a common issue with vegetation management to mitigate fire risk is how easy it is to dispose of any removed vegetation [13].

8.3.4   Wildfire-specific considerations

Fire is relatively unique as a hazard in that although it can disrupt key social values it also plays an integral ecological role in many valued ecosystems. This fact can complicate land and fire management decisions but also appears to come into play with mitigation decisions. Multiple studies show that homeowners more readily adopt fire mitigation practices that are in line with local ecological needs [7a] and that understanding the ecological benefits of fire is often a more important consideration in acceptance of prescribed fire than recognition of its role in reducing fire risk [33].

Fire also is a somewhat unique hazard in that as a landscape scale process, effective mitigation activities generally need to take place across property lines: It is a shared risk. Studies have found that the level of fire hazard and management activities on adjacent lands can be an important consideration in mitigation decisions, albeit in an inconsistent manner that is similar to that of experience. In some cases lack of mitigation on adjacent lands can deter mitigation and in other cases landowners choose to mitigate more to compensate or find ways to work together with the adjacent landowners to mitigate the fire risk [7a].

8.3.5   Nonwildfire considerations

Finally, a weakness in many discussions about how to improve mitigation and preparedness is that the discussion occurs in what might be called a hazard-specific vacuum. As with the societal level, individual response to a single hazard is influenced by a range of factors external to the hazard itself. Individuals have other risks (e.g., driving to work) and concerns (economic, social, etc.) to worry about and must make tradeoffs in how they will respond to wildfire with other considerations in their daily lives. The need to secure daily livelihoods tends to be mentally more salient than risk perception related to a specific natural hazard [21]. Hence, efforts to increase preparedness that do not actively take into account potential competing interests are less likely to lead to increased wildfire preparedness [22]. It also can be useful to identify complimentary interests; in many cases individuals have implemented fire mitigation measures not to mitigate their wildfire risk but because they confer other benefits.

8.3.6   Evacuation decisions

Natural hazards research has extensively studied how individuals learn of and then respond to a hazard event. When an event occurs, it disrupts normal life and increases uncertainty; many actions individuals take are efforts to reduce that uncertainty and regain or maintain a sense of normality.

When considering information provision for evacuations, a belief that appears to inform decisions by emergency responders during an event is that panic is a common public response to an imminent threat. However, research clearly demonstrates that actual panic (irrational, nonadaptive, or antisocial behavior) in response to natural hazards, including wildfire, is extremely rare [23,24]. Instead, evidence indicates that although there may be heightened anxiety, fear, and more rapid action (all rational responses to impending danger), individuals tend to respond to an imminent threat by first engaging in gathering more information to determine the best course of action and then proceeding to act in a manner congruent to their situation. These actions often include helping behavior (informing neighbors, helping others evacuate) which is the inverse of panic (e.g., prosocial rather than antisocial behavior). For example, during the 2016 Fort McMurray Horse River wildfire in Canada, a number of residents who evacuated left in a vehicle of a neighbor or friend or someone they did not know before the evacuation [24]. Research has shown that behaviors seen as panic by outside observers in reality are a rational response from the actual individual's perspective [25].

Descriptions of "I panicked' appear to be less about describing irrational behavior than recognition of a moment when there is a shift to more focused and rapid thought processes. The continued belief in the panic myth may in part occur because it is an easy way to explain poor outcomes, driving toward flames to try to rescue someone is seen as panic if there is loss of life but labeled as heroic if they succeed in saving a life [25]. Scholars also argue that the media focuses on panic because of its inherent drama and that emergency response organizations focus on it because it reinforces their central role in the command and control structure [23]. However, assuming panic can be problematic as hesitance to provide warnings out of concern about causing 'panic' can lead to worse outcomes as individuals are not provided with timely information that could help them make the safest decision for their situation.

Research into warnings has identified a number of characteristics of effective warnings including that they are from a credible source; consistent in content and tone (e.g., not indicate that things are terrible but everything is under control); accurate and clear; and provide specific information about what people should do and in what timeframe [26]. Once aware of a threat, individuals tend to seek information from multiple sources to confirm and validate the initial information, make sense of the situation, reduce uncertainty, and identify their best courses of action. Official warnings are a critical information source with a clear connection to increased evacuation. Individuals also have been shown to pay attention to environmental cues (e.g., smoke, flames) as well as information from and the behavior of those around them (social cues); both types of cues have been found in studies to be associated with evacuation decisions, although the decision might not always be to evacuate [27,28]. For instance, a study of wildfire evacuation decisions found that while all respondents relied on official warnings to make a decision, and those who most relied on them were more likely to leave early, the majority of respondents also relied on environmental cues and that greater reliance on environmental cues was associated with individuals being more likely to wait and see how conditions played out rather than immediately evacuate [28]. A study of the Fort McMurray evacuation found that social cues led people to decide to just carry on with their day instead of to prepare to evacuate [24].

Self-efficacy beliefs also come into play in evacuation decisions with concerns about ability to evacuate or limited evacuation options inhibiting evacuation, items that are particularly relevant for wildfires. The response efficacy of an action, whether evacuating or not evacuating, in protecting key values (life and property) also has been shown to be influential in relation to wildfires [15]. McCaffrey et al. [28] found that respondents who felt more strongly that evacuation was an effective protective action were more likely to leave early rather than wait to see what happens, while those who had a stronger belief in the efficacy of staying and defending were much more likely to stay. Notably, the study also found an indication that those who saw mitigation actions as effective were less likely to leave early as opposed to waiting to see how the fire evolved. The study also found that risk attitudes underlay different decisions with those who were more generally risk tolerant more likely to stay and defend and those who were more financially risk tolerant more likely to leave early [28]. Other situational factors that studies have found can influence when and whether individuals evacuate include the time of day, whether all family members are present, and the presence of children, the elderly, or animals (both pets and livestock).

8.4 Diffusion of innovations

The field of "diffusion of innovations" works to understand the process by which a new idea or technology is communicated and adopted. Three key aspects of this long-standing field provide useful insights into mitigation and preparedness: preventive innovations, how the attributes of a new practice influence its adoption, and the role of change agents.

8.4.1 Preventive innovations

Hazard mitigation is a particular type of innovation: preventive. Most innovations are adopted in the expectation that it will in some way improve one's life through improved knowledge or increased income or comfort. In constrast, preventive innovations are actions adopted primarily to *potentially* protect one's current lifestyle. As preventive innovations do little to decrease uncertainty, they tend to have a slow adoption rate as the rewards of adoption "are often delayed in time, are relatively intangible, and the unwanted consequence may not occur anyway" [29].

8.4.2 Characteristics that influence adoption of new practices

The attributes of an innovation are important because the risks and benefits of adopting it are unclear. Several characteristics of a new practice or tool contribute to how much uncertainty is involved in the cost—benefit calculation surrounding its adoption. Rogers [3] identifies five, often interconnected, characteristics of an innovation that play a role in its rate of adoption:

- Relative advantage—the degree an innovation is seen as superior, in economic or social terms, to existing practice. Perceived relative advantage is a key predictor of adoption rates.
- Compatibility—how well the innovation fits with the lifestyle, needs, experience, and values of the adopter.
- Trialability—how easy it is to test the innovation in a limited manner. A successful trial decreases uncertainty around the innovation's usefulness and increases likelihood of full adoption.
- Observability—how easy it is for others to see the benefits of the innovation. Seeing an innovation adopted by peers tends to influence adoption more than receiving formal information.
- Complexity—how difficult the innovation is to understand and use.

In general, the first four items are positively related to an innovation's adoption rate, whereas complexity is negatively related [29].

8.4.3 Change agents

Change agents are an example of a mesoscale element that diffusion of innovation has identified as playing a particularly important role in whether or not an innovation is adopted. A change agent is someone who provides "a communication link between a resource system of some kind and a client system" [3]. With wildfire such change agents or 'champions' can be fire chiefs, political leaders, forestry workers or community members [30]. The role of a change agent, who may hold a professional position but can also be less formally trained, is to provide information, create interest in, and support the adoption of an innovation by a target population. Factors that facilitate a change agent's effectiveness include: whether the change agent's attitude and the innovation itself are directed toward meeting the client's needs; frequency of contact with clients; whether the agent is of the same peer group as the client; and the degree that the change agent is seen as credible and encourages the client's ability to understand and evaluate the innovation [3].

8.5 Risk and crisis communication

Research has also examined how to effectively communicate about natural hazards. Risk communication has mainly focused on how to provide information about a potential hazard and mitigation options, while crisis communication has focused on how to provide information during an actual event. Over time, research in these two areas has increasingly overlapped as work demonstrated how communication during one phase in the disaster process (mitigation, preparedness, response, and recovery) can influence outcomes at another phase. An assessment of common characteristics of effective communication across this risk and crisis literature as well as wildfire social science research identified five key considerations for effective risk and crisis communication: (1) use of interactive processes and dialog; (2) use credible sources, especially appropriate authority figures; (3) take the local social context into account; (4) provide honest, timely, accurate, and reliable information; and (5) communicate about hazard response during all stages—before, during, and after an event [31]. This section elaborates on the first three of these items which have been shown to also be important considerations in the larger context of public response to wildfires.

8.5.1 Interactive processes

Social marketing and adult learning research have both shown that interactive processes are a critical part of efforts to shift norms and behavior. Interaction allows all parties to ask questions, clarify misperceptions (both of emergency responders and of different stakeholder groups), and identify how the topic is relevant for their particular situation and key concerns or barriers that might need to be addressed [32]. Wildfire studies have frequently found that social interactions and use of interactive processes are a key dynamic in increased preparedness with a homeowner preference for one-on-one interactions, with agency personnel as well as with neighbors and community leaders, to learn about how best to mitigate their fire risk [33]. Wildfire studies have also shown that interactive information sources are likely to be seen as more useful and trustworthy and that agency outreach efforts, particularly personal relationships with agency personnel, can influence assessment of agency activities and whether individuals adopt protective measures [34]. During a fire, interactive communication is particularly important as affected individuals seek to decrease the uncertainty of their situation and regain a sense of control [35].

For preventive innovations, interpersonal communication networks, especially via peer networks and champions, can be particularly effective in creating localized incentives to adopt [29]. A number of wildfire studies have shown how both peer-to-peer interactions and efforts that connect fire agency staff with community members can be influential in motivating adoption of mitigation measures and that outreach programs can be a key part of fostering such interactions. Such programs can help build the social networks and relationships that facilitate information sharing and the social learning that often underlies proactive mitigation and preparedness efforts [1,14,36−39].

8.5.2 Trust

Interactive processes are also critical to building trust which is one of the most consistent dynamics found to shape public wildfire response. Two key aspects of trust are credibility and competence. Credibility is important in how much attention is paid to information, while beliefs about individual competence underlie acceptance of various land management practices, with trust in a manager's ability to implement a practice shown to be a key factor influencing acceptance of both thinning and prescribed fire practices [40,41]. Credibility of the information source or message provider is particularly important with preventive innovations. If the source is seen to have ulterior motives or to be contradicting past practices, it is likely to be given short shrift. Transparency of communication is critical as, particularly during an event when there is limited time to build a relationship, it can act as an indicator of trustworthiness.

8.5.3 Local context

Finally, studies have shown that efforts that actively take local knowledge and experience into account are more likely to be effective as they can better address local considerations that may shape the hazard and its outcomes. Communication and outreach efforts that incorporate local knowledge are more likely to be seen as relevant and trusted, positively effecting preparedness efforts [42]. Whether local knowledge and values are considered in management decisions also has been found to influence views of agency management decisions, with views of management actions, particularly response during wildfires, trending toward more negative when local knowledge and resources have not been taken into account in management decisions [1].

8.6   Conclusion

Wildfires have several characteristics that make them a particularly complex hazard to assess and manage, particularly in relation to effective mitigation and communication. As a critical ecological process in many ecosystems, too narrow a focus on removing fire from a system can be counterproductive in the long term. As a physical process, fire behavior can vary substantially at local spatial and temporal scales (e.g., vegetation type, topography, wind shifts), further complicating the ability to determine the most effective ways to respond to a specific event. Nor is the human relationship with wildfire simple, fire has always been an integral part of human lives and few areas in the world do not have potential for wildfires given the right conditions. Changing climate conditions are likely to exacerbate this potential in many locations. Given these factors, it is difficult to predict which dynamics will be most critical in creating the degree and type of fire risk for any given location. Therefore, care needs to be taken about overgeneralizing the global fire "problem" as no single approach will work in all locations and inaccurate assessments of the main drivers of wildfire risk in an area are likely to lead to ineffective solutions. Instead, minimizing harmful consequences of future wildfires will require careful consideration of how a range of factors, from the national to the local scale, may inform the fire risk and likely human response in a given location.

However, a number of social science research fields can provide guidance for identifying critical elements to consider in developing programs and plans to increase wildfire preparedness and mitigation in a specific country or region. Natural hazard research provides a framework for how societies and individuals perceive and respond to the wildfire hazard with specific insights on how scale, risk interpretation, experience, and views of self-efficacy and response efficacy can influence adoption of protective actions before and during a wildfire. Diffusion of innovations provides additional insights into processes that may influence adoption of fire mitigation measures, and research related to risk and crisis communication can help identify key considerations in effective outreach efforts. Research specific to wildfires provides specific examples of how key variables may come into play in individual decisions to undertake mitigation measures on their property or when and whether to evacuate.

Not all wildfires with harmful consequences occur in extreme conditions, nor will every extreme wildfire lead to significant social impacts. However, the potential for increased extreme fire behavior that overwhelms response capacity means that a focus on understanding and utilizing the full range of mitigation options becomes even more critical. Ultimately, understanding human response to wildfires that is based on empirical evidence rather than nurtured narratives will only become more critical as more communities are affected by fire.

References

[1] S.M. McCaffrey, Community wildfire preparedness: a global state-of-the-knowledge summary of social science research, Curr. For. Reports 1 (2015) 81–90.

[2] R.J. Eiser, A. Bostrom, I. Burton, D.M. Johnston, J. McClure, D. Paton, J. van der Pligt, M.P. White, Risk interpretation and action: a conceptual framework for responses to natural hazards, Int. J. Disaster Risk Reduct. (2012), https://doi.org/10.1016/j.ijdrr.2012.05.002.

[3] E.M. Rogers, Diffusion of Innovations, third ed., 1995 doi:citeulike-article-id:126680.

[4] B. Mitchell, Geography and Resource Analysis, second ed., Longman Group, Singapore, 1993.

[5] K. Hewitt, Regions of Risk: A Geographical Introduction to Disasters, 1997, https://doi.org/10.1016/S0143-6228(97)00049-0.

[6] I. Burton, R.W. Kates, G.F. White, The Environment as Hazard, second ed., The Guilford Press, New York, 1993.

[7] P.H. Rossi, J.D. Wright, E. Weber-Burdin, Natural Hazards and Public Choice: The State and Local Politics of Hazard Mitigation, Academic Press, New York, 1982, https://doi.org/10.2307/2578380;

[7a] E. Toman, M. Stidham, S. McCaffrey, B. Shindler, Social Science at the Wildland-Urban Interface: A Compendium of Research Results to Create Fire-Adapted Communities, Gen. Tech. Rep. NRS-111 (2013).

[8] M.R. Coughlan, Wildland arson as clandestine resource management: a space–time permutation analysis and classification of informal fire management regimes in Georgia, USA, Environ. Manage. (2016), https://doi.org/10.1007/s00267-016-0669-3.

[9] R. Palm, Natural Hazards: An Integrative Framework for Research and Planninge, Johns Hopkins University Press, Baltimore, 1990.

[10] A. Maranghides, W. Mell, A case study of a community affected by the witch and guejito wildland fires, Fire Technol. 47 (2010) 379–420, https://doi.org/10.1007/s10694-010-0164-y.

[11] A. Westhaver, Why Some Homes Survived: Learning from the Fort McMurray Wildland/Urban Interface Fire Disaster, Toronto, Canada, 2017. ICLR research paper series – number 56.

[12] S. McCaffrey, Thinking of wildfire as a natural hazard, Soc. Nat. Resour. 17 (2004), https://doi.org/10.1080/08941920490452445.

[13] E. Toman, M. Stidham, S. McCaffrey, B. Shindler, Social science at the wildland-urban interface: a compendium of research results to create fire-adapted communities, Gen. Tech. Rep. NRS-111 (2013).

[14] H. Brenkert-Smith, K.L. Dickinson, P.A. Champ, N. Flores, Social amplification of wildfire risk: the role of social interactions and information sources, Risk Anal. 33 (2013) 800–817, https://doi.org/10.1111/j.1539-6924.2012.01917.x.

[15] I.M. McNeill, P.D. Dunlop, J.B. Heath, T.C. Skinner, D.L. Morrison, Expecting the unexpected: predicting physiological and psychological wildfire preparedness from perceived risk, responsibility, and obstacles, Risk Anal. 33 (2013) 1829–1843, https://doi.org/10.1111/risa.12037.

[16] T.K. McGee, B.L. McFarlane, J. Varghese, An examination of the influence of hazard experience on wildfire risk perceptions and adoption of mitigation measures, Soc. Nat. Resour. 22 (2009) 308–323, https://doi.org/10.1080/08941920801910765.

[17] I. Burton, R.W. Kates, G.F. White, Hazard, response, and choice, in: Environ. As Hazard, 1978.

[18] J.H. Sims, D.D. Baumann, Educational programs and human response to natural hazards, Environ. Behav. (1983), https://doi.org/10.1177/0013916583152003;

[18a] K. Tierney, Disaster Response: Research Findings and their Implications for Resilience Measures, CARRI Research Report 6, Oak Ridge, TN, 2009.

[19] R.L. Dillon, C.H. Tinsley, W.J. Burns, Near-misses and future disaster preparedness, Risk Anal. (2014), https://doi.org/10.1111/risa.12209.

[20] J. Whittaker, K. Haynes, J. Handmer, J. McLennan, Community safety during the 2009 Australian "Black Saturday" bushfires: an analysis of household preparedness and response, Int. J. Wildl. Fire. 22 (2013) 841, https://doi.org/10.1071/WF12010.

[21] G. Wachinger, O. Renn, C. Begg, C. Kuhlicke, The risk perception paradox-implications for governance and communication of natural hazards, Risk Anal. (2013), https://doi.org/10.1111/j.1539-6924.2012.01942.x.

[22] C. Eriksen, N. Gill, Bushfire and everyday life: examining the awareness-action 'gap' in changing rural landscapes, Geoforum 41 (2010) 814–825, https://doi.org/10.1016/j.geoforum.2010.05.004.

[23] K. Tierney, Disaster beliefs and institutional interests: eecycling disaster myths in the aftermat of 9–11, Res. Soc. Probl. Public Policy (2003), https://doi.org/10.1016/S0196-1152(03)11004-6.

[24] T.K. McGee, Preparedness and Experiences of Evacuees from the 2016 Fort McMurray Horse River Wildfire, Fire, 2019, https://doi.org/10.3390/fire2010013.

[25] R.F. Fahy, G. Proulx, L. Aiman, Panic or not in fire: clarifying the misconception, in: Fire Mater, 2012, https://doi.org/10.1002/fam.1083.

[26] D.S. Mileti, J.H. Sorensen, Communication of Emergency Public Warnings: A Social Science Perspective and State-of-The-Art Assessment, 1990, https://doi.org/10.2172/6137387.

[27] M.K. Lindell, R.W. Perry, The protective action decision model: theoretical modifications and additional evidence, Risk Anal. 32 (2012) 616–632, https://doi.org/10.1111/j.1539-6924.2011.01647.x.

[28] S. McCaffrey, R. Wilson, A. Konar, Should I stay or should I go now? or should I wait and see? Influences on wildfire evacuation decisions, Risk Anal. 38 (2018) 1390–1404, https://doi.org/10.1111/risa.12944.

[29] E.M. Rogers, Diffusion of preventive innovations, Addict. Beyond Behav. 27 (2002) 989–993, https://doi.org/10.1016/S0306-4603(02)00300-3.

[30] L.M.M. Labossière, T.K. McGee, Innovative wildfire mitigation by municipal governments: two case studies in Western Canada, Int. J. Disaster Risk Reduct (2017), https://doi.org/10.1016/j.ijdrr.2017.03.009.

[31] T.A. Steelman, S. McCaffrey, Best practices in risk and crisis communication: implications for natural hazards management, Nat. Hazards 65 (2013), https://doi.org/10.1007/s11069-012-0386-z.

[32] M.C. Monroe, L. Pennisi, S. McCaffrey, D. Mileti, Social Science to Improve Fuels Management: A Synthesis of Research Related to Communicating with the Public on Fuels Management Efforts, U.S. Department of Agriculture, Forest Service, North Central Research Station, St. Paul, MN, 2006, p. 42.

[33] S.M. McCaffrey, C.S. Olsen, Research Perspectives on the Public and Fire Management: A Synthesis of Current Social Science on Eight Essential Questions, 2012, https://doi.org/10.2737/NRS-GTR-104.

[34] T.A. Steelman, S.M. McCaffrey, A.-L.K. Velez, J.A. Briefel, What information do people use, trust, and find useful during a disaster? Evidence from five large wildfires, Nat. Hazards 76 (2014) 615−634, https://doi.org/10.1007/s11069-014-1512-x.

[35] J.B. McCaffrey, S.A.L. Velez, Difference in information needs for wildfire evacuees and non-evacuees, Int. J. Mass Emerg. Disasters 31 (2013) 4−24.

[36] T.K. McGee, Public engagement in neighbourhood level wildfire mitigation and preparedness: case studies from Canada, the US and Australia, J. Environ. Manage. (2011), https://doi.org/10.1016/j.jenvman.2011.05.017.

[37] T. Prior, C. Eriksen, Wildfire preparedness, community cohesion and social-ecological systems, Glob. Environ. Chang. 23 (2013) 1575−1586, https://doi.org/10.1016/j.gloenvcha.2013.09.016.

[38] P. Fairbrother, M. Tyler, A. Hart, B. Mees, R. Phillips, J. Stratford, K. Toh, Creating "community"? Preparing for bushfire in rural Victoria, Rural Sociol 78 (2013) 186−209, https://doi.org/10.1111/ruso.12006.

[39] M. Stidham, S. McCaffrey, E. Toman, B. Shindler, Policy tools to encourage community-level defensible space in the United States: a tale of six communities, J. Rural Stud. 35 (2014), https://doi.org/10.1016/j.jrurstud.2014.04.006.

[40] E. Toman, M. Stidham, B. Shindler, S. McCaffrey, Reducing fuels in the wildland urban interface: community perceptions of agency fuels treatments, Int. J. Wildl. Fire. 20 (2011) 340−349.

[41] B. Shindler, R. Gordon, M.W. Brunson, C. Olsen, Public perceptions of sagebrush ecosystem management in the Great Basin, Rangel. Ecol. Manag. 64 (2011) 335−343, https://doi.org/10.2111/REM-D-10-00012.1.

[42] A. Christianson, T.K. Mcgee, L. L'Hirondelle, The influence of culture on wildfire mitigation at Peavine Métis settlement, Alberta, Canada, Soc. Nat. Resour. 27 (2014) 931−947, https://doi.org/10.1080/08941920.2014.905886.

Resident and community recovery after wildfires

Tara K. McGee[1], Sarah McCaffrey[2], Fantina Tedim[3,4]
[1]Department of Earth and Atmospheric Sciences, University of Alberta, Edmonton, AB, Canada; [2]Rocky Mountain Research Station, USDA Forest Service, Fort Collins, CO, United States; [3]Faculty of Arts and Humanities, University of Porto, Porto, Portugal; [4]Charles Darwin University, Darwin, NWT, Australia

9.1 Introduction

Recovery after a wildfire is a process, both at the community or larger scale and for individuals. The United Nations Office for Disaster Risk Reduction (UNISDR) defines recovery as

> *The restoring or improving of livelihoods and health, as well as economic, physical, social, cultural and environmental assets, systems and activities, of a disaster-affected community or society, aligning with the principles of sustainable development and 'build back better', to avoid or reduce future disaster risk.*
>
> UNISDR [1].

Despite the growing wildfire social science literature and increasing impacts of wildfires worldwide, most social science wildfire research has focused on pre-fire mitigation and preparedness. While recent years have seen an increase in research that has focused on during fire dynamics such as evacuation decision-making, there remains little research specific to wildfire recovery [2].

This chapter draws on existing wildfire social science literature to examine the recovery of people and communities after wildfires. First, models of disaster recovery are presented. Then the recovery of people is examined, including the honeymoon period of increased social cohesion immediately after a wildfire, disillusionment as residents deal with challenges including insurance and rebuilding, distress that residents experience and adjustments they make during the recovery process, and increased preparedness that may occur during the recovery process. Community recovery after a wildfire is then examined, including aid provided after a wildfire, rebuilding, and building back better.

9.2 Disaster recovery frameworks

In the US, the Federal Emergency Management Agency (FEMA) and the Substance Abuse and Mental Health Services Administration (SAMHSA) have developed a model which identifies three phases (Fig. 9.1) to show the emotional highs and lows

Extreme Wildfire Events and Disasters. https://doi.org/10.1016/B978-0-12-815721-3.00009-6

Figure 9.1 FEMA/SAMHSA phases of disaster collective reactions model. *SAMHSA*, Substance Abuse and Mental Health Services Administration.
Source: https://www.samhsa.gov/programs-campaigns/dtac/recovering-disasters/phases-disaster

experienced by people after the disaster and initial relief period: honeymoon, disillusionment, and reconstruction [3]. The honeymoon phase is where people experience emotional highs as they receive disaster assistance, community members bond as they help each other, and people are optimistic that everything will return to normal quickly. This model indicates that this phase typically lasts a few weeks. The disillusionment phase involves emotional lows as people deal with insurance and other challenges, and social networks that were stronger during the honeymoon phase may become divided, and community conflict may occur. The reconstruction phase involves a feeling of recovery where people adjust to a new "normal" while continuing to grieve losses. The model indicates that the reconstruction phase often begins around the 1 year anniversary of the disaster. The three phases of the FEMA/SAMHSA model [3] provide a valuable framework for examining wildfire recovery of residents, discussed in Section 9.3.

Chang has developed a framework of indicators of community and regional recovery: regaining predisaster conditions and attaining a new normality (returning to a stable state) [4]. More recently, Thomalla et al [5] distinguish four different approaches to community recovery, drawing from the international disaster literature:

(1) Early restoration (ER) which involves replacing lost assets and getting lives back to normal as quickly as possible [5];
(2) Linking Relief, Rehabilitation, and Development (LRRD) which focuses on reducing gaps between humanitarian aid and development cooperation [5];
(3) Build Back Better (BBB) which focuses on improving predisaster conditions, which may include more emphasis on environmental sustainability, stronger buildings, and other hazard mitigation, to name a few [5]; and

(4) Empower local communities (ELC), which involves reducing vulnerabilities and root causes of disasters to empower communities [5].

The existing wildfire recovery literature provides insights into the ER and BBB approaches, which are discussed in Section 9.4.

9.3 Wildfire recovery: Residents

Although wildfire recovery studies are limited, they provide evidence for all three of the FEMA/SAMHSA phases.

9.3.1 Honeymoon period

The community cohesion and emotional highs that characterize the honeymoon period early in the recovery process have been found after wildfires [6–8]. The 2002 Rodeo-Chediski fire in the US that burned 189,541 ha and destroyed 426 buildings was the worst wildfire in Arizona's history at that time. Carroll et al [6] conducted interviews 5–6 months after the fire and found evidence that residents and organizations "pulled together," with people sharing food and supplies and providing transportation and information. This increased social cohesion continued after people returned to their homes with neighbors providing shelter and social support to each other [6]. It therefore appears that the honeymoon period can extend well beyond the short phase identified in the FEMA/SAMHSA model.

9.3.2 Disillusionment

In contrast to the honeymoon period where residents experience emotional highs, emotional lows predominate in the disillusionment phase. Pujadas-Botey and Kulig [9] conducted research 4–7 months after the 2011 Slave Lake wildfire in Alberta, Canada, which burned 22,000 ha, caused the evacuation of more than 10,000 residents, and destroyed 333 single-family homes, 11 multi-family homes, six apartment buildings, three churches, 10 businesses, and part of the Government Center in the Town of Slave Lake, and 56 properties outside the town of Slave Lake. Residents in the Town of Slave Lake received little or no warning before the wildfire entered town [10], so many had very little if any time to prepare before they had to leave their home and community. These researchers found that during this phase of the recovery process 4–7 months after the wildfire, parents experienced high levels of stress due to constant concerns about the future of their families, survivors guilt, and disagreements with local authorities about being able to clean up their burned property. Families in their study underwent many adjustments, including re-evaluating life goals and priorities, establishing new routines, changes in attitudes, changes in interactions within the family and community, and new values and perceptions [9].

During the disillusionment phase, community conflicts may occur, and people may attach blame for the impacts of the wildfire [6]. Conflict often focuses on perceived inadequacies of firefighting effort and aid provided to residents. For example, Rodriguez-Mendez et al [11] studied the 1994 Tyee fire in the State of Washington, US, which burned 54,632 ha and destroyed 35 homes and cabins in the small rural town of Entiat. The researchers conducted interviews 1 month after the fire and found that some residents in Entiat blamed the federal government for their land management and firefighting strategy. Edgeley and Paveglio [7] recently conducted a study 1 year after the 2014 Carlton Complex fire in Washington State, US, which burned 103,599 ha, destroyed 256 houses, and caused other impacts to infrastructure. These researchers identified conflict over the firefighting efforts, with some participants unhappy with the Department of Natural Resources' fire management response. Interestingly, the authors report that a group of residents filed a lawsuit against the Department, claiming that the disaster was preventable [7]. However, these researchers found that some residents affected by the Carlton Complex fire were not critical of the government's firefighting response, causing conflict amongst residents. In addition, FEMA was blamed for not providing Individual Assistance during the recovery process [7], which includes services for temporary housing, home repair and replacement, unemployment insurance, legal services, income tax credits, and crisis counseling [12]. Carroll et al [13] also found that there was conflict around the distribution of helping resources from agencies including FEMA and the Red Cross when they studied six wildfire case studies in the western US.

The disillusionment phase also involves challenges associated with insurance. Many people who lose a home in a wildfire are under insured, and some are not insured at all [14]. For example, Mockrin et al. [15] found that lack of or inadequate insurance posed challenges to rebuilding after three wildfires in Colorado: The Fourmile Canyon Fire (2010) which destroyed 168 homes and burned 2501 ha; High Park Fire (2012) which caused one fatality and destroyed 259 structures; and the Waldo Canyon Fire (2012) which caused two fatalities, 347 homes were destroyed, and 30,000 people were evacuated. However, even if residents are insured adequately, dealing with insurance companies can cause distress during the disillusionment phase after a wildfire. For example, McGee [16] surveyed residents 1 month after the Horse River (Fort McMurray) wildfire and found that dealing with insurance was one of the challenges faced by survey respondents once they returned to Fort McMurray.

9.3.3   Reconstruction

In the reconstruction process, residents affected by a wildfire may continue to experience distress and have setbacks as they adjust to the "new normal." A few studies indicate that distress can continue several years after wildfires have occurred. Kirsch et al [17] examined the public health impacts of 2011 Bastrop County Complex Fire in Texas, US, which was the most destructive wildland urban interface fire in Texas history. The wildfire caused the death of two people, destroyed 1645 homes, and burned 13,786 ha of land. The researchers found that in 2015, 3.5 years after the wildfire, households exposed to the wildfire were still significantly more likely to report symptoms of depression and higher stress than those who were not exposed.

Pfitzer et al. [18] completed research 3 years after the 2009 Black Saturday wildfires in Victoria, Australia, to examine the psychosocial adjustment of residents who had serious burn injuries that required complex medical treatments. The researchers found that one-third of the 13 participants suffered high to very high levels of distress, and 58% fulfilled some or all criteria for posttraumatic stress disorder. Participants also experienced significantly impaired physical health functioning. Although some participants experienced little distress 3 years after the wildfires and others had a decline in stress levels over time, some patients had high levels of distress throughout the 3-year period or experienced an increase in distress later in the recovery process. Researchers have also found that residents can experience psychological distress due to damage to the natural environment after a wildfire. One year after the 2011 Wallow Fire in Arizona, US, Eisenman et al. [19] found that higher loss of solace from the landscape (solastalgia) and adverse impact from the fire were associated with significant psychological distress.

The increased social cohesion of the honeymoon phase may continue during the reconstruction phase. In a study 5 years after the Rodeo-Chediski fire, Carroll et al [20] found there was still evidence of local residents and organizations continuing to support each other 5 years after the fire, clearly indicating that the increased social cohesion that occurs in the honeymoon phase can continue for a long time. In contrast, researchers in Australia found that the period of increased social cohesion may not last. In a study after wildfires in East Gippsland, Australia, Whittaker et al [21] found that social cohesion increased among residents and rural landholders and that divisions in the community that existed before the wildfires were broken down after the wildfires, with people sharing donated goods and working together to clean up after the wildfire. However, the researchers found that after an initial period of increased cohesion, the social divisions gradually re-established as people began to recover.

Although there are mixed findings regarding whether residents will implement recommended mitigation measures to their property as part of the reconstruction process [22], there is some evidence that people may undertake measures to increase their preparedness so that they can respond effectively during a wildfire. In the study after the 2011 Bastrop County Complex fire in Texas, US, Kirsch et al [17] found that 4 years after the wildfire, more than half of respondents were more prepared and had a 3-day supply of food and water, family meeting place, evacuation route, copies of personal documents, and plans for pets and/or livestock. Those in the area during the 2011 fire were more likely to be more prepared than those who were not in the area at the time when the fire occurred.

9.4 Wildfire recovery: Community

9.4.1 Early Restoration

Evidence to date indicates that most approaches to wildfire recovery tend to use on the ER approach, which focuses on replacing lost assets and getting lives back to normal as quickly as possible after a disaster [5]. For residents, the ER approach enables them to return to their "new normal" quickly in the reconstruction phase. At the community level, this ER approach fits the need for government authorities to be seen to be taking immediate action after a disaster [5]. This strong desire to rebuild quickly was apparent after the 2016 Fort McMurray Horse River wildfire where large "Together we will rebuild" billboards were set up by the Regional Municipality of Wood Buffalo on the highway into Fort McMurray shortly after the fire. Similarly, after the 2003 Canberra (Australia) wildfires, MacKenzie [23] examined the rebuilding process and found that planners who were involved in the rebuilding process responded to

> *social and political pressures from elected officials and senior bureaucrats to return*
> *the community to a stable state as soon as possible. Many designers and residents*
> *complained that things were happening too slowly.*
>
> *Mackenzie [23], p. 351*

Municipal governments may try to streamline the rebuilding process after a wildfire to make it easier and faster for residents to rebuild. In their study of the 2010 Fourmile Canyon Fire in Colorado US, Mockrin et al. [24] found that Boulder County tried to speed up the rebuilding process by hiring a recovery coordinator for 2 years, and waiving the requirement for site-plan review if the homeowner's new building was no more than 530 square feet larger than the destroyed home and if they applied for a building permit within 2 years of the fire. After other fires in the Colorado Front Range between 2010 and 2012, Mockrin et al. [15] found that Larimer County reduced building permit fees for those who were under- and uninsured; and Colorado Springs simplified site-plan review by allowing the use of previous plans and master plans, and reduced fees for site-plan review and utility reconnection.

Despite pressure from governments and homeowners, and efforts to streamline the rebuilding process, rebuilding after wildfires may be slow. Alexandre et al. [25] examined rebuilding and new building development after wildfires across the US between 2000 and 2005 and found low rates of rebuilding, with only 25% of burned homes rebuilt within 5 years [25]. After the 2010 Fourmile Canyon Fire in Colorado, Mockrin et al. [24] found that 34 months after the fire, only 30% of those who had lost homes had rebuilt and were living in the home, 20% were in the process of rebuilding, and 50% were not yet rebuilding. Similarly, in Canberra, Australia, MacKenzie [23] found that 40% of homes destroyed in the 2003 wildfires were rebuilt 3 years later.

Although rebuilding has been found to be slow, Alexandre et al. [25] found high rates of new housing development within fire perimeters 5 years after US fires between 2000 and 2005. Interestingly, the researchers found geographic differences in housing development rates, with higher rates inside the fire perimeter in some states and lower development rates inside fire perimeters in others. In the states of Kentucky and West Virginia, the number of buildings within the fire perimeters increased even though there was a decline in housing within the surrounding counties. In contrast, in the states of California, Arizona, Colorado, Wisconsin, and most of Utah, there were lower development rates within fire perimeters than in surrounding counties.

The evidence to date indicates that most approaches to wildfire recovery tend to focus on ER. While this approach benefits residents and governments that wish to return to normal as quickly as possible, it may miss the opportunity to rebuild in a way that will reduce future wildfire risks. For example, after the 2003 Canberra (Australia) wildfires, MacKenzie [23] interviewed design professionals involved in rebuilding homes after the fires. He found that their resident clients who needed to rebuild after the Canberra fires started out the process by wanting to build a safer home, but the designer encouraged them instead to build larger homes to maximize their investment or meet a desire for the ideal home.

9.4.2 Building back better

Following the 2003 Canberra fires, MacKenzie [23] found that some planners recognized that building back as quickly as possible may not lead to the best outcome. As a development assessment officer said:

> *So I think that whole issue of slowing down the whole decision making process [was important], and I can understand why it happened so quickly because they were desperate to normalise their life again […] I think the ones who worked through a very deliberate process ended up with a better outcome than the ones that moved very quickly.*
>
> *Mackenzie [23]*

BBB is a guiding principle of the Sendai framework for Disaster Risk Reduction 2015–20 [26]. As stated in the Sendai framework,

> *In the post-disaster recovery, rehabilitation and reconstruction phase, it is crucial to prevent the creation of and to reduce disaster risk by 'Building Back Better' and increasing public education and awareness of disaster risk.*
>
> *UNISDR [26], p. 14*

BBB provides an opportunity to reduce the vulnerability of communities to future wildfires by implementing mitigation and preparedness activities during the recovery process. Construction standards can be used to ensure that homes are rebuilt to be more resistant to wildfires. One of the most comprehensive construction standards is the Australian Standard AS3959—Construction of Buildings in Bushfire-prone Areas, which was developed in 1999 with revisions in 2009 and 2011. This standard prescribes the minimum level of construction required for new houses and extensions to reduce the risk of ignition from a wildfire while the fire front passes. In the US, California Public Resources Code 4290 and 4291 set out requirements for vegetation management within 100 feet of structures, and minimum fire resistant building codes for new structures.

A small group of researchers has studied BBB after wildfires. After the 2009 Black Saturday fires in the State of Victoria, Australia, Mannakkara et al [27] examined measures implemented by governments to facilitate rebuilding that is more resistant to wildfires, including categorizing property into high-, medium-, and low-risk areas; a revision to the Australian Building Code to include more stringent design and construction specifications; and stricter implementation of the wildfire management overlay to ensure that homes were designed and constructed appropriately. In the US, Mockrin et al. [15] examined how fires between 2010 and 2012 encouraged building back better in Boulder, Larimer, and Colorado Springs Counties in the state of Colorado. While all three counties had WUI regulations before the wildfires, the researchers found that after wildfires Boulder and Larimer counties did not lessen requirements to facilitate rebuilding, while Colorado Springs strengthened their regulations by encouraging home location changes if they reduced fire hazard, improved access, or reduced land-use impacts.

9.5   Conclusion

Research shows that residents affected by an EWE or smaller wildfire follow the steps in the recovery process identified in the FEMA/SAMHSA model. Studies after wildfires clearly identify a honeymoon phase where residents experience stronger social cohesion, with mixed findings regarding how long this honeymoon period lasts. The disillusionment phase, including community conflict, blaming, and dealing with insurance is also apparent in studies of recovery after wildfires. During the reconstruction phase, many residents continue to experience distress as they try to return to their "new normal." In some instances the increased social cohesion in the honeymoon phase may continue throughout the reconstruction phase. There is evidence that some residents will increase their preparedness after experiencing a wildfire.

At the community level, wildfire recovery studies indicate that there is strong emphasis on rebuilding quickly after a wildfire, in line with the ER approach identified by Thomalla et al. [5]. However, there can be significant delays in rebuilding after wildfires. The ER approach serves residents who wish to return to normal as quickly as possible and political leaders who want to be seen to be acting quickly after a disaster; however, this can limit opportunities to reduce vulnerability in the event of a future wildfire. BBB is advocated by the UN as part of the Sendai Framework for Disaster Risk Reduction. In the context of wildfire recovery, BBB can include a variety of activities including building codes and construction standards for buildings constructed after a wildfire, vegetation management requirements, to name a few.

Additional research is needed in several areas. Further study is needed to examine how long the period of increased social cohesion lasts after a wildfire. More research is also needed to examine residents' efforts to increase their preparedness during the recovery process. Finally, additional research is needed to examine rebuilding during the recovery process to reduce vulnerability in the event of a future wildfire.

References

[1] UNISDR, 2017. https://www.unisdr.org/we/inform/terminology.

[2] S.M. McCaffrey, Community wildfire preparedness: a global state-of-the-knowledge summary of social science research, Curr. For. Rep. 1 (2) (2015) 81−90.

[3] FEMA/SAMHSA (undated) Phases of Disaster. https://www.samhsa.gov/programs-campaigns/dtac/recovering-disasters/phases-disaster [accessed 21 March 2019].

[4] S.E. Chang, Urban disaster recovery: a measurement framework and its application to the 1995 Kobe earthquake, Disasters 34 (2) (2010) 303−327, https://doi.org/10.1111/j.0361-3666.2009.01130.x.

[5] F. Thomalla, L. Lebel, M. Boyland, D. Marks, H. Kimkong, S.B. Tan, A. Nugrohu, Long-term recovery narratives following major disasters in Southeast Asia, Reg. Environ. Chang. 18 (2018) 1211−1222. https://doi.org/10.1007/s10113-017-1260-z.

[6] M.S. Carroll, P.J. Cohn, D.N. Seesholtz, L. Higgins, Fire as a galvanizing and fragmenting influence on communities: the case of the Rodeo-Chediski Fire, Soc. Nat. Resour. 18 (4) (2005) 301−320.

[7] C.M. Edgeley, T.B. Paveglio, Community recovery and assistance following large wild-fires: the case of the Carlton Complex Fire, Int. J. Disaster Risk Reduct. 25 (2017) 137−146.

[8] J.C. Kulig, D.S. Edge, I. Townsend, B. Reimer, N. Lightfoot, Impacts of wildfires: the aftermath at individual and community level, Aust. J. Emerg. Manag. 28 (3) (2013) 29−34.

[9] A. Pujadas-Botey, J.C. Kulig, Family functioning following wildfires: recovering from the 2011 Slave Lake fires, J. Child Fam. Stud. 23 (2014) 1471−1483.

[10] KPMG Lesser Slave Lake Regional Urban Interface Wildfire - Lessons Learned Final Report, November 6, 2012. https://open.alberta.ca/dataset/8b69f242-0b66-4cd4-bdf3-944de68f3ae1/resource/beac1cb7-767f-4883-8686-9682beae772f/download/6520642-2013-lessons-learned-final-report.pdf.

[11] S. Rodriguez-Mendez, M.S. Carroll, K.A. Blatner, A.J. Findley, G.B. Walker, S.E. Daniels, Smoke on the Hill: a comparative study of wildfire and two communities, WJAF 18 (1) (2003) 60−70.

[12] K. O'Donovan, Disaster Recovery Service Delivery: Toward a Theory of Simultaneous Government and Voluntary Sector Failures, Administration & Society, 2015, pp. 1−20.

[13] M.S. Carroll, L.L. Higgins, P.J. Cohn, J. Burchfield, Community wildfire events as a source of social conflict, Rural Sociol. 71 (2) (2009) 261−280, https://doi.org/10.1526/003601106777789701.

[14] K. Booth, B. Tranter, When disaster strikes: under-insurance in Australian households, Urban Stud. 55 (14) (2017) 3135−3150.

[15] M.H. Mockrin, S.I. Stewart, V.C. Radeloff, R.B. Hammer, Recovery and adaptation after wildfire on the Colorado front Range (2010−12), Int. J. Wildland Fire 25 (2016) 1144−1155.

[16] T.K. McGee, Residents' experiences of the 2016 Fort McMurray Wildfire, Alberta. In Advances in Forest Fire Research, D.X. Viegas (ed.). Chapter 6 − Socio Economic Issues 2018. https://doi.org/10.14195/978-989-26-16-506_129.

[17] K.R. Kirsch, B.A. Feldt, D.F. Zane, T. Haywood, R.W. Jones, J.A. Horney, Longitudinal community assessment for public health emergency response to wildfire, Bastrop County, Texas, Health Secur. 14 (2) (2016) 93−104, https://doi.org/10.1089/hs.2015.0060.

[18] B. Pfitzer, L.J. Katona, S.J. Lee, M. O'Donnell, H. Cleland, J. Wasiak, S. Ellen, Three years after Black Saturday: long-term psychosocial adjustment of burns patients as a result of a major bushfire (2016), J. Burn Care Res. 37 (3) (2016) e244−e253.

[19] D. Eisenman, S. McCaffrey, I. Donatello, G. Marshal, An ecosystems and vulnerable populations perspective on solastalgia and psychological distress after a wildfire, Eco-Health 12 (2015) 602−610, https://doi.org/10.1007/s10393-015-1052-1.

[20] M.S. Carroll, T. Paveglio, P.J. Jakes, L.L. Higgins, Nontribal community recovery from wildfire five years later: the case of the rodeo-Chediski fire, Soc. Nat. Resour. 24 (7) (2011) 672−687, https://doi.org/10.1080/08941921003681055.

[21] J. Whittaker, J. Handmer, D. Mercer, Vulnerability to bushfires in rural Australia: a case study from East Gippsland, Victoria, J. Rural Stud. 28 (2) (2012) 161−173. https://doi.org/10.1016/j.jrurstud.2011.11.002.

[22] T. McGee, C. Eriksen, Defensive actions and people preparedness, in: S.L. Manzello (Ed.), Encyclopedia of Wildfires and Wildland-Urban Interface Fires, Springer, 2018. https://doi.org/10.1007/978-3-319-51727-8_93-1.

[23] A. Mackenzie, Planning for the redevelopment after a fire event, Int. J. Disaster Resilience Built Environ. 8 (4) (2017) 344−356. https://doi.org/10.1108/IJDRBE-03-2016-0008.

[24] M.H. Mockrin, S.I. Stewart, V.C. Radeloff, R.B. Hammer, P.M. Alexandre, Adapting to wildfire: rebuilding after home loss, Soc. Nat. Resour. 28 (8) (2015) 839−856, https://doi.org/10.1080/08941920.2015.1014596.

[25] P.M. Alexandre, M.H. Mockrin, S.I. Stewart, R.B. Hammer, V.C. Radeloff, Rebuilding and new housing development after wildfire, Int. J. Wildland Fire 24 (2015) 138−149. https://doi.org/10.1071/WF13197.

[26] UNISDR, Sendai Framework for Disaster Risk Reduction 2015-2030, United Nations, Geneva Switzerland, 2015. https://www.unisdr.org/files/43291_sendaiframeworkfordrren.pdf.

[27] S. Mannakkara, S. Wilkinson, R. Potangaroa, Build back better: implementation in Victorian bushfire reconstruction, Disasters 38 (2) (2014) 267−290, https://doi.org/10.1111/disa.12041.

Part Four

How to Cope with the Problem of Extreme Wildfires and Disasters

Wildfire policies contribution to foster extreme wildfires

Paulo M. Fernandes[1], Giuseppe Mariano Delogu[2], Vittorio Leone[3], Davide Ascoli[4]

[1]Centre for the Research and Technology of Agro-Environmental and Biological Sciences (CITAB), University of Trás-os-Montes and Alto Douro, Vila Real, Portugal; [2]Former Chief Corpo Forestale e di Vigilanza Ambientale (CFVA), Autonomous Region of Sardegna, Italy; [3]Faculty of Agriculture, University of Basilicata (retired), Potenza, Italy; [4]DISAFA Department, University of Torino, Grugliasco, Italy

10.1 Introduction

Opposite worldviews and discourses about fire management policies are vehemently debated and often lead to conflict [1−3]. Historically, and in general, development of fire regulation policies parallels the maturity stage of forest resources management [4]: Low value or local use corresponds to indifference, i.e., a lack of policy; limited efforts to control fire are associated to developing management and abundance of forest resources; intensive (and concerned with scarcity) management, namely when focused on a few main resources, compels to aggressive fire suppression; finally, if substantial demands exist for manifold and rival uses, fire suppression is expected to give way to more integrated fire management. Nonetheless, change and evolution in fire policies is driven by extreme wildfire events (EWEs) and seasons, and in this respect, Fernandes et al [5] have provided a number of examples from around the world. Reactive changes in fire policies can suffer from selective filtering of information [6] and foster misperceptions and inconsistency [7].

Federal forest-fire exclusion policies in the U.S. had their inception in the late 19th century and were put in the form of national law in 1914 [8], after the devastating fire season of 1910 during which 78 firefighters were killed and two million hectares of Forest Service land were burned [9]. A fire control policy was deemed as a necessary condition to develop forestry and was molded by the practices and system of values of European silviculture, where fire is viewed solely as a damaging disturbance [9]. Hence, fire exclusion policies largely ignore environmental and social issues and do not recognize the role of fire in ecosystems, despite substantial controversy and debate at the time the policy was shaping up [8,9]. Enforcement of the policy involved considerable social coercion through influence or force in both the U.S [10] and Europe [11], with impacts on people livelihoods wherever cultural (traditional) burning was in place [12−15].

Policies variously known as fire exclusion, suppression, or control are based on avoiding human-caused ignitions and adopting effective presuppression measures

Extreme Wildfire Events and Disasters. https://doi.org/10.1016/B978-0-12-815721-3.00010-2

and hard-hitting suppression whenever fire occurs. *"Only You Can Prevent Forest Fires,"* the message conveyed by Smokey the Bear in the U.S. was instrumental and quite successful as part of the fire exclusion policy [9,16]. Firefighting operations under this policy are guided by rigid principles, regardless of the resources at risk and the conditions under which fire occurs, seeking to minimize burned area under the assumption that larger fires equate to higher losses; to attain a contained fire status by 10:00 a.m. of the following day was one of such principles [17].

Over time, increased fuel hazard and decreased forest health in U.S. forests resulting from fire exclusion, development of fire ecology science, and the need to reduce fire suppression costs demanded a change in federal fire policy. Such change began with Florida's Osceola National Forest adopting prescribed fire in 1943, but expansion of the practice to Western federal lands did not occur until late 1960s [8]. Only in mid 1970s was the 10:00 a.m. policy abandoned [9]. Fire management (or integrated fire management) was embraced and defined as the integration of all fire-related information and activities in obedience to forest and land management goals [4,17–19]. Fire management seeks to minimize the difference between the negative and positive effects of fire, not area burned [20]. Fire management planning follows a hierarchy, from the global objectives of resources management, to the specific aims of fire management, to the formulation of strategies, tactics, and actions [18]. Integrated fire management is further distinguished from fire control by

- zoning to guide fuel treatments and a priori define how a wildfire will be responded to;
- fuel management in general, and prescribed burning in particular, as a relevant component;
- wildfire suppression operations commensurate with its characteristics and potential damage, thus variable in effort and timing (fast and aggressive, modified, or delayed);
- deliberate and planned response to a given fire, weighing the consequences of the decision and including monitoring and no-suppression options;
- regular policy assessment based on results of follow-up and cost-effectiveness.

As per its initial concept, integrated fire management expanded the scope of the former policy on ecological and socioeconomic rationale grounds. As expected, fire management developed in different countries and regions in different moments, at different paces, and to different depths.

The Mediterranean corner of Western Australia was ahead of its time, as it never practiced a full-exclusion policy, implemented fire management zoning as early as the 1920s, and decided for large-scale prescribed burning in 1954 [21]. Fire management in Canada evolved in parallel with the U.S., being well known for the modified fire suppression policy in parts of the country [22]. The South African programs are more focused in biodiversity than in hazard reduction and include a strong adaptive management component [23,24]. Involvement with local communities and reinstatement or regulation of their traditional burning practices are a recent trademark of Northern Australia programs [25] and emergent efforts in South America [26]. Finally, the extent to which integrated fire management has advanced in southern Europe can be assessed by the level of prescribed burning adoption, from prohibited (Greece) or still experimental (Italy) to locally (Portugal, Spain) or regionally significant (France) [27].

Increasingly encompassing and holistic views of fire management and governance have been proposed more recently, further considering and evaluating the negative and positive sides of fire and its multiple and complex contexts [28] in the frame of a coupled human and natural systems [29,30]. However, rigid responses to wildfire occurrence and fire exclusion are still largely prevalent and detract from a sustainable human coexistence with fire [31], even when an official policy entailing a more progressive, balanced or nuanced approach is in place [8,32]. The remainder of this chapter addresses the consequences of command and control fire policies and why they perpetuate and introduces basic principles that should guide policy development and implementation.

10.2 Shaping wildfire disasters through misguided policy

Wildfire ignition control programs can substantially succeed, both in terms of adoption by the population [33] and fire incidence decrease [34]. However, not all causes of fire are preventable, and even the most successful policy will not be able to significantly avoid those ignitions that will become a problem [35,36]. Causes of ignition are greatly emphasized in detriment of the factors that govern fire spread, deterring the complete understanding of the phenomenon. In addition, significant conflicts can arise because decision-makers lean to command and control responses [37] and will often prohibit or severely constrain traditional burning practices against the will and interests of local communities and without preceding educational programs [11]. Ecosystem services delivered through the traditional use of fire, which are made more relevant by climate change [37] but are seldom recognized [12], have declined, and such trend will continue if more land is further abandoned in the Mediterranean Basin [15,38].

The effectiveness of fire control operations is contingent on fire behavior magnitude, namely on its rate of spread and fireline intensity [39]. The ability to constrain fire size is then physically limited by fire weather and fuel quantity, structure, and condition, regardless of the amount and extinction capacity of firefighting resources. This implies that while wildfires occurring under relatively mild fire weather conditions are easily tackled, those driven by strong winds and very dry fuels will run unchecked until the fire environment changes [40,41]. Consequently, fire control enforcement implies that the majority of fires burning under mild conditions will be timely suppressed. However, under extreme conditions, a few fires will account for most of the burned area [42] and, over time, decreased vegetation patchiness and increased fuel accumulation due to less frequent but larger fires will create more hazardous fuel conditions over broader spatial scales [43]. Additional factors can conflate with fire exclusion to shift the fire regime, namely large-scale afforestation and climate change [14]. Control of the fire regime through fuel patterns varies across ecosystems and vegetation types, is dependent on treatment scale and spatial arrangement, and is better revealed by fire severity than by area burned, but is well supported by research carried out worldwide [44].

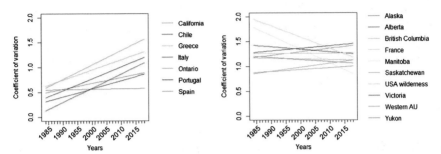

Figure 10.1 Trends in the coefficient of variation (CV) of annual burned area calculated using a moving window of 7 years for 1984−2018 data series: high fire suppression territories (left panel) versus low fire suppression territories (right panel). CV increases over time indicate increased interannual variability in burned area.

Exacerbation of the wildfire problem through fire suppression is known as the *wildfire paradox* [45] or the *firefighting trap* [46] and is a positive feedback outcome of shifting the distribution of fire behavior to only the most severe through misguided management [47]. The fire paradox can be detected by increases in the interannual variability in burned area in a given territory subjected to fire suppression. Fig. 10.1 shows linear regression trends in the coefficient of variation calculated using a moving window of 7 years for several series of yearly burned area in different territories subjected to high or low fire suppression polices from 1984 to 2018 (Table 10.1). The contexts leading to low fire suppression vary: Southwestern Australia forests are characterized by a large prescribed burning effort [48]; naturally ignited fires in Western U.S. wildernesses are not suppressed [49]; a natural fire regime also dominates Canadian provinces and Alaska, and either modified fire suppression policies are implemented or the effect of fire suppression is made irrelevant by remoteness and scarce resources [50]; somewhat in between the later and the former are Victoria, Australia [51], and France, with the largest prescribed burning program in Southern Europe and reinstatement of best-practice traditional fire use within a regulated context [27].

Mean slope coefficients for the trends in area burned variability in Fig. 10.1 are 0.019 and −0.004 in territories characterized by high and low fire suppression, respectively. This indicates that where fuel management is residual full fire suppression policies lead to more variable fire years over time by being more effective at controlling fires during favorable fire weather years, which in turn increases landscape-level fuel hazard. Spain does not conform to this general trend, which possibly reflects the exceedingly high budget allocated to fire suppression [13]. Some caution is advised in interpreting Fig. 10.1 literally, as other factors are involved and can influence the results, namely changes in climate and land use and variability within country or region. Table 10.1 reports the sources of series of yearly burned area used for the trend analysis in Fig. 10.1.

Awareness of the wildfire paradox in the U.S. has grown over the last decades, as the annual area burned and the incidence of large and severe fires increased, despite

Table 10.1 Series of yearly burned area used for the analyses in Fig. 10.1. The high fire suppression category includes southern Europe countries, California, Chile, and Ontario. The low category includes fire regimes either dominated or influenced by management-ignited fires and/or naturally ignited fires.

Region(s)	Data source	Suppression level	Data link
Alaska	Alaska dept. Natural resources	Low	http://forestry.alaska.gov/firestats/
Alberta, British Columbia, Manitoba, Saskatchewan, Yukon	Canadian National Fire Database	Low	http://cwfis.cfs.nrcan.gc.ca/ha/nfdb
[a]California	CALFire	High	http://cdfdata.fire.ca.gov/
Chile	CONAF	High	http://www.conaf.cl/incendios-forestales/incendios-forestales-en-chile/estadisticas-historicas/
France	European Forest Fires Information System	Low	http://effis.jrc.ec.europa.eu
Greece, Italy, Portugal, Spain	European Forest Fires Information System	High	http://effis.jrc.ec.europa.eu
Ontario	Canadian National Fire Database	High	http://cwfis.cfs.nrcan.gc.ca/ha/nfdb
[b]Western U.S. wildernesses	Parks et al. [49]	Low	https://www.fs.fed.us/rm/pubs_journals/2015/rmrs_2015_parks_s001
Victoria	[c]Country fire Authority	Low	—
SW Western Australia	[d]Dept. Biodiversity, Conservation and Attractions	Low	—

[a]Nonfederal land only.
[b]Frank Church-River of No Return, Crown of the Continent Ecosystem, Gila, and Aldo Leopold wildernesses.
[c]Supplied by Alen Slijepcevic.
[d]Supplied by Lachlan McCaw.

escalating (and unsustainable) expenditures in fire suppression that delay fire only to make it worse [8,16,52]; area burned and fire suppression costs in the U.S. have increased by factors of four and 10 since 1985, respectively (Fig. 10.2). In the words of Ingalsbee [32] this stems from "*anti-ecological human beliefs* (e.g., *the goal of controlling nature in the guise of attempted fire exclusion) (...) that conflict with the reality of living on a pyrogenic plane*t." However, and although current wildfire severity is higher than historically, a recent Western U.S. study found a burned area deficit of one order of magnitude [53]. This suggests that the relevance of the fire paradox can be overstated regarding fire extent, i.e., the positive feedback is lower than generally thought, and that the contemporary increase in burned area is mainly driven by climate change.

The combination of rural abandonment, afforestation policies, and fire exclusion is creating landscapes in southern Europe that have been shown to foster large wildfires [54–59]. Nonetheless, the area burned in Mediterranean Basin countries has decreased by two-thirds from 1985 to 2011 [60]. Similarly, and after the peaks observed in the 1970–1980, regional and country-level studies indicate sustained decreases in area burned since 1990s [61–63]. These changes can be attributed to a decline in the number of fires and strong investment in fire suppression, which effectively control fire size in years wild mild fire weather, but record-size mega fires, EWEs, and catastrophic

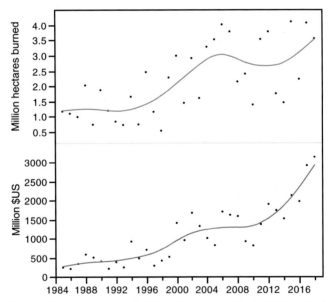

Figure 10.2 Observations and spline smoothing of area burned in the U.S., including all private, state, and federal lands, and federal fire suppression costs, not adjusted for inflation (1985–2018)

Data from the National Interagency Fire Center (https://www.nifc.gov/fireInfo/fireInfo_statistics.html).

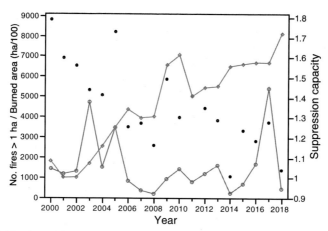

Figure 10.3 Number of fires, burned area, and suppression capacity in Portugal (2000–18), respectively, black dots and red and blue lines.

seasons were observed in Portugal in 2003 (and then again in 2017) and in Greece in 2007 and 2018.

Portugal is the country most affected by large fires in southern Europe, and the results of the current policy are depicted in Fig. 10.3, for which we used official data and estimated the relative suppression capacity per year by combining the type and amount of available firefighting resources with their standard productivity rates and fireline intensity limits for effective fire control [64]. A decreasing trend in the number of fires and increasing fire suppression capacity, nearly doubling in 2 decades, could not avoid the catastrophic year of 2017, both in burned surface and number of human fatalities [65].

In France, the contemporary decrease in area burned [61] led the authors to consider fire suppression as a strategy able to counteract the effects of global change on fire activity, although they failed to acknowledge the leverage exerted by prescribed burning and questioned whether the results are sustainable. System dynamics–based simulations of the effects of fire suppression in Portugal indicated that ∼30 years are needed for the fire paradox to reveal itself, but >50 years are needed for its full expression [46]. Simulation modeling studies predict higher fire intensity in Mediterranean climate regions under a fire exclusion policy [66].

10.3 The perpetuation of misguided fire policies

Catastrophic fire events and seasons are usually followed by blame games and reactive and short-lived bursts of activity by governments and organizations, which may include inquiries and their subsequent recommendations [36,67]. Instead, wildfire disasters can, or could, serve as tipping points for change through societal pressure [47]. Pyrocrises are opportunities for changing paradigms and reform policies. And yet, more often than not, the response to fire disasters consists of instinctive enlargement of fire suppression capacity, thus addressing not the root of the problem but its symptoms [68]. This expanded ability to respond to wildfire will seem successful, and therefore rational, when in fact it will worsen the problem by reinforcing the firefighting trap, because of the positive feedback loops caused by the suppression—fuel—fire interactions previously described [46,47,69,70].

A conceptual model of the factors driving response to wildfire has been proposed by Calkin et al. [47]. Social and political expectations and pressures are sufficiently strong to constrain the options available to fire managers and agencies and increase their risk aversion in relation to practices supported by policy but that deviate from the full-suppression paradigm. Announcing or using expanded fire suppression capability is appealing and reassuring to the public and mass media, especially when wildland—urban interfaces are at risk [46,69]; displays of firefighting aircraft are particularly popular, but often are merely choreographic ("*CNN drops*") [71] and expectations regarding their effectiveness are unrealistic. Public perceptions and the associated sociopolitical pressures are in part the result of lack of understanding and acceptance of risk management concepts [47]. Additional explanations for the firefighting trap persistence are the focus on short-term results, difficulty in identifying and claiming damage avoidance by prevention measures, and an economic-motivated interest of the firefighting establishment to maintain the status quo [46].

Fire management is also conditioned by external societal demands in the form of multiple, competing, and sometimes conflicting goals for resources management and external environmental stressors such as climate change and biotic risks [72]. Both tend to aggravate fire risk and reinforce fire suppression, with value assigned to the former implying poor contemplation of the long-term impacts of fire management on resource values.

Higher external demands for fire suppression resources are a disincentive for reforming fire management agencies and further increase risk by deviating resources from forest management and fuel treatments [73]. However, endogenous pressures on the fire management strategy can be equally important [74], namely because addressing multiple goals is difficult to achieve by large natural resource management organizations [75]. Consecutive and frequent changes in the organization of land management agencies and their objectives and strategies are a pattern common to southern Europe countries, e.g., Portugal [76], and to Australia [51,77], which obviously weakens performance and hinders the outline and accomplishment of long-term goals.

Improvement of fire presuppression and suppression activities is also a goal of sustainable fire management policies. Conflicts and potential dysfunction are likely to arise whenever responsibilities for the single problem of wildfire are dispersed among multiple agencies, especially when acting in the same territory and following different policies or conflicting interests and goals. Such problems are known to occur in Australia and California, with northern Europe countries facing similar challenges as fire activity increases [78]. A critical issue in southern Europe has been the separation between forest and fire management, consistent with the increasingly dominant but misguided political perception of wildfire as solely an emergency to be addressed by a civil protection strategy. Contemporary conservationist and esthetic visions of the forest promoted by environmental organizations add to the difficulty in mitigating fuel hazard. In Portugal, Greece, and Italy, structural or poorly trained volunteer firefighters were assigned with wildfire suppression tasks without incorporating the know-how that resided in the Forest Service [76] and that is commonly requested to professionals by firefighting agencies worldwide. Less resources for land management and loss of fire suppression duties to emergency management agencies disrupts forest and fire management, thus contributing to increase fuel hazard, decrease fire management skills, and intensify the firefighting trap. Capture by the firefighting trap implies that the impacts from severe fire events and seasons will be mistakenly reduced to a matter of lacking firefighting resources and inadequate preparedness and organization.

10.4 Transforming fire management policies

The previous sections made manifest the need to evolve toward more effective and sustainable policies, such that integrated fire management can become the norm rather than the exception. Public, institutional, and political understanding and recognition of the consequences of current suppression-centered policies are crucial and would allow changing expectations and accepting alternatives [9], contributing to alter managerial incentive structures and increase awareness of the long-term impacts of decisions [47]. Increased understanding of the role of fire in ecosystems and of the need for managing unplanned fires [52,79−81] would greatly benefit. The scientific community is pivotal in this respect, through more targeted R&D [82] and by actively engaging with the society [77].

Fire management is inherently complex, comprising various partial solutions and involving trade-offs that depend on the existing stakeholders [67]. Complexity in environmental problems and disasters is best handled by integrative governance, which comprises cooperative planning and deliberative processes and is facilitated by inclusive, participatory, and reflexive practices and mechanisms [2], enabling adaptive strategies to assist in the processes of learning and adapting public policies [83]. The adopted approaches should embrace

1. a risk-based framework as a collective scientific basis for fire management decision support and to justify decisions, build trust, and share responsibility [47,70,72,78,80,84,85].
2. formulation and planning accounting for cross-sectional dynamics, namely, fire-land management and, more generally, as a social-ecological problem [29,31,70]. In addition, fire management plans should be consistent and compatible across agencies, clear and comprehensive, and spatially and temporally scalable [86].
3. flexible and specific approaches, with a bottom-up component to address the local natural and social contexts, namely, risk and traditional fire knowledge and practices [11,31,47,70].

10.5 Conclusion

Extreme fire events should be properly understood and recognized as the outcome, at least in part, of misguided or outdated fire policies. Mega fire occurrence points to approaches (or lack thereof) to land and fire management that are incongruent with the socio-ecological environment and the factors shaping the fire regime. Fire management in human-dominated landscapes is undoubtedly complex and, citing Gill [67], should be tackled as a "multistakeholder, multivariable, multiscale problem."

Unfortunately, either governments ignore the root causes of catastrophic fires and concentrate on the symptoms, or more progressive policies are barely operationalized because of a variety of institutional, socioeconomic, and cultural impediments. The former is particularly relevant in Europe, where a more encompassing and integrated approach is not even politically and publicly debated and fire management rests largely on ignition and fire control strategies, with all the shortcomings that have been presented in this chapter.

Acknowledgments

The authors thank Lachlan McCaw and Alen Slijepcevic for the wildfire area statistics for SW Australia and Victoria, respectively.

References

[1] M. González-Hidalgo, I. Otero, G. Kallis, Seeing beyond the smoke: the political ecology of fire in Horta de Sant Joan (Catalonia), Environmental Planning A 46 (2014) 1014–1031.

[2] S. Ruane, Using a worldview lens to examine complex policy issues: a historical review of bushfire management in the South West of Australia, Local Environ. 23 (2018) 777–795.

[3] J. Whittaker, D. Mercer, The Victorian Bushfires of 2002–03 and the politics of blame: a discourse analysis, Aust. Geogr. 35 (2004) 259–287.

[4] J.E. Lotan, Integrating fire management into land-use planning: a multiple-use management research, development, and applications program, Environ. Manag. 3 (1979) 7–14.

[5] P.M. Fernandes, N. Guiomar, P. Mateus, T. Oliveira, On the reactive nature of forest fire-related legislation in Portugal: a comment on Mourão and Martinho (2016), Land Use Policy 60 (2017) 12–15.

[6] G. Busenberg, Wildfire management in the United States: the evolution of a policy failure, Rev. Policy Res. 21 (2004) 145–156.

[7] S. Jensen, G. McPherson, Living with Fire: Fire Ecology and Policy for the Twenty-First Century, University of California Press, Berkeley, CA, 2008.

[8] S.L. Stephens, L.W. Ruth, Federal forest-fire policy in the United States, Ecol. Appl. 15 (2005) 532–542.

[9] G.H. Donovan, T.C. Brown, Be careful what you wish for: the legacy of Smokey Bear, Front. Ecol. Environ. 5 (2007) 73–79.

[10] D. Twidwell, C.L. Wonkka, H.-H. Wang, W.E. Grant, C.R. Allen, S.D. Fuhlendorf, A.S. Garmestani, D.G. Angeler, C.A. Taylor, U.P. Kreuter, W.E. Rogers, Coerced resilience in fire management, J. Environ. Manag. 240 (2019) 368–373.

[11] Y. Birot, Living with Wildfires: What Science Can Tell Us — A Contribution to the Science-Policy Dialogue, Discussion Paper - European Forest Institute (EFI), Joensu, 2009.

[12] F.E. Putz, Are rednecks the unsung heroes of ecosystem management? Wild Earth 13 (2003) 10−15.

[13] F. Seijo, R. Gray, Pre-industrial anthropogenic fire regimes in transition: the case of Spain and its implications for fire governance in Mediterranean type biomes, Hum. Ecol. Rev. 19 (2012) 58−69.

[14] P.M. Fernandes, C. Loureiro, N. Guiomar, G.B. Pezzatti, F.T. Manso, L. Lopes, The dynamics and drivers of fuel and fire in the Portuguese public forest, J. Environ. Manag. 146 (2014) 373−382.

[15] D. Ascoli, G. Bovio, Prescribed burning in Italy: issues, advances and challenges, iFor. Biogeosci. For. 6 (2013) 79.

[16] J. Minor, G.A. Boyce, Smokey Bear and the pyropolitics of United States forest governance, Political Geogr. 62 (2018) 79−93.

[17] C. Chandler, P. Cheney, P. Thomas, L. Trabaud, D. Williams, Fire in Forestry, John Wiley & Sons, Inc., New York, 1983.

[18] L.T. Egging, R.J. Barney, Fire management: a component of land management planning, Environ. Manag. 3 (1979) 15−20.

[19] W. Fischer, Fire management techniques for the 1980's, Ames Forester 66 (1980) 23−28.

[20] S. Simard, Fire severity, changing scales, and how things hang together, Int. J. Wildland Fire 1 (1991) 23−34.

[21] W.L. McCaw, N.D. Burrows, Fire management, in: The Jarrah Forest, Springer, Dordrecht, 1989, pp. 317−334.

[22] B.J. Stocks, D.L. Martell, Forest fire management expenditures in Canada: 1970−2013, For. Chron. 92 (2016) 298−306.

[23] B. W Van Wilgen, N. Govender, H.C. Biggs, D. Ntsala, X.N. Funda, Response of savanna fire regimes to changing fire-management policies in a large African national park, Conserv. Biol. 18 (2004) 1533−1540.

[24] B.W. Van Wilgen, Fire management in species-rich Cape fynbos shrublands, Front. Ecol. Environ. 11 (2013) e35−e44.

[25] J. Russell-Smith, G.D. Cook, P.M. Cooke, A.C. Edwards, M. Lendrum, C. Meyer, P.J. Whitehead, Managing fire regimes in north Australian savannas: applying Aboriginal approaches to contemporary global problems, Front. Ecol. Environ. 11 (2013) e55−e63.

[26] J. Mistry, I.B. Schmidt, L. Eloy, B. Bilbao, New perspectives in fire management in South American savannas: the importance of intercultural governance, Ambio 48 (2019) 172−179.

[27] P.M. Fernandes, G.M. Davies, D. Ascoli, C. Fernández, F. Moreira, E. Rigolot, C.R. Stoof, J.A. Vega, D. Molina, Prescribed burning in southern Europe: developing fire management in a dynamic landscape, Front. Ecol. Environ. 11 (2013) e4−e14.

[28] Global Fire Initiative, Living with fire: sustaining ecosystems & livelihoods through integrated fire management, in: R.L. Myers (Ed.), The Nature Conservancy, Tallahassee, Florida, 2006.

[29] T. Steelman, U.S. wildfire governance as social-ecological problem, Ecol. Soc. 21 (2016). https://doi.org/10.5751/ES-08681-210403.

[30] F. Tedim, V. Leone, G. Xanthopoulos, A wildfire risk management concept based on a social-ecological approach in the European Union: fire Smart Territory, Int. J. Disaster Risk Reduct. 18 (2016) 138−153.

[31] M.A. Moritz, E. Batllori, R.A. Bradstock, A.M. Gill, J. Handmer, P.F. Hessburg, J. Leonard, S. McCaffrey, D.C. Odion, T. Schoennagel, A.D. Syphard, Learning to coexist with wildfire, Nature 515 (2014) 58−66.

[32] T. Ingalsbee, Whither the paradigm shift? large wildland fires and the wildfire paradox offer opportunities for a new paradigm of ecological fire management, Int. J. Wildland Fire 26 (2017) 557−561.

[33] A. Jarrett, J. Gan, C. Johnson, I.A. Munn, Landowner awareness and adoption of wildfire programs in the Southern United States, J. For. 107 (2009) 113−118.

[34] J.P. Prestemon, D.T. Butry, K.L. Abt, R. Sutphen, Net benefits of wildfire prevention education efforts, For. Sci. 56 (2010) 181−192.

[35] C.M. Countryman, Can Southern California Wildland Conflagrations Be Stopped? General Technical Note PSW-7, USDA Forest Service, Pacific Southwest Forest and Range Experiment Station, Berkeley CA, 1974.

[36] A.M. Gill, S.L. Stephens, G.J. Cary, The worldwide "wildfire" problem, Ecol. Appl. 23 (2013) 438−454.

[37] M. Huffman, The many elements of traditional fire knowledge: synthesis, classification, and aids to cross-cultural problem solving in fire-dependent systems around the world, Ecol. Soc. 18 (2013). https://doi.org/ES-05843-180403.

[38] M.R. Coughlan, Errakina: Pastoral fire use and landscape memory in the Basque region of the French Western Pyrenees, J. Ethnobiol. 33 (2013) 86−104.

[39] K.G. Hirsch, D.L. Martell, A review of initial attack fire crew productivity and effectiveness, Int. J. Wildland Fire 6 (1996) 199−215.

[40] P.M. Fernandes, A.P. Pacheco, R. Almeida, J. Claro, The role of fire-suppression force in limiting the spread of extremely large forest fires in Portugal, Eur. J. For. Res. 135 (2016) 253−262.

[41] M. Finney, I.C. Grenfell, C.W. McHugh, Modeling containment of large wildfires using generalized linear mixed-model analysis, For. Sci. 55 (2009) 249−255.

[42] D. Strauss, L. Bednar, R. Mees, Do one percent of the forest fires cause ninety-nine percent of the damage? For. Sci. 35 (1989) 319−328.

[43] R.A. Minnich, Y.H. Chou, Wildland fire patch dynamics in the chaparral of Southern California and Northern Baja California, Int. J. Wildland Fire 7 (1997) 221−248.

[44] P.M. Fernandes, Empirical support for the use of prescribed burning as a fuel treatment, Curr. For. Rep. 1 (2015) 118−127.

[45] S.F. Arno, J.K. Brown, Overcoming the paradox in managing wildland fire in western wildlands, in: University of Montana, Montana Forest and Conservation Experiment Station, Missoula, MT, 1991, pp. 40−46.

[46] R.D. Collins, R. de Neufville, J. Claro, T. Oliveira, A.P. Pacheco, Forest fire management to avoid unintended consequences: a case study of Portugal using system dynamics, J. Environ. Manag. 130 (2013) 1−9.

[47] D.E. Calkin, M.P. Thompson, M.A. Finney, Negative consequences of positive feedbacks in US wildfire management, For. Ecosyst. 2 (2015) 9.

[48] N. Burrows, L. McCaw, Prescribed burning in Southwestern Australian forests, Front. Ecol. Environ. 11 (2013) e25−e34.

[49] S.A. Parks, L.M. Holsinger, C. Miller, C.R. Nelson, Wildland fire as a self-regulating mechanism: the role of previous burns and weather in limiting fire progression, Ecol. Appl. 25 (2015) 1478−1492.

[50] S.R.J. Bridge, K. Miyanishi, E.A. Johnson, A critical evaluation of fire suppression effects in the boreal forest of Ontario, For. Sci. 51 (2005) 41−50.

[51] P.M. Attiwill, M.A. Adams, Mega-fires, inquiries and politics in the eucalypt forests of Victoria, South Eastern Australia, For. Ecol. Manag. 294 (2013) 45−53.

[52] S.L. Stephens, J.K. Agee, P.Z. Fulé, M.P. North, W.H. Romme, T.W. Swetnam, M.G. Turner, Managing forests and fire in changing climates, Science 342 (2013) 41−42.

[53] R.D. Haugo, B.S. Kellogg, C.A. Cansler, C.A. Kolden, K.B. Kemp, J.C. Robertson, K.L. Metlen, N.M. Vaillant, C.M. Restaino, The missing fire: quantifying human exclusion of wildfire in Pacific Northwest forests, USA, Ecosphere 10 (2019) e02702.

[54] T. Curt, L. Borgniet, C. Bouillon, Wildfire frequency varies with the size and shape of fuel types in southeastern France: implications for environmental management, J. Environ. Manag. 117 (2013) 150−161.

[55] A. Duane, L. Kelly, K. Giljohann, E. Batllori, M. McCarthy, L. Brotons, Disentangling the influence of past fires on subsequent fires in Mediterranean landscapes, Ecosystems (2019), https://doi.org/10.1007/s10021-019-00340-6.

[56] P.M. Fernandes, T. Monteiro-Henriques, N. Guiomar, C. Loureiro, A.M.G. Barros, Bottom-up variables govern large-fire size in Portugal, Ecosystems 19 (2016) 1362−1375.

[57] P.M. Fernandes, C. Loureiro, M. Magalhães, P. Ferreira, M. Fernandes, Fuel age, weather and burn probability in Portugal, Int. J. Wildland Fire 21 (2012) 380−384.

[58] F. Lloret, E. Calvo, X. Pons, R. Díaz-Delgado, Wildfires and landscape patterns in the Eastern Iberian Peninsula, Landsc. Ecol. 17 (2002) 745−759.

[59] L. Loepfe, J. Martinez-Vilalta, J. Oliveres, J. Piñol, F. Lloret, Feedbacks between fuel reduction and landscape homogenisation determine fire regimes in three Mediterranean areas, For. Ecol. Manag. 259 (2010) 2366−2374.

[60] M. Turco, J. Bedia, F.D. Liberto, P. Fiorucci, J. von Hardenberg, N. Koutsias, M.-C. Llasat, F. Xystrakis, A. Provenzale, Decreasing fires in Mediterranean Europe, PLoS One 11 (2016) e0150663.

[61] T. Curt, T. Frejaville, Wildfire policy in Mediterranean France: how far is it efficient and sustainable? Risk Anal. 38 (2018) 472−488.

[62] A. Ganteaume, R. Barbero, Contrasting large fire activity in the French Mediterranean, Nat. Hazards Earth Syst. Sci. 19 (2019) 1055−1066.

[63] M.V. Moreno, M. Conedera, E. Chuvieco, G.B. Pezzatti, Fire regime changes and major driving forces in Spain from 1968 to 2010, Environ. Sci. Policy 37 (2014) 11−22.

[64] M.P. Plucinski, Fighting flames and forging firelines: wildfire suppression effectiveness at the fire edge, Curr. For. Rep. 5 (2019) 1−19.

[65] S. Gómez-González, F. Ojeda, P.M. Fernandes, Portugal and Chile: longing for sustainable forestry while rising from the ashes, Environ. Sci. Policy 81 (2018) 104−107.

[66] J. Piñol, M. Castellnou, K.J. Beven, Conditioning uncertainty in ecological models: assessing the impact of fire management strategies, Ecol. Model. 207 (2007) 34−44.

[67] A.M. Gill, Landscape fires as social disasters: an overview of 'the bushfire problem, Environ. Hazards 6 (2005) 65−80.

[68] S. Ellis, P. Kanowski, R. Whelan, National Inquiry on Bushfire Mitigation and Management, Faculty of Science - Papers (Archive), 2004. https://ro.uow.edu.au/scipapers/4.

[69] I. Otero, J.Ø. Nielsen, Coexisting with wildfire? Achievements and challenges for a radical social-ecological transformation in Catalonia (Spain), Geoforum 85 (2017) 234−246.

[70] IUFRO, in: F.-N. Robinne, J. Burns, P. Kant, B. de Groot, M.D. Flannigan, M. Kleine, D.M. Wotton (Eds.), Global Fire Challenges in a Warming World, Occasional Paper No. 32, IUFRO, Vienna, 2018.

[71] J. Cart, B. Boxall, Air Tanker Drops in Wildfires Are Often Just for Show, 2008. Latimes.com. https://www.latimes.com/local/la-me-wildfires29-2008jul29-story.html.

[72] D.C. Calkin, M.A. Finney, A.A. Ager, M.P. Thompson, K.M. Gebert, Progress towards and barriers to implementation of a risk framework for US federal wildland fire policy and decision making, For. Policy Econ. 13 (2011) 378−389.

[73] M.P. North, S.L. Stephens, B.M. Collins, J.K. Agee, G. Aplet, J.F. Franklin, P.Z. Fulé, Reform forest fire management, Science 349 (2015) 1280−1281.

[74] T.A. Steelman, S.M. McCaffrey, What Is limiting more flexible fire management—public or agency pressure? J. For. 109 (2011) 454–461.

[75] A.M. Petty, C. Isendahl, H. Brenkert-Smith, D.J. Goldstein, J.M. Rhemtulla, S.A. Rahman, T.C. Kumasi, Applying historical ecology to natural resource management institutions: lessons from two case studies of landscape fire management, Glob. Environ. Chang. 31 (2015) 1–10.

[76] P. Mateus, P.M. Fernandes, Forest fires in Portugal: dynamics, causes and policies, in: F. Reboredo (Ed.), Forest Context and Policies in Portugal, Springer International Publishing, 2014, pp. 97–115.

[77] P.M. Attiwill, M.A. Adams, Harnessing forest ecological sciences in the service of stewardship and sustainability: a perspective from 'down-under, For. Ecol. Manag. 256 (2008) 1636–1645.

[78] R. Gazzard, J. McMorrow, J. Aylen, Wildfire policy and management in England: an evolving response from Fire and Rescue Services, forestry and cross-sector groups, Philos. Trans. R. Soc. Biol. Sci. 371 (2016) 20150341.

[79] P.M. Fernandes, Fire-smart management of forest landscapes in the Mediterranean basin under global change, Landsc. Urban Plan. 110 (2013) 175–182.

[80] M.A. Moritz, S.L. Stephens, Fire and sustainability: considerations for California's altered future climate, Clim. Change 87 (2008) 265–271.

[81] A. Regos, N. Aquilué, J. Retana, M. De Cáceres, L. Brotons, Using unplanned fires to help suppressing future large fires in Mediterranean forests, PLoS One 9 (2014) e94906.

[82] M.R. Huffman, Making a world of difference in fire and climate change, Fire Ecol. 10 (2014) 90–101.

[83] G.J. Busenberg, Adaptive policy design for the management of wildfire hazards, Am. Behav. Sci. 48 (2004) 314–326.

[84] J. O'Laughlin, Policy issues relevant to risk assessments, balancing risks, and the National Fire Plan: needs and opportunities, For. Ecol. Manag. 211 (2005) 3–14.

[85] M.P. Thompson, D.G. MacGregor, C.J. Dunn, D.E. Calkin, J. Phipps, Rethinking the wildland fire management system, J. For. 116 (2018) 382–390.

[86] M.D. Meyer, S.L. Roberts, R. Wills, M. Brooks, E.M. Winford, Principles of effective USA federal fire management plans, Fire Ecol. 11 (2015) 59–83.

Fire Smart Territory as an innovative approach to wildfire risk reduction

Vittorio Leone[1], Fantina Tedim[2,3], Gavriil Xanthopoulos[4]

[1]Faculty of Agriculture, University of Basilicata (retired), Potenza, Italy; [2]Faculty of Arts and Humanities, University of Porto, Porto, Portugal; [3]Charles Darwin University, Darwin, NWT, Australia; [4]Hellenic Agricultural Organization "Demeter", Institute of Mediterranean Forest Ecosystems, Athens, Greece

11.1 The wildfire paradoxes

Wildfires are a very complex issue as it is affected by a multitude of environmental, social, economic, political, and even psychological factors. This complexity is being brought up in scientific literature with an increasing frequency in the last few years, in an effort to explain shortcomings and failures of fire management policies that under a simplistic approach in the past seemed very obvious and certain to work. However, paradoxically, they have not worked. The money and effort internationally devoted to fire management have not brought a long-term solution to the problem [1]. The international literature, trying to explain this general shortcomings, has identified and reported a series of paradoxes (with the current meaning of absurd statement or proposition) and metaphors, which are influencing the wildfire problems with unexpected and contradictory outcomes. The most common among them are reported in the following paragraphs:

The *"wildfire paradox"*: Surprisingly, wildfire suppression to eliminate large and damaging wildfires induces the inevitable occurrence of these fires; it deals with the unexpected effect of a generalized and progressive excess fuels buildup, which periodically can feed impressive fires, as it releases in a very short time huge amounts of energy accumulated in the forest biomass [2,3].

The *"firefighting trap"* metaphor refers to a shortsighted way of problem-solving: Dealing with problems ("fires" in business language), as they arise, but failing to address the underlying cause, thereby increasing the chance that the same problem will crop up in the future [4,5]. This kind of intervention is self-perpetuating the wildfire problem.

The *"safe development paradox"* says that in trying to make hazardous areas safer, governments can in fact increase the potential for catastrophic property damages and economic loss [6]. In short, it is the perverse consequence of the typical reaction to protect the areas with stronger firefighting instead of accommodating the probability of fire impact and preparing for it regarding safety of people and properties. The eventual firefighting failures are contributing to a deepening of wildfire problems in the socioecologic system [7].

Extreme Wildfire Events and Disasters. https://doi.org/10.1016/B978-0-12-815721-3.00011-4

The *"local government paradox"* is that while citizens bear the brunt of human suffering and financial loss in disasters, local officials do not pay sufficient attention to policies to limit vulnerability [6]. Vulnerability has been substituted by resilience in many political approaches [8], but it makes sense to reduce vulnerability because it can happen to reduce also poverty, bad life conditions, and increase economic activities.

In addition to the aforementioned paradoxes, we would like to propose what we label as the *"set-aside paradox."* It borrows its name from an incentive scheme introduced by the European Economic Community (EEC) in 1988 [Regulation (EEC) 1272/88], to reduce agricultural surpluses produced in Europe under the guaranteed price system of the Common Agricultural Policy (CAP) and to deliver environmental benefits to agricultural ecosystems and wildlife by reducing the damage induced by the intensification of agriculture. Farmers were requested to leave a proportion of their land out of intensive production, making it a *"set-aside"* land. This policy contributed to decrease the extension of agriculture lands in Europe, decrease the landscape mosaic, and increase the continuity of fuels. In our opinion, the narrow and limited sectoral analysis of the processes with direct intervention on the landscape, which did not recognize that fire is a component of it, can be named the *"set-aside paradox"*. This is typically present in areas where alternative activities, not belonging to the primary sector, emerge and gradually substitute the traditional ones. For instance, in rural areas where depopulation and land abandonment is strong, tourism is often based on the exploitation of natural resources as a way to create some economic alternative. Because of the tourism, some fuel management and reduction activities can be seen unfavorably; their avoidance can influence (increase) future fire occurrence, which in turn can induce the feeling that there are no conditions of safety in the countryside, with a consequent reduction or abandonment of visits to rural areas. This happened in Portugal, for instance, in the world-known Geopark of Arouca: In 2016, because of the increasing tourist presence, the Geopark management did not allow Forest Services and firefighters to do more prescribed burning interventions, afraid of the impact of this activity on tourism. Although prescribed burning is useful for reducing the conditions of hazard, it is also disturbing the contemplation of the landscape with the evident burned patches it creates; in the case of Arouca the biomass that was not treated because of the nonapplication of this fuel reduction technique burned in a wildfire a few months later, resulting in the largest fire (27,440 ha) that ever occurred in the area so far [9]. Fire as a natural process is ignored instead of being accommodated in the local dynamics of socioecological systems [7].

Without solving these paradoxes, the wildfire problem will continue to escalate.

11.2 Wildfires: An unsolved problem

As mentioned previously, the problem of wildfires is far from being solved at global level, as the tragic events in Australia in 2009, in the United States in 2017 and 2018, in Greece in 2018, and in Portugal in 2017 show. To solve the wildfire problem does not mean to eliminate fire from Earth, which is utopic and not desirable; as sustained by Pausas and Keeley [10], there are benefits for contemporary societies from living in a flammable world as fire provides supporting, provisioning, and regulating cultural ecosystem services.

To solve the wildfire problem means to accept the beneficial use of fire, to reduce some causes and motivations of fires as there are alternative actions to fire use and some behaviors cannot be acceptable by society, to minimize the appearance of extreme wildfire events (EWEs [11]), and to avoid the occurrence of wildfire disasters. Although EWEs represent a minority of events, they are responsible for the majority of losses and damage. By definition, they overwhelm any capacity of control, exceeding the commonly accepted threshold of intensity of 10,000 kWm^{-1} (see Chapter 1). Fire intensity is not the object of a standard assessment, but there is evidence of wildfires that have exceeded the aforementioned threshold by far. EWEs that empirically demonstrated the incapacity to control them are the *Black Saturday* Kilmore East and Murrundindi fires in 2009 in Victoria, Australia, with intensity estimated at 90,000 kWm^{-1} [12] or 150,000 kWm^{-1} [13], and the Pedrógão Grande and Sertã fires in 2017 in Portugal, with 60,000 and 90,000 kWm^{-1}, respectively [14,15]. We do not know if 150,000 kWm^{-1} is the highest possible value that can be reached on Earth or if it is possible to experience even higher values.

The aforementioned fire intensity values put in evidence that suppression is only able to control a very small fraction of the possible fire scenarios that can occur around the world (Fig. 11.1). What is crucial is to understand that the current approach to cope with fires is not adequate to deal with EWEs and to avoid the occurrence of wildfire disasters. Thus, suppression is not the solution for all types of fires, as people usually think. For Calkin et al. [16], the increase of losses is not a wildfire control problem, but rather a home ignition problem. EWEs are mainly a problem of land management, rural development, spatial planning, and political priorities [12,17].

Most wildfire disasters are related to events affecting the wildland–urban interface (WUI). Proposing alternatives is crucial because when EWEs affect an area without

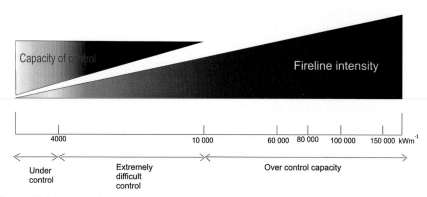

Figure 11.1 The maximum control capacity of fires is very reduced if compared with potential intensity based on existing records.

people, it is not a big issue for the public and also for many ecosystems that are adapted to high intensity fires. However, when they affect WUI areas, they may turn into large-scale disasters if people are not prepared for this type of threat. In short, people need learning to *"coexist"* with wildfires, and this means to be prepared for the time that the flames will reach their settlement, which according to statistics is likely to occur rather quickly in the foreseeable future.

Proposing alternatives on how to achieve this is even more crucial, considering that WUI areas are expanding around the world. As an example, in the United States, the WUI grew from 1990 to 2010 in terms of both new houses (from 30.8 to 43.4 million; 41% growth) and land area (from 581,000 to 770,000 km^2; 33% growth), making it the fastest-growing land-use type; approximately one in three houses and one in 10 hectares in the United States are now in the WUI, where, within the perimeter of wildfires (1990−2015), there were 286,000 houses in 2015, compared with 177,000 in 1990 [18].

The expansion of the WUI can be a result of urban-sprawl related with people who want to live near open space and in close contact with nature. Significant housing growth occurs in suburban and rural areas, especially in and near forests [19]. This growth is the result of the demand of new styles of life but also in many cases of the availability of cheaper settlement conditions. WUI expansion can also be a consequence of the increase in wildland vegetation within and near previously developed areas, resulting from accelerated processes of land abandonment that occur in many countries, and is expected to continue to increase in the future. Farmland abandonment predominantly occurred in developed countries in Europe, the United States, Australia, and Japan. It has been largely reported in mountainous areas of China, Latin America, and Southeast Asia [20]. In the period 2015−30, more than 20 million hectares of agricultural land in the European Union are under high potential risk of abandonment [21]. In southern Europe fire risk associated with land abandonment [22] is a major environmental concern.

11.3 Communities and wildfires: How to reduce losses?

As a response to the multiplication of wildfires in WUI, different initiatives—*Fire Smart Communities, Firewise Communities, Fire-Adapted Communities, Community Wildfire Protection Plans (CWPP)*—at various scales of space (house, group of houses, community), became rather common in North America to enhance the resistance to fires. The assumption that sooner or later a fire could occur leads to the proposal of different solutions, among which a different and strategically placed array of barriers to its potential spread and an emphasis toward reduction of losses.

The goal of the *Fire Smart Communities* approach is to make sure that a house can survive a wildfire with little or no damage (https://www.firesmartcanada.ca/what-is-firesmart). It applies to single homes and is implemented in Canada. It is an important tool to reduce losses, although there is a lack of commitment to self-protection [23].

The *Firewise Community* program is oriented toward communities that take appropriate measures to become more resistant to wildfire structural damage (http://www.firewise.org/about.aspx). It applies to single homes, neighborhoods, and communities in the United States.

The *Fire-Adapted Community* program is focused on increasing awareness and promoting actions of residents regarding infrastructure, buildings, landscaping, and the surrounding ecosystem. It lessens the need for extensive protection actions and enables the community to safely accept fire as a part of the surrounding landscape [24–26]. It applies to communities and the surrounding environment in the United States. As communities in WUI areas encompass strong social diversity and use different ways to prepare, respond, and recover from fires [27], their fire adaptation is different from one another. As there is no checklist of measures, there is no entity that certifies that any given community is fire adapted. In addition, there is the need for continual reevaluation and adjustment (https://fireadapted.org/).

Another initiative is the *Community Wildfire Protection Plans (CWPP)*. It puts emphasis on a community plan to reduce the risk of wildfire that identifies strategic sites and methods for fuel reduction. In the United States, communities with the CWPP have priority for Federal grants (https://sccfiresafe.org/). CWPPs can be incorporated into local wildfire risk management programs [28].

All these initiatives mainly operate in WUI areas and modify the characteristics of buildings and vegetal landscape surroundings to achieve (1) protection of structures through passive defensible space around them; (2) elimination of flammable vegetation in the home ignition zone, adjacent to structures, to limit ember sources and direct flame contact; (3) use of fire-resistant building materials to make buildings safer and fire-resistant; (4) reduction of the amount of fuel in nearby areas; and (5) removal of trees and brush where needed and use of suitable fire-resistant plants for firewise landscaping [25].

Building codes, protection zone requirements, standards for subdivision design, and land-use restrictions complement the aforementioned programs. They include prevention of building in very specific, highly dangerous locations, making roads easily accessible to firefighting crews and securing adequate water supply and road access [25–27].

In the United States, Canada, Australia, and other fire-prone countries that have large areas covered by fire-dependent ecosystems with significant wildfire hazard, and an increasing number of houses built in highly hazardous landscapes, at-risk residents are encouraged to adopt two different types of behaviors in case of fire: (A) Evacuation or (B) to stay and defend their properties ("*Shelter in Place*" for the United States; "*Leave Early or Stay and Defend*" or simply "*Stay or Go*" for Australia). The decision to stay or to leave demands a clear understanding of fire dynamics and of the strong physical and psychological requirements for remaining in place, with the risk that individuals may decide to leave at the last minute, the worst and least safe option available. Whereas the *Stay or Go* process is an active one, if the homeowners are not well prepared and experienced on actively protecting their home before, during (from within the house), and after the fire front passes through, they should not stay [28]. An increasing number of homeowners refuse to evacuate, as observed in many cases also in Europe. However, recent major disasters make civil protection authorities move to promote evacuation. A clear shift took place in Australia after the 2009 fires in Victoria with 173 fatalities. The result of that experience and the studies that followed have been a major shift in policy across the country where fire Authorities once advised communities to *"Prepare, stay, and defend or leave early."* This message placed an emphasis on residents staying to defend property in even the direst circumstances. *"Prepare, act, survive"* is now the community message adopted nationwide. This emphasizes the importance of evacuation [29]. In Greece, after the disastrous wildfire of July 2018 in eastern Attica with 102 fatalities, the authorities started ordering evacuations even for medium intensity fires. Without specific criteria, good advance planning, and population preparation, it soon became evident that this reaction often increased the risk of damages and accidents (Fig. 11.2).

11.4   Fire Smart Territory as a model to *"thrive with fire"*

11.4.1   The advantages of territory scale

The programs and approaches presented in the previous section mainly cover limited portions of landscape or are limited to single assets or communities and their immediate surrounding spaces. These actions respond to fire by creating a patchwork with the purpose to increase the resistance of buildings to fire; thus they reduce the influence of

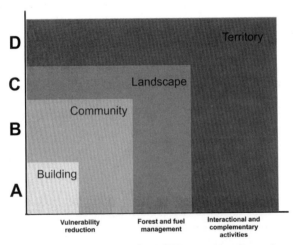

Figure 11.2 Scale of intervention and scope from different approaches and programs following the perspective of *"living with fire"* (A-*"Fire Smart: community protection,"* *"Firewise Communities,"* *"Fire-adapted communities"*; B-*"Community-based fire management,"* C-*"Integrated fire management,"* *"Fire smart landscape management"*; D- *"Fire Smart Territory."*

fires entering the treated properties, by modifying behavior and severity of fires, but they have no general effect. Another approach, distinct from the initiatives described in the previous section, is the concept of Fire Smart Territory (FST) [31,32]. FST is different because it operates at the level of territory, i.e., a portion of geographical space, closely *"interconnected with society on different spatial, temporal and social scales"* ([33], p.133). The concept of territory encompasses society, conflicts, tensions, activities, and relationships; it is a socially produced reality, defined by relations of power that reflect the dynamic interdependence of social–ecological system.

Landscape, the visible portion of living space, does not give the key to the invisible, i.e., the dynamic of the territory [34]; the latter is closely related to "land" or "terrain" but is more than them. Addressing the wildfire problem using a territory perspective is helpful because the relational system is just as important as the material realm, if not more [34]. Territory accommodates the dynamics and power relations among different groups with different and sometimes conflicting interests, which interact in different ways in a certain geographic space [33,35,36]. These relations of power although framed by political and economic factors and different spatial scales must be considered under a perspective centered on everyday social life [37].

Territory is continually "produced" by spatial strategies that transform, and yet are constrained by, social/spatial relations. States and other actors operate within the territories they inhabit, and inherit, from previous attempts at political, economic, strategic, legal, and technical transformation; as they seek to transform them in the present moment, the context for future operations is created [39].

A territorial approach helps to identify where fire is necessary to accomplish land management and conservation goals (e.g., pasture renovation) and where it is possible to use alternatives to avoid fire use (e.g., burning domestic garbage is not necessary if adequate landfills or garbage treatment facilities are available).

11.4.2   Fire Smart Territory: The definition

FST has been defined as *"a fire-prone territory in which the management of economic and social activities, promoting sustainable development, is aimed at risk reduction and conservation of natural values and ecosystem services* (e.g., biodiversity conservation, soil protection, protection of water resources). *It is accomplished by cognizant, well trained and empowered communities able to decide and implement the objectives and practices for preventing, controlling and utilizing fires"* [35]. Here we refine and integrate this definition as follows: *FST is a territory with a shared governance model, in which empowered communities with high levels of knowledge and skills are able to decide and manage wildfire risk to keep it very low, through economic and social activities that not only can contain (in the end eliminate) wildfire hazard but promote the benefits of fire use.* FST promotes sustainable development, adaptation or transformation to climate change, enhance the quality of life of the citizens, and a harmonized relationship with wildfire.

The reduction of wildfire risk (i.e., hazard, and vulnerability that integrates exposure) can and must thus be integrated in the local development process [42].

FST goes beyond the conventional views of risk assessment and management. FST is mainly focused on the reduction of the occurrence of unwanted fires, on the modification of potential fire behavior characteristics, on improving the building conditions and creating conditions to suppress any fire. FST accommodates the social diversity, dynamic, and pressures. In an FST, an EWE can arrive, but it should abate and loose its extreme behavior characteristics there and become a fire within control capacity. At this point, we think that a risk−benefit analysis is the best way to balance the hazardous face of the fire with its beneficial one. In the assessment of risk, FST theorizes that the EWEs cannot be solved by a checklist of even theoretically adequate procedures, but by operating/acting on several fields (see Chapter 13, in this book) before the fire outbreak, to enhance physical and psychological capabilities to cope during the fire, knowing that the firefighting resources will likely be overwhelmed and will be unable to achieve control.

11.4.3 FST components

In planning and developing an FST, it is required to address three main components. First, it is needed to build knowledge and capability [43,44] to:

(1) Understand the fire problem and its characteristics. Human activity plays a significant role in determining the occurrence, frequency, intensity, and extent of landscape fires [45], but usually this is underestimated. Scientists estimate that humans are responsible for about 90% of biomass burning, with only a small percentage of natural fires contributing to the total amount of vegetation burned [46].

Concerning the causes, a frequent way to address them is through the criminalization of fire and through criminal laws. This could be adequate if all the fires were arson, but in fact it is not true. By the end of the 1980s, the European Commission recognized the need for a pan-European research effort in the field of forest fires, aiming to improve our understanding and to manage them better. However, regarding fire causes the main effort was in the direction of harmonizing terminology and definition [47]. Little research was devoted to gain a better knowledge of fire causes and specially to the human factor. This precludes improvements in fire prevention based on modification of behaviors because we do not know why or who started the majority of fires and "unknown cause" is the prevailing class in many wildfire statistics at national level [46,48]. As there are many ecosystems that need fire to regenerate, in some cases people need fire to accomplish certain land management activities because there are no alternatives. Restrictive regulations on the use of fire have not solved the problem. Actually, they have destroyed the traditional ecological knowledge (TEK) of fire. Also owing to the diffuse criminalization of fire use, traditional burning, as a viable cultural practice, is in continued decline.

Monitoring the effectiveness in wildfire management, an in-depth knowledge of fire metrics (fire intensity, fire magnitude, rate of spread, spotting activity and distance, radiant heat at different distance from the flames, fire severity, and fatalities and losses) is needed, to go beyond the simplistic characterization of wildfire activity based on the traditional metrics of the number of fires and burned area. Lack of such knowledge constitutes a barrier to effectively adopt new fire management approaches to minimize the wildfire problem. There is need to understand the complex relationships between fire intensity, fire severity, ecosystem response [49,50], and human systems preparedness to a phenomenon that we cannot eliminate.

(2) Anticipate what can happen by interpreting how the wildfire physical properties and fire behavior can affect individuals and interact within the area. People must know that it is possible to act efficiently on abating the hazard but also how to reduce fire consequences. Psychological preparedness (anticipating sources of anxiety, identifying distressing thoughts and emotions that may exacerbate anxiety, and developing stress management strategies) can help overcome this barrier to anticipation [51].

(3) Understand what needs to be done to decrease the wildfire hazard and vulnerability and increase the capacity to respond when hazardous events occur. Several authors (e.g., [43]) have been proposing functional preparedness categories, i.e., comprising structural, survival, planning, psychological, community capacity building, livelihood, community agency, and evacuation preparedness. Considering wildfires, it is necessary to add hazard modification (reduction of the number of ignitions and abating fire behavior) to the mentioned functional categories. Literature puts in evidence that there are different levels of preparedness and in general wildfire preparedness is low [42,52,53].

Second, it is needed that FST adopts a shared governance model and an adaptive management process to continuously adapt to the existing conditions. It relies on communities and on their connate, sometimes residual capabilities to manipulate fire in a knowledgeable way, but demands changing the antifire bias, which severely hampers using it as a beneficial and economic tool. If these capabilities have already been eroded, it is imperative to rebuild them. The goal of FST is to reframe and adapt the relationship between society and territory and to reduce wildfire damage by reestablishing a fire management wisdom and making the judicious use of fire a legal and widely accepted practice rather than continuing to promote the failed policy of fire exclusion. The ultimate goal of FST is minimizing public exposure to risk and preventing property and environmental damage at territory level. As demonstrated in numerous countries throughout the world, minimization must be applied to the whole territory, not just on selected or privileged portions of it [35].

Third, it is needed to develop a system of certification for areas organized as FST [36] to support and compensate individuals, communities and organizations for their efforts to reduce and maintain wildfire risk at low levels, as well as for their contribution for sustainable development. In fact, FST by itself already represents an advantage, but adding some financial benefits or offering priority in access to grants, could increase the implementation of FST.

11.4.4 FST: SWOT analysis

"*Living with fire*" has been proposed by many authors [54−56] and may mean different things. "*Living with fire*" can be seen with a fatalistic connotation [57] and a way of "*adapting to*" wildfire [56]. Several authors thus prefer "*coexist with fire*" [7,30,57−62]. However, "*coexist with fire*" can also have a negative connotation.

Table 11.1 SWOT analysis of FST.

Strengths	Weaknesses
Proposes social solutions for the social construct of the wildfire problem	*Requires political acceptance and commitment to implement it*
Creates a synergy among prevention practices and primary activities	*Must cope with the self-reinforcing mechanisms of the policy network which supports the maintenance of policy errors through time*
Reduces the occurrence on unwanted fires and losses through the introduction and valorization of economic activities and ecosystems services instead of just activities that represent a cost (fuel management)	*Resistance and difficulty to change the operational monopoly of fire management and governance model*
Conciliates interests from different actors through the establishment of agreements	*Requires conciliation and harmonization of different actors' interests*
Partial or modular implementation, with a gradient of leverage	*Requires to overcome the antifire bias*
Changes the top-down suppression approach by an interactive approach that takes advantage of both bottom-up and top-down strengths	*Requires a change in the culture of organizations*
Based on shared governance	
Builds trust between citizens and organizations	
Recovers and emphasizes TEK and TFK (traditional wisdom and knowledge in manipulating fire), avoiding clandestine fire setting	
Makes suppression activities easier and more efficacious	
Reinvents traditional practices in agriculture, just differently planned and scheduled	
Connects wildfire risk reduction with sustainable development and safety promotion	
Proposes a system of certification with added value to access and benefit of funding opportunities	
Revitalizes rural areas	
Reduces the fire risk in the WUI	
Introduces and maintains wildfire risk reduction in the daily life of people without additional costs, so easier to maintain through the time	
Opportunities	**Threats**
Represents a feasible solution to EWEs	*Resistance of organizations to shift their culture*
Easily adapts to the changing scenarios imposed by climate change	*Lack of acceptance by all the involved communities*

Source: Modified from [26].

As an alternative we propose *"thrive with fire"* (see Chapter 13), a term with a highly positive connotation that frames a more beneficial, successful, proactive, long-term, safer, and more cost-effective approach, which better translates the rationale of FST.

SWOT analysis puts in evidence the original features of FST and the main constraints that hamper its full efficaciousness. Strengths clearly prevail; since they are present and actual (whereas Opportunities and Threats are future or expected events), they can help in evaluating the feasibility of the model and its implementation. Weaknesses mainly involve problems of governance, acceptance at different levels, difficulty to overcome the strong, established antifire bias, and to gradually abandon the suppression model, which is the preferred option at many political and operational levels. In terms of governance, it means to gradually pass from a situation of passive expectation of institutional intervention in case of a fire event, to a situation where local communities are immediate and efficacious actors of an interactive approach that takes advantage of both bottom-up and top-down strengths in the implementation of activities before, during and after the fire (Table 11.1).

11.5 Conclusion

The concept of FST, as a holistic planning philosophy, acting at territory level on both the physical space and on the communities living in that space framed by power relations, is innovative and promising. From one side, it presents an alternative to fuel management that is a very expensive activity, proposing the introduction or maintenance of productive activities (e.g., conservation and production agriculture, grazing) that break landscape continuity and connectivity, assuring a blanket reduction of fuel load, and contributing to decrease the vulnerability of the exposed local communities and, consequently, losses and damage. On the other side, the perspective of acting on concerned communities by empowering them can be obtained through training, recovery of traditional wisdom and knowledge, and enhancing their resilience. This passes through reduction of vulnerability and return of rural communities to their natural role of *guardian of the territory,* agents of transformation, and, when possible, actors of first and prompt attack on fire events of attenuated intensity; it surely represents a social solution to a socially constructed phenomenon. This opportunity can also help boosting local rural economies and enhancing rural prosperity, as new jobs' opportunities and increase of growth potential [63].

Not hiding difficulties, FST appears as a feasible solution able to cope with the increasing menace of EWEs, where also the more advanced technology appears inadequate and inefficacious. Reducing the risk of wildfire by the introduction or maintenance of agriculture and grazing (mixing traditional with modern models of sustainable agriculture) should be considered with priority within the measures of prevention against unwanted and uncontrollable wildfires diffusion and spread. FST not only can be effective in containing wildfires but also contributes to the implementation of the United Nations Sustainable Development Goals.

References

[1] European Commission, Forest Fires Sparking Firesmart Policies in the EU, European Commission, Luxembourg, 2018, https://doi.org/10.2777/248004.

[2] S.F. Arno, J.K. Brown, Overcoming the paradox in managing wildland fire, West. Wildlands 17 (1991) 40–46. https://www.frames.gov/catalog/34743.

[3] P.M. Fernandes, M.M. Fernandes, C. Loureiro, Post-fire live residuals of maritime pine plantations in Portugal: structure, burn severity, and fire recurrence, For. Ecol. Manage. 347 (2015) 170–179, https://doi.org/10.1016/J.FORECO.2015.03.023.

[4] R.D. Collins, R. de Neufville, J. Claro, T. Oliveira, A.P. Pacheco, Forest fire management to avoid unintended consequences: a case study of Portugal using system dynamics, J. Environ. Manag. 130 (2013) 1–9, https://doi.org/10.1016/J.JENVMAN.2013.08.033.

[5] J. Chu, Study Finds More Spending on Fire Suppression May Lead to Bigger Fires, MIT News, 2013. http://news.mit.edu/2013/forest-fire-management-1120.

[6] R.J. Burby, Hurricane Katrina and the paradoxes of government disaster policy: bringing about wise governmental decisions for hazardous areas, Ann. Am. Acad. Pol. Soc. Sci. 604 (2006) 171–191, https://doi.org/10.1177/0002716205284676.

[7] M.A. Moritz, E. Batllori, R.A. Bradstock, A.M. Gill, J. Handmer, P.F. Hessburg, J. Leonard, S. McCaffrey, D.C. Odion, T. Schoennagel, A.D. Syphard, Learning to coexist with wildfire, Nature 515 (2014) 58–66, https://doi.org/10.1038/nature13946.

[8] S.L. Cutter, L. Barnes, M. Berry, C. Burton, E. Evans, E. Tate, J. Webb, A place-based model for understanding community resilience to natural disasters, Glob. Environ. Chang. 18 (2008) 598–606, https://doi.org/10.1016/j.gloenvcha.2008.07.013.

[9] A. Teodoro, A. Amaral, A. Teodoro, A. Amaral, A statistical and spatial analysis of Portuguese forest fires in summer 2016 considering landsat 8 and sentinel 2A data, Environments 6 (2019) 36, https://doi.org/10.3390/environments6030036.

[10] J.G. Pausas, J.E. Keeley, Wildfires as an ecosystem service, Front. Ecol. Environ. 17 (2019) 289–295, https://doi.org/10.1002/fee.2044.

[11] F. Tedim, V. Leone, M. Amraoui, C. Bouillon, M. Coughlan, G. Delogu, P. Fernandes, C. Ferreira, S. McCaffrey, T. McGee, J. Parente, D. Paton, M. Pereira, L. Ribeiro, D. Viegas, G. Xanthopoulos, Defining extreme wildfire events: difficulties, challenges, and impacts, Fire 1 (2018) 9, https://doi.org/10.3390/fire1010009.

[12] M.G. Cruz, A.L. Sullivan, J.S. Gould, N.C. Sims, A.J. Bannister, J.J. Hollis, R.J. Hurley, Anatomy of a catastrophic wildfire: the Black Saturday Kilmore East fire in Victoria, Australia, For. Ecol. Manage. 284 (2012) 269–285, https://doi.org/10.1016/J.FORECO.2012.02.035.

[13] K. Tolhurst, Report on the Physical Nature of the Victorian Occurring on 7th February 2009 Fires, Victoria, 2009, http://royalcommission.vic.gov.au/getdoc/5905c7bb-48f1-4d1d-a819-bb2477c084c1/EXP.003.001.0017.pdf.

[14] CTI, Análise e apuramento dos factos relativos aos incêndios que ocorreram em Pedrógão Grande, Castanheira de Pêra, Ansião, Alvaiázere, Figueiró dos Vinhos, Arganil, Góis, Penela, Pampilhosa da Serra, Oleiros e Sertã entre 17 e 24 de junho de 2017, 2017. https://www.parlamento.pt./Documents/2017/Outubro/RelatórioCTI_VF .pdf.

[15] CTI, Avaliação dos Incêndios ocorridos entre 14 e 16 de outubro de 2017 em Portugal Continental. Relatório Final, 2018. https://www.parlamento.pt./Documents/2018/Marco/RelatorioCTI190318N.pdf.

[16] D.E. Calkin, J.D. Cohen, M.A. Finney, M.P. Thompson, How risk management can prevent future wildfire disasters in the wildland-urban interface, Proc. Natl. Acad. Sci. U.S.A. 111 (2014) 746–751, https://doi.org/10.1073/pnas.1315088111.

[17] J.M. Martínez Navarro, La gestión territorial del riesgo antrópico de ignición forestal en Castilla-La Mancha, 2018. https://ruidera.uclm.es/xmlui/handle/10578/16612.

[18] V.C. Radeloff, D.P. Helmers, H.A. Kramer, M.H. Mockrin, P.M. Alexandre, A. Bar-Massada, V. Butsic, T.J. Hawbaker, S. Martinuzzi, A.D. Syphard, S.I. Stewart, Rapid growth of the US wildland-urban interface raises wildfire risk, Proc. Natl. Acad. Sci. U.S.A. 115 (2018) 3314−3319, https://doi.org/10.1073/pnas.1718850115.

[19] S.I. Stewart, V.C. Radeloff, R.B. Hammer, T.J. Hawbaker, Defining the wildland−urban interface, J. For. 105 (2007) 201−207, https://doi.org/10.1093/jof/105.4.201.

[20] S. Li, X. Li, Global understanding of farmland abandonment: a review and prospects, J. Geogr. Sci. 27 (2017) 1123−1150, https://doi.org/10.1007/s11442-017-1426-0.

[21] D.X. Viegas, A. Simeoni, G. Xanthopoulos, C. Rossa, L. Ribeiro, L. Pita, D. Stipanicev, Recent Forest Fire Related Accidents in Europe, 2009, https://doi.org/10.2788/50781.

[22] J. JM Terres, L. Nisini, E. Anguiano, Assessing the Risk of Farmland Abandonment in the EU, Luxembourg, 2013. doi:LB-NA-25783-EN-N.

[23] M. Ergibi, Awareness and Adoption of Firesmart Canada: Barriers and Incentives, 2018. https://harvest.usask.ca/handle/10388/8498.

[24] S. McCaffrey, Community wildfire preparedness: A global state-of-the-knowledge summary of Social Science Research, Curr. For. Reports 1 (2015) 81−90, https://doi.org/10.1007/s40725-015-0015-7.

[25] E. Toman, M. Stidham, S. McCaffrey, B. Shindler, Social science at the wildland-urban interface: A compendium of research results to create fire-adapted communities, 2013.

[26] FAC Self-Assessment Tool, FAC - Fire Adapt. Communities, 2014. https://fireadaptednetwork.org/resources/fac-assessment-tool/.

[27] M. Carroll, T. Paveglio, Using community archetypes to better understand differential community adaptation to wildfire risk, Philos. Trans. R. Soc. B Biol. Sci. 371 (2016), https://doi.org/10.1098/rstb.2015.0344.

[28] J.D. Absher, J.J. Vaske, C.L. Peterson, Community wildfire protection plans in Colorado, J. For. 116 (2017) 25−31, https://doi.org/10.5849/jof.2016-053R3.

[29] CFA, Landscaping for Bushfire: Garden Design and Plant Selection, 2011.

[30] F. Tedim, V. Leone, Enhancing resilience to wildfire disasters: from the "war against fire" to "coexist with fire Disaster resilience: an integrated approach, in: D. Paton, D. Johnston (Eds.), Resil. An Integr. Approach, Charles C Thomas, Publisher, 2017, pp. 362−383, 362−383.

[31] C. Kousky, S. Olmstead, R. Sedjo, In harm's way: homeowner behavior and wildland fire policy, in: K.M. Bradshaw, D. Lueck (Eds.), Wildfire Policy Law Econ. Perspect., Taylor and Francis for RFF Press, New York, U. S., 2011.

[32] S. McCaffrey, Applying Australia's stay or go approach in the U.S, Would it work? 2 (2017). http://idahofirewise.org/wp-content/uploads/2017/04/stayXorXgoXinXtheXus.pdf.

[33] G. Xanthopoulos, C. Bushey, C. Arnol, D. Caballero, Characteristics of wildland−urban interface areas in Mediterranean Europe, North America and Australia and differences between them, in: G. Boustras, N. Boukas (Eds.), Proc. 1st Int. Conf. Saf. Cris. Manag. Constr. Tour. SME Sect. (1st CoSaCM), Brown Walker Press, Boca Raton, Florida., USA, Nicosia, Cyprus, 2012, pp. 702−734.

[34] F. Tedim, O conceito de "fire smart territory": contributo para a mudança de perspetiva na gestão dos incêndios florestais em Portugal, in: Geogr. Paisag. e Riscos. Livro Homenagem Ao Prof. Doutor António Pedrosa, Imprensa da Universidade de Coimbra, Coimbra, Portugal, 2016, pp. 251−283.

[35] F. Tedim, G. Xanthopoulos, V. Leone, Forest Fires in Europe: Facts and Challenges, Wildfire Hazards, Risks and Disasters, 2015, pp. 77−99, https://doi.org/10.1016/B978-0-12-410434-1.00005-1.

[36] F. Tedim, V. Leone, G. Xanthopoulos, A wildfire risk management concept based on a social-ecological approach in the European Union: fire Smart Territory, Int. J. Disaster Risk Reduct 18 (2016) 138−153, https://doi.org/10.1016/J.IJDRR.2016.06.005.

[37] C. Raffestin, Géographie économique du pouvoir, Libr. Tech., Paris, 1980.

[38] C. Raffestin, Space, territory, and territoriality, Environ. Plan. Soc. Space 30 (2012) 121−141, https://doi.org/10.1068/d21311.

[39] S. Elden, Territory/territoriality, in: Wiley Blackwell Encycl. Urban Reg. Stud., Wiley, 2019, pp. 1−11, https://doi.org/10.1002/9781118568446.eurs0339.

[40] C. Raffestin, From the territory to the landscape: the image as a tool for discovery, in: Nat. Policies Landsc. Policies, Springer International Publishing, Cham, 2015, pp. 93−101, https://doi.org/10.1007/978-3-319-05410-0_10.

[41] F.R. Klauser, Thinking through territoriality: introducing claude Raffestin to Anglophone sociospatial theory, Environ. Plan. Soc. Space 30 (2012) 106−120, https://doi.org/10.1068/d20711.

[42] D. Paton, F. Tedim, Wildfire and Community: Facilitating Preparedness and Resilience, Charles C Thomas Publisher, 2012.

[43] D. Paton, Disaster risk reduction: psychological perspectives on preparedness, Aust. J. Psychol. (2018), https://doi.org/10.1111/ajpy.12237.

[44] UNDRR, What Is Disaster Risk Reduction?, 2017. https://www.unisdr.org/who-we-are/what-is-drr.

[45] M.R. Coughlan, A.M. Petty, Linking humans and fire: a proposal for a transdisciplinary fire ecology, Int. J. Wildland Fire 21 (2012) 477, https://doi.org/10.1071/WF11048.

[46] V. Leone, N. Koutsias, J. Martínez, C. Vega-García, B. Allgöwer, R. Lovreglio, The Human Factor in Fire Danger Assessment, in: 2003: pp. 143−196. doi:10.1142/9789812791177_0006.

[47] A. Camia, T. Durrant, J. San-Miguel-Ayanz, Harmonized Classification Scheme of Fire Causes in the EU Adopted for the European Fire Database of EFFIS, Executive, Publications Office of the European Union, Luxembourg, 2013.

[48] F. Tedim, V. Leone, F. Gutierres, F.J.M. Correia, C. Magalhães, Evidences about wildfire causes in the northern region of Portugal, in: L. Lourenço, F. Tedim, C. Ferreira (Eds.), Incêndios Florestais. Em Busca Um Novo Paradig. - II Diálogo Entre Ciência e Util. -, NICIF - Núcleo de Investigação Científica de Incêndios Florestais, Faculdade de Letras da Universidade de Coimbra, Coimbra, 2019, pp. 59−92, https://doi.org/10.34037/978-972-8330-25-5_2.

[49] J.E. Keeley, T. Brennan, A.H. Pfaff, Fire severity and ecosytem responses following crown fires in California shrublands, Ecol. Appl. 18 (2008) 1530−1546, https://doi.org/10.1890/07-0836.1.

[50] J.E. Keeley, W.J. Bond, R.A. Bradstock, J.G. Pausas, P.W. Rundel, Fire in Mediterranean Ecosystems: Ecology, Evolution and Management, Cambridge University Press, 2011.

[51] S.A. Morrissey, J.P. Reser, Evaluating the effectiveness of psychological preparedness advice in community cyclone preparedness materials, Aust. J. Emerg. Manag. 18 (2003) 46.

[52] J.S. Becker, D. Paton, D.M. Johnston, K.R. Ronan, J. McClure, The role of prior experience in informing and motivating earthquake preparedness, Int. J. Disaster Risk Reduct. 22 (2017) 179−193, https://doi.org/10.1016/J.IJDRR.2017.03.006.

[53] M. Marti, M. Stauffacher, J. Matthes, S. Wiemer, Communicating earthquake prepared-ness: the influence of induced mood, perceived risk, and gain or loss frames on home-owners' Attitudes toward general precautionary measures for earthquakes, Risk Anal. 38 (2018) 710–723, https://doi.org/10.1111/risa.12875.

[54] Y. Birot, Living with Wildfires: What Science Can Tell Us-A Contribution to the Science-Policy Dialogue, European Forest Institute, 2009.

[55] A.M.S. Smith, C.A. Kolden, T.B. Paveglio, M.A. Cochrane, D.M. Bowman, M.A. Moritz, A.D. Kliskey, L. Alessa, A.T. Hudak, C.M. Hoffman, J.A. Lutz, L.P. Queen, S.J. Goetz, P.E. Higuera, L. Boschetti, M. Flannigan, K.M. Yedinak, A.C. Watts, E.K. Strand, J.W. van Wagtendonk, J.W. Anderson, B.J. Stocks, J.T. Abatzoglou, The science of firescapes: Achieving fire-Resilient communities, Bioscience 66 (2016) 130–146, https://doi.org/10.1093/biosci/biv182.

[56] H. Brenkert-Smith, J.R. Meldrum, P.A. Champ, C.M. Barth, Where you stand depends on where you sit: qualitative inquiry into notions of fire adaptation, Ecol. Soc. 22 7 (2017) 22, 7.

[57] A.M. Gill, S.L. Stephens, G.J. Cary, The worldwide "wildfire" problem, Ecol. Appl. 23 (2013) 438–454.

[58] R.L. Olson, D.N. Bengston, L.A. DeVaney, T.A.C. Thompson, Wildland Fire Manage-ment Futures: Insights from a Foresight Panel, 2015, https://doi.org/10.2737/NRS-GTR-152.

[59] D. Paton, P.T. Buergelt, F. Tedim, S.M. McCaffrey, Wildfires: international perspectives on their sociale ecological implications, in: D. Paton, P.T. Buergelt, S.M. McCaffrey, F. Tedim (Eds.), Wildfire Hazards, Risks, and Disasters, Elsevier, Oxford, UK, 2015, pp. 1–14.

[60] M.A. Moritz, S.G. Knowles, Coexisting with wildfire: promoting the right kind of fire–and smarter development–is safer and more cost-effective than fighting a losing battle, Am. Sci. 104 (2016) 220–228. https://go.galegroup.com/ps/anonymous?id=GALE%7CA458803770&sid=googleScholar&v=2.1&it=r&linkaccess=abs&issn=00030996&p=AONE&sw=w.

[61] R. Fazio, Coexist with Fire, Jonh Muir Proj, Earth Isl. Inst., 2015. http://johnmuirproject.org/2015/08/coexist-with-fire/.

[62] A.B. Guy, Coexist or Perish, Wildfire Analysis Says, Berkeley News, 2014. https://news.berkeley.edu/2014/11/05/coexist-or-perish-new-wildfire-analysis-says/.

[63] European Commission, The future of food and farming - communication on the common agricultural policy post-2020, Eur. Community (2017). https://europa.eu/rapid/press-release_MEMO-17-4842_en.htm.

How to create a change in wildfire policies

12

Vittorio Leone[1], Fantina Tedim[2, 3]
[1]Faculty of Agriculture, University of Basilicata (retired), Potenza, Italy; [2]Faculty of Arts and Humanities, University of Porto, Porto, Portugal; [3]Charles Darwin University, Darwin, NWT, Australia

12.1 Introduction

In 1933 an international survey on forest fire problems was carried out by the Rome International Institute of Agriculture (IIA, established in 1905 as an international organization for food and agriculture, after 1945 merged into the FAO). Questionnaires were sent to all countries, scientific institutions, and national experts; the IIA received back only 72 replies [1]. Countries that currently distinguish for ferocious fires (e.g., Portugal, Greece, Spain, Chile, Russia) were absent, whereas US was present with 27 replies, followed by 13 replies by British Colonies, Dominions, and Protectorates. In the section illustrating *"Prevention and control measures,"* a suppression-centric position is clearly evident in all countries, together with the interest in technical means at all scales (from individual hand tools to big machines) and prevention infrastructures (water-filling points, lookout towers, firebreaks, road system). The use of water with hose and pumps is strongly advocated. Nowhere is present the ecological role of fire. The solution of the problem is reductively in terms of extinguishing flames, in analogy with other fires, irrespective of causes, trends, geography of risk, and interactions among social and ecological systems across multiple spatial and temporal scales [2].

12.2 The origins of suppression policy

Peshtigo fire occurred on October 8th, 1871, the same day of the Great Chicago Fire. It annihilated the town of Peshtigo, in Wisconsin, and 16 smaller surrounding communities [3–5]. The fire destroyed 6216 km^2, killing between 1200 and 2400 people, with damages for more than $5 million (https://fhsarchives.wordpress.com/2011/10/08/october-8-1871-peshtigo-wisconsin-is-consumed-by-fire/9). Peshtigo fire is one of the first examples of wildland–urban interface (WUI) fire [5]. And it was not the first awesome event in the US. Other extraordinary fires occurred from 1825 to 1868 [6] in Maine and Oregon, the biggest one being the Minamichi Fire of October 1825, with 1,214,575 ha burned and 160 lives lost. After other ferocious and massive wildfires (Michigan, 1881; Hinckley, Minnesota, 1894) [5], a fire suppression movement began in the late 1890s. It was based on the philosophy, advocated by early

Extreme Wildfire Events and Disasters. https://doi.org/10.1016/B978-0-12-815721-3.00012-6

conservationists [7] that fire destroys forests [8] and that excluding fire would help to conserve forest values. The movement was led by Gifford Pinchot, first America's trained forester [9], chief of the USDA Division of Forestry in 1898, successively the father and first chief of USDA Forest Service (established in 1905). He shared the fire exclusion theories and precepts of the National School of Water Resources and Forestry of Nancy, France, where he had been trained in 1889−90 [9]. In the European Academies of Forestry, fire in the forests was demonized [10] and considered the main threat to timber production, together with goat grazing; it was seen as an expression of social disorder [11] and blamed as primitive [12]. Accordingly, the opinion of Pinchot was very explicit: *"Of all the foes that attack the woodlands of North America, no other is so terrible as fire"* ([13], p.41).

In 1910, a series of wildfires known as " Big Blowup " burned about 1.2 million ha in Montana, Idaho, and Washington in only 2 days; at least 85 people died, most of them being firefighters. This event once again suggested that the devastation of forest could have been prevented just by enough men and equipment and that only total wildfire suppression could prevent a similar event from occurring again [14]. For US Forest Service wildfire exclusion became thus a priority over all other duties and activities [14]. This was the basis of the policy of total exclusion of fires that developed in the following years, together with the Forestry School of Cornell University established in 1898 [9] and the Yale Forestry School established in 1900 (https://environment.yale.edu/about/history/). In 1902 and 1903 other forestry schools were founded in Michigan, Main, Minnesota, and Pennsylvania [14]. Forestry was still in its infancy in the US, without academic heritage of fire or fire management [12], and there was a need to find forestry schools with a different mindset from the European ones.

Thus, since early 1900s, the US Forest Service followed a policy of intolerance and all-out fire suppression, i.e., to stamp out wildfires before they could grow big enough to do real damage [15,16]; its rationale was *"The first measure necessary for the successful practice of forestry is protection from forest fires"* ([6], p.6). Suppression was fostered by the Forest Fires Emergency Act (1908), stipulating that in fire emergencies the Forest Service could direct any available funds toward suppression, and Congress would later reimburse those expenses [17].

In line with the approach advocated by Pinchot *"to replace with carefully gathered facts the vague general notions that now exist about forest fires"* ([18], p. 4), in 1915 H.S. Graves, the second chief of Forest Service, established the Branch of Research in the Forest Service, as an instrument of help in a better management of forests and not only to address the needs for wildfire research [12].

The suppression policy, having the objectives of preventing fires through early finding and suppressing a fire as quickly as possible [7] but controlling the costs [19], progressively gained urgency until the adoption of the Silcox zero-tolerance directive of *"10 a.m. policy"* in 1935 [20,21].This demanded that all wildfires, regardless of how remote, be brought under control by 10 a.m. the morning after ignition, i.e., within the first *"burn period,"* the interval between 2:00 p.m. and 6:00 p.m. when fires most actively burn [20,21].

In Canada a similar approach to wildfire control was emerging, based on fighting fires with hand tools, and developing infrastructures to detect and facilitate access to fire, integrated with removal of causes. The measures to protect forests from fires were: "(…) *the removal by education or legislation adequately enforced of the causes of the fires, the organization of a patrol to find and extinguish such fires which will inevitably start, and the improvement and organization of the forested area so as to render most efficient the efforts of fire fighters and minimize the chance of any fires getting beyond control*" (McMillan, 1911 cited by Murphy [64], p.23).

When wildfire problems suddenly surged in Europe, in the 1960s, favored by deep changes in post-WWII society and economy, the most immediate and generalized reply was to mount a suppression organization, following the time-tested American model of organization, popularized by the FAO manual by Show and Clark [22].

The suppression-centric approach was thus perpetuated, but without the long preparatory phase that preceded it in the US and without the successive shifts that characterized the period after WWII. It is the model currently adopted worldwide by the majority of countries.

In Australia a series of disastrous bushfires (the term equivalent to wildfires) started in 1851 with 6 February Black Thursday in Victoria State, which affected 5,000,000 ha and one million sheep and had a toll of 12 fatalities. All chronologies of bushfires start with this event, because of lack of comprehensive recorded information in the early times of European settlement in Australia. Until 2009, 42 years with major and deadly events, involving fatalities and million hectares burnt, have swept Australia, of which 31 in the state of Victoria (particularly near Melbourne) that is an outstanding hotbed for bushfire activity. Between 1901 and 2011, 825 people lost their lives in more than 260 bushfires. Of those killed, 92 were firefighters [23].

More than 80% of the deaths were in January and February, and 61% happened in Victoria; 65% of all the people killed in bushfires across that 110-year period died in just 9 days, including Black Saturday in 2009, when 173 people died in Australia's worst natural disaster [24]. Throughout Australia, the initial response to the outbreak of a bushfire is generally the task of the agency with responsibility for the land on which the fire occurs. Forest services and other government agencies actively participate, but their responsibility is mainly exercised in public lands, a minor surface of Australia's extent. In the other lands, volunteer brigades and individual owners operate in bushfires. Response is usually managed at a district level using locally deployed resources. Fires that escape initial suppression attempts will then subsequently attract support from neighboring districts coordinated at the state level and other agencies with firefighting resources protection. Ground crews generally form the primary type of suppression force, if water-bombing aircraft are not immediately available for early attack on fires [25].

12.3   Wildfire research and policies

12.3.1   Origins of wildfire research

At a difference from forestry and rational silviculture, which was developed in Germany and France in the 19th century, forest fire research and the resulting fire science took origin in the US, in the second half of 19th century, in the aftermath of the Peshtigo fire. *"Forest fire consciousness was a development which accompanied recognition of the place of forestry on the part of the nation. For 150 years the United States considered the forest a liability instead of an asset because it screened the hostile Indians and was in possession of soil needed for crops. This viewpoint gradually changed until Congress passed the Timber Culture Act of 1873; forests then became assets and not liabilities in the Great Plains region of the United States"* ([23], p.8).

Regarding the wildfire problem, as it existed in the US at the beginning of 20th century, the incipit of Folweiler and Brown's book entitled *"Fire in the forests of the United States"* ([24], p.3) says *"The forest fire problem as it exists in the United States today is essentially a new factor in forestry tradition and a new subject in forest literature. Scientific European forestry almost ignores the control of fire as an important part of the forester's job and gives scant attention to what happens to silvicultural systems when fire intervenes. Professional forestry training in this country, influenced strongly by European traditions, began in the pattern of intensive training in scientific systems of silviculture, in mensuration, in dendrology, and botany, with only an introduction to forest fire control as an innovation peculiar to the United States. The effect of this has been that many young foresters have found themselves on a job of which four-fifths was protection of the forest from fire, but with their training in inverse ratio."*

The prominence of wildfires fostered an effort in producing the knowledge necessary to cope with a challenging problem, never explored by forestry. Since 1879, activities of forest fire research are recorded in the US, but a 1904 memorandum confirms in 1899 the timeline of the beginning of fire science research in the Forest Service [18].

Research activity was instrumental to provide knowledge tools to the policy of total fire exclusion, strongly advocated by the veterans of the 1910 fire, which successively served as the Chief of Forest Service [7]. Research progressed and technical manuals (e.g., The operational manuals [7,26]) aided in establishing sound basements to suppression activities. In the following years wildfire research in the US developed in an exponential way, first around three areas of interest to improve safeguard of forested areas (fire control, fire behavior, and fire effects). After the WWII, interest was directed to the militarization of firefighting, through the capitalization of surplus military means, available technologies, techniques, and trained war veterans [18]. The establishment of dedicated fire laboratories and the activity of the Research Stations created a true web in the establishment of a wildfire science.

As a matter of fact, to confirm that *"forest fire problem of the United States was unique and that European forestry had nothing to offer in terms of solutions"* ([12], p.31), the first textbooks on fire were by Folweiler [27,28] and the first FAO manual on forest fire control by Show and Clarke [22], this latter spreading the American model of fighting forest fires on a global scale.

In Canada, forest fire research officially started in 1920s, its originator being J.G. Wrigth [29]. Goals for Canadian fire research might be summarized in six categories (fire behavior, fire suppression, fire management systems, fire ecology, prescribed fire, and fire economics) [29]. Main products are fire hazard rating to indicate the predisposition of forested areas to fire [30], culminated with the Canadian Forest Fire Weather Index (FWI) System in 1970 and the Canadian Forest Fire Behavior Prediction System, a systematic method for assessing wildland fire behavior potential, released to fire management agencies in 1984. The FWI is by far considered the most performing fire danger index at world level [31,32]. Wildfire research in Canada has always held to its main obligation of producing practical output of value in fire and forest management [29]. The responsibility for forest land management, fire protection, and use rests with other federal provincial and territorial government agencies. Research products must therefore be introduced to user agencies and accepted on the basis of merit and performance. Acceptance is enhanced by involving user agencies in both identifying the research need and participating in the research and development process [30].

In Australia, since 1926, forest fire research is carried out under the umbrella of the Commonwealth Scientific and Industrial Research Organization (CSIRO), an Australian Government corporate entity and Australia's National Science Research Agency. CSIRO bushfire research is improving the understanding of fire and technologies and strategies to save lives and limit damage, namely, on advancing fire spread prediction and bushfire suppression systems. Working with state land management, rural fire agencies, and other research agencies, CSIRO scientists apply knowledge of bushfire dynamics to real events and help predict risks. Bushfire research in Australia focused on (1) the development of fire data analysis tools; (2) understanding and predicting bushfire behavior; (3) the impact of bushfires on infrastructure; (4) ecological responses to fire; (5) the impact of climate change on bushfire risk; and (6) pollutants and greenhouse gases as a result of bushfires. Among the works by CSIRO fire scientists, we can mention: (1) models for predicting the rate of fire spread, for operational use in prescribed burning and wildfire suppression in different Australian vegetation types; (2) bushfires impact on houses and other buildings; (3) laboratory investigations of building performance under fire exposure conditions; (4) a comprehensive house vulnerability assessment tool for assessing built assets in bushfire-prone areas; and (5) assessment of bushfire danger conditions [McArthur Mk5 Forest Fire Danger Meter; Grassland Fire Danger Meter; Grassland Fire Spread Meter; Fire Spread Meter for Northern Australia (https://www.csiro.au/en/About/We-are-CSIRO)]. Research is carried out also by Bushfire and Natural Hazards CRC and Universities.

We just mention here the former Union Soviet Socialist Republics (USRR) but for reminding the resolutions of the Council of Ministers of the USRR, in January 1973, in terms of strengthening of the means of conservation and improvement of the utilization of natural resources [33]. We mention also the Leningrad Research Institute of Forestry and the Far-eastern Scientific Research Institute of Forestry, the experimental laboratory of the Central Air Force (Avialesookhrana) base, the Wood Pyrology Laboratory of the Forest and Timber Institute of the Siberian branch of the Soviet Academy of Science, and the Arkhangelsk Institute of Forest and Wood Chemistry [33]. Many institutions are currently involved with the functioning of the boreal forest, threatened by wildfires annually burning an average of eight million ha [34].

The European Union (EU) has, since the early 1990s, considered forest fires (the official term for wildfires) as a common issue with a cross-border nature. The EU thus paid attention on research about forest fires, financing a number of projects that permitted fire science in Europe to reach a respectable level at a world scale. This adds to operative measures, such as the European Forest Fire Information System (EFFIS) [35] that includes fire danger rating by the FWI; active fire detection; burnt area mapping; land cover damage assessment; appraisal of Natura 2000 sites affected; forest fire emissions; potential soil erosion; and estimates of economic losses caused by forest fires in Europe (http://ies.jrc.ec.europa.eu/forest-fire-information). Research on forest fires in the EU started under the Framework Programmes for Research and Technological Development (FP2 1989–92). Over the past 2 decades, 56 research projects received EU funding, worth more than €103 million (https://ec.europa.eu/info/research-and-innovation/research-area/environment/forest-fires_en). The EU selectively financed proposals having operative issues, such as fire science (6); fire prevention (21), fire detection (4), fire suppression (20), postfire recovery (6), and integrated fire management (5) [36].

Among the most important projects, we mention: (1) FIRE PARADOX (https://cordis.europa.eu/project/rcn/79792/factsheet/en), distinguished for a multinational and multidisciplinary research consortium of 34 participants partners, many of which not belonging to the EU (Tunisia, Morocco, Argentina, South Africa, and Russian Federation). Under the slogan *Combating fire with fire*, FIRE PARADOX proposed *"learning to live with fire"* viewing it as a means to prevent and fight fires by reducing the available fuel. It proposed the integration and regulation of the use of fire as a fuel-management and suppression tool [36], in fire prevention and fighting strategies [37], including the use of prescribed burning emphasized by other projects (STEP 087, FIRETORCH, FUME); (2) FUME (https://cordis.europa.eu/project/rcn/94659/factsheet/en), by understanding of the causes underlying past changes in fire regime and with projections of future climate and other socioeconomic factors, the project aimed at presenting the likely impacts on the vegetation, landscapes, and fire regime (http://www.fumeproject.eu/); and (3) AF3, ADVANCED FOREST FIRE FIGHTING (https://cordis.europa.eu/project/rcn/185483/factsheet/en), the most expensive project ever funded by the EU in forest fires domain, focused on the implementation of a system to disperse extinguishing materials from high altitude by aircrafts and helicopters in any condition; innovative countermeasures with the fast buildup of defensive lines; early detection and monitoring; integration and deployment of satellites, airplanes, UAVs, and both mobile and stationary ground systems for the early detection of fire and for monitoring the propagation of smoke and toxic clouds; and integrated crisis management to perform overall coordination of all firefighting missions. In the EU about 400 scientific articles about forest fires are published every year, placing it immediately after the US [38]. Despite the impressive amount of forest wildfire research at world level, yearly made available to institutions, both governmental and private, to scientists, to private owners, the agreement is general that *"high quality wildland fire science is available but that translating this science into actionable policy initiatives is difficult"* ([33], p. 16). Some exceptions must be made: For instance, research results permitted to the suppression-centered system of the US to gradually and more closely adapt to science finding.

12.3.2   Changes in US wildfire policies

First changes in wildfire policy may be dated to around 1970, by recognizing the limits of full suppression, the ecological role of fire [39], and that, conversely, total suppression was producing forests with high fire hazard, and more prone to high-severity wildfires [40]. From 1977, the US Forest Service changed its policy from fire control to fire management [39]. The policy shift was also partly due to an increased role in the fire management responsibilities of other federal agencies [39], such as Bureau of Land Management (BLM) and National Parks Service (NPS). Private organizations too appeared, such as the Tall Timbers Research Station, an independent, privately run preserve established in Florida in 1958 to study fire on the landscape [39]; during the 1960s it was an alternative forum, for promulgating ideas about fire ecology and prescribed fire [39,41]. The term *prescribed fire* was proposed in 1942 by Raymond Conarro as a compromise between state-sanctioned suppression, and *laissez-faire* folk burning [11]. Planned or prescribed natural fires thus became a formal part of fire management policy. In 1978 there was a shift in suppression policy, allowing some natural fires to burn in specific locations, under previously identified conditions (prescribed natural fires [PNFs] [16]). In 1995 a revision of federal fire policy recognized the role of fire in ecological systems, calling for the implementation of fuel reduction programs by prescribed burning, to reduce the likelihood of catastrophic fire events [40]. A National Interagency Prescribed Fire Training Center was established in 1998 [12]. In 2017 an estimated 4,568,887 ha was treated in the US with prescribed fire (Melvin, 2018), 20% of which in agricultural terrains and 80% in forests.

The most active states to implement prescribed burning are the South-East ones, traditionally characterized by the use of light fires as prevention tools. The 2018 report by the National Association of State Foresters [42] enumerates the top impediments limiting the use of prescribed burning, among which weather, liability/insurance, and capacity are the most selected ones, accounting for 74% of responses. Private institutions such as the Nature Conservancy are very active in carrying out prescribed burning activity, almost with the same acreage of US Forest Service [12]. In the 1980s, attention was devoted to the WUI, on problems of evacuation and on the adoption of measures to make firewise communities. In 1998, the Joint Fire Science Program (JFSP) was created, supporting the use of fire and fuels treatment with the goals of reducing severe wildland fires and improving ecosystem health [12].

In 2000, the National Fire Plan was implemented, allocating funding for fire science and for the treatment of fuels (for example, through thinning or prescribed burning) [12]. The National Fire Plan (2000), the Western Governors Association 10-Year Comprehensive Plan (2001), and the Healthy Forest Restoration Act (2003) represent the shift from a policy of complete fire suppression to an alternative one, including restoration of fire-adapted ecosystems, reduction of fuels, and economic assistance to rural communities [39].

In 2009 the Federal Land Assistance Management and Enhancement Act mandated that federal agencies work with stakeholders to develop a National Cohesive Wildfire-Management Strategy (2014) [43], a strategic push to work collaboratively among all stakeholders and across all landscapes, using best science, to make progress toward the three goals of (1) resilient landscapes; (2) fire-adapted communities; and (3) safe and effective wildfire response. The cohesive vision can be summarized as follows: To safely and effectively extinguish fire, when needed; use fire where allowable; manage our natural resources; and as a Nation, live with wildland fire [43].

In the US, Canada, and Australia, results of research (inspired by the state forestry agency that have institutional hegemony and funding, and thus the capacity of steering fire research) have thus been exploited in terms of innovative wildfire policy.

On the contrary, where the *"culture of suppression"* [16] dominates, the scenario is totally different: Coping with wildfires consists of costly centralized, decentralized, or regional suppression organizations, endowed with updated technology, mainly relying on 4WD and aerial means, to ensure a systematic, rapid, hard-hitting initial attack on all fire ignitions. For all EU countries, forest firefighting represents a yearly budget of € 2.2 billion [36]. Goals, strategies, tactics, and human and technical resources are aimed and directed to put out flames, thus reductively interpreting the phenomenon as a mere combustion. Modest attention is paid to the socioecological pathology of wildfire and the necessity of a different risk governance system [2] and to an integrated fire management. This approach emphasizes the old suppression-centered strategy, without evolving toward new perspectives, but increasing the gap with other countries where a more prevention-oriented research can produce relevant changes. Wildfires are still mainly considered a civil protection or civil defense issue.

Where wildfire research is not inspired by the agency/organization/service that needs solutions for problems to be handled, it is mainly a task of a more general forestry research. Not always, forestry research institutions have specialized sections for wildfires, and departments that have special interest and experience in fire science topic not always see their results transformed in precepts; rather often research does not translate into effects.

12.3.3   The resistance to change: The example of Italy

We propose Italy as paradigmatic of a suppression model, and an example of the problems arising from such option, when activity is not properly assisted by research or worst, when availability of research results is almost refused.

Before WWII, wildfires in Italy were rather rare and occasional events, in the predominantly rural scenario of the country. After WWII, an accelerated change of scenario started, with a gradual massive migration from the countryside and consequent abandonment of land (about three million ha) and of many activities in the woods, such as the right to freely collect dry wood. In some areas this change was accompanied by the reduction of grazing in common lands and woods and resulted in an increase of fuel load.

Fires started increasing in number after WWII in Italy: In the period 1964−70 it ranged from 1158 to 3444 (1968) events and burned area passed from 8588 ha in 1964 to 42,966 in 1967. From 1971 to 1976 the number ranged from 2358 in 1972 to 5681 in 1973 and the total burned area from 27,303 ha in 1972 to 108,838 ha in 1973 [44].

In 2008−2017 a total of 42,748 events were recorded for Italy, with a burned area of 380,739 ha, of which 35.8% is nonforested area. Events with size more than 500 ha (a megafire, according the EU standard; [45]) are 52, with a total area of 52,491 ha. Values confirm that 0.1% of events are responsible for 10.8% of burned area. The cumulated values of percentage of both number and burned area, per size classes, confirm that a small amount of events participate to the total and that since the present situation is almost out of control, as relatively "mild" events, determine difficulty of containment.

In long-term EU statistics (1980−2017) [46], Italy is currently third in the EU countries after Portugal and Spain. In the observation period, burned areas were more than 100,000 ha for 17 times and more than 200,000 ha for 4 times, with a maximum of 229,850 ha in 1981. This means the gradual rise of a rather complex phenomenon.

The event historically marking the beginning of wildfire awareness in Italy was the National Forest Fire Conference of Bergamo, in June 1967, where for the first time Italian authorities and representatives of CFS (Corpo Forestale dello Stato, the State Forestry Authority) declared concern for the increasing number of fires [47]. In the Proceedings of the Conference, a vague knowledge of the phenomenon is evident, conversely opposed to the attractive power of technical means (tracks, 4WD, and mainly aerial means, helicopters, retardants) in the effort to delineate an efficient system of control based on suppression [47]. Hoffman [48], the most outstanding expert of CFS, after putting in evidence the lack of scientific research in the wildfire domain, states that the future of fire control is in aerial means [47].

Not surprisingly, in 1971, Italy started building up a powerful state fleet of Canadair 215, reaching the top of 19 some years ago and now stabilized on 15, mainly Canadair 415. The fleet was built without any assessment of cost and benefits derived by the use of a water bomber, considering that for fire intensities exceeding 3000 kWm^{-1}, water bombing becomes ineffective and stopping the advance of the fire front is impossible [49−51].

To contain the growing number of fires, but without awareness of fire origin and motives, the concerned authorities in the 1970s unwisely integrated fire suppression crews with workers seasonally hired for plantation activities. This represented a new, unexpected job opportunity in summer season, when agriculture demands a minimum of activity and therefore created the possible vicious circle of arson fires, motivated by burning, extinguishing, and replanting, called "*forest fire industry*" [10,50].

In 1975 the Italian Law 47 for the control of forest fires established the creation of an office for the study and control of forest fires. There is scarce evidence of the activity of the office, called the National Center for Research and Experimentation about Forest Fires, acting at the Ministry of Agriculture; in the early 1980s it was surely involved in the experimentation for the use of Mc Arthur forest fire index. The use of fire danger index was made with contrasting results, never entering in the routine activity, but stimulated the creation of new indexes, now currently used in Italy. As forest fire fighting is an institutional task of Regions since 2000, at regional level, different danger indexes are used. More than on the integration of danger assessment as instrumental to the efficaciousness of the firefighting activity, it seems that the goal is to demonstrate the availability of regional indexes.

The maximum discrepancy is in the domain of prescribed burning. In Italy this practice, largely adopted in other countries as an ecologically efficacious and cheap preventive tool, as demonstrated by the FIRE PARADOX project, is allowed by regional laws in six Regions, included in nine Regional Fire Management Plans and in six Fire Management Plans of National Parks [52]. However, its implementation finds difficulties and insidious opposition, where lack of knowledge and uncertainty in predicting wildfire behavior [53], vested interests, deeply rooted opinions, fears and traditions, inadequate information dissemination, administration constraints and, sometimes, simply resistance to change, all act as barriers to the acceptance [36]. Impediments to prescribed burning are represented by limitations in terms of (1) land (extent of WUI; forests, shrub lands and rangelands in mountain territories); (2) management policies focused on suppression; (3) culture (Loss of expertise in traditional fire knowledge by the rural population; public opinion accustomed to fire control policies; poor knowledge of the ecological role of fire); and (4) legislation (Absence of a national legislative framework; heterogeneity of the legislation at the regional level; confusion between prescribed burning and backfire; lack of clear authorization processes and definition of liability; lack of certification programs) [52].

Similarly, also the use of backfire and in general of suppression fires [54] is rather rare in Italy, not only because of lack of clear authorization processes and definition of liability but also because the general policy of strict rules concerning the use of fire is gradually obliterating the residual traditional fire knowledge of rural people. The criminalization of such use drastically reduces the use of the only available tool to contain EWEs. Proactive use of fire as a management tool to optimize biomass reduction is thus limited, because of liability and casualty risks and little tolerance for management errors [36].

Attitudinal barriers are present in Italy, similarly to the US *"legacy of* 'suppression bias' *afflicting land management agencies"* ([16], p.10). Agency personnel trained in fire suppression perceive themselves as professionals whose job it is to put fires out, not to let them burn. For them, the idea of letting a natural fire burn may be anathema [16]. Conversely, these barriers are often accompanied, in Italy, by emphasis on the supremacy of practical experience on the ground over the results of research and their implementation. It is a very dangerous bias, when climate change could allow the occurrence of EWEs, and the relative success of current practice on mild and easy to control fires could be dramatically overwhelmed by extreme behavior of future events.

Italy epitomizes the diffuse situation where the threat of fire, initially contained by suppression crews, has gradually increased in number and burned area, but mainly in extreme fire behavior manifestations. Insisting on suppression, despite the increase in costs, has not been accompanied by the necessary changes in policy, strategy, and tactics supported by better knowledge of facts, which only a dedicated research activity can provide. Firefighting continues to face fire as a light seasonal inconvenience, with an emergency approach, focusing and devoting all resources in the phase of the quick and efficacious response to hazard occurrence. It is difficult to describe points of acceptance of, and compliance with, the many recommendations made by the EU, regarding forest fires and the importance of research [36]. In Italy, fire management and its associated governance are making scarce use of science-based findings and innovations, with emphasis on suppression, and marginal and local activities of prevention, without attention to problems of vulnerability and conversely, to the necessity to foster resilience of people and assets.

This is a dangerous option, ignoring the existence of EWEs and wildfire disasters. The incumbent risk for the country is to be suddenly obliged to cope with unexpected EWEs, with an emergency organization that is neither able nor prepared to handle with the values of intensity and behavior of such fires.

12.4   How to create a change in wildfire policies

The current entrenched wildfire policies centered on suppression are maintained and protected by powerful interests. The barriers to change are rooted in misunderstanding of wildfire problem, in the governance system, in the culture of organizations, and in the top-down relationship between firefighting institutions and citizens. The increasing threat represented by EWEs, unpredictable in their behavior and largely exceeding any control capacity, requires a policy change to cope with the new scenarios, mainly when they affect the WUI.

Recent wildfire disasters in Australia, California, Portugal, and Greece and also other extreme events that did not capture the attention of the international media, such as the ones occurred in 2017 and 2019 in New South Wales [55] and Tasmania [45] show the need to enhance adaptation to wildfires, as climate change is highly contributing to fire-prone environment and fosters the occurrence of EWEs.

The change cannot be reached without promoting two critical steps: awareness of the wildfire nature and recognition of the limits of suppression activities (and making citizens aware of it).

12.4.1   Awareness of the wildfire nature: The pyrometrics

Any earthquake, at world level, is currently identified and labeled by the amount of released energy; people have learnt to currently discuss about it in this sense, not in terms of number of events and surface impacted. Wildfires, on the contrary, are still presented in postfire metrics such as number, area burned, and, sometimes severity, that do not capture the nature of fire but the result of the interaction between fire behavior and the characteristics of the area affected. This notwithstanding the research work starting in the 1950s by pioneers in the US such as Byram [56], Rothermel, Alexander, and others [20], and Canada [29] about the physical properties of fire and its quantitative analysis.

Wildfire behavior pyrometrics [57], such as intensity, rate of spread, height or length of flames, and magnitude of a fire (that have a crucial role to understand the operational limits of suppression activities, the threats fires represent to societies and ecosystems, the levels of death and destruction) are almost unknown by people and totally absent in the common statistic reports, whereas they are a normal presence in scientific literature. People (citizens and professionals) must be fully aware of the physical properties of wildfire events, to understand that EWEs create a complex emergency, that cannot be managed with the same tools and procedures of the normal fires, and that a fire of category 1 (in the classification by Tedim et al. [58]) puts different threats than one of category 5 greater (see chapter 1).

Understanding wildfires in terms of physical properties is important because *"The old way of fighting fires by sending firefighters — that's gone"* [59]. Under the influence of the undoubted climate change, more days of extreme heat, longer heat waves [60] and more frequent droughts are expected, conditions that favor the occurrence of EWEs. In risk and crisis communication activities there is a strong need of objective and well-understood quantitative information. The threats fire puts are different depending on fire category. The lack of knowledge about the metrics does not allow to understand fire identity and consequently to distinguish its double role: beneficial when it is a tool of land management and fuel reduction, and negative when it happens out of its natural regime and is caused by anthropic actions. In addition, it is not enough to say to citizens that they are in an area at risk. It is crucial to make them aware of how the fire hazard looks like, how they can cope with it, what is their vulnerability, what are the physical and mental capabilities to deal with EWEs. An appropriate knowledge of fire dynamics could help collectivity to better understand and critically evaluate the intrinsic risk of personal or political options to cover the rural space with a continuous blanket of inflammable vegetation, sometimes in close contact with WUI border and assets embodied in it, as it happened in some cases (e.g., Portugal, Spain, South Africa) for *Eucalyptus* industrial plantations. People living in the area often have scarce or no awareness of their risk situation and blindly rely on the intervention of aerial means in case of wildfires. Risk is over their shoulders, without sharing the benefits of plantation presence.

12.4.2   Recognition of the limits of suppression activities and making citizens aware of it

Persisting in ignoring the limit of the suppression activities or overemphasizing them with deceptive assurances of efficiency of the defensive apparatus is not only confused understanding but also an unacceptable behavior, mainly if it is the result of the *optimism bias*, the mistaken belief that our chances of experiencing negative events are lower and our chances of experiencing positive events are higher than those of our peers [61].

People must be aware that when fires exceed the threshold of 1000 ha in size, it is expected that they exhibit characters of extreme fire behavior [62]; so, they cannot remain unaware that the upper threshold of the current control capacity (commonly accepted in $10,000 \, \text{kWm}^{-1}$) covers a small fraction (about 7%) of the possible values of intensity. For instance, in October 2017 wildfires in Portugal, intensity went over $50,000 \, \text{kW m}^{-1}$ in the majority of involved municipalities [63].

12.5   Conclusions

Despite the impressive amount of wildfire research yearly made available to institutions, at world level the agreement is general that translating this science into actionable policy initiatives is difficult. Science can advise, but not decide; thus additional research does not seem able to cope with the challenges presented by wildfires. In addition, in many countries there is some reluctance to accept the results of research, and changes of policy suggested and supported by research are sometimes impeded by resistance, lack of information, vested interests, culture, and personal and political convenience. This maintains the all-out suppression model, which offers visible and tangible results, consistent with the duration of political office of policy-makers.

There are two ways to promote changes that are able to reduce the negative impacts of wildfires and avoid the occurrence of disasters: (1) More objective knowledge and full awareness by political deciders of the incoming scenario fostered by the climate change, and of the new challenge to cope with EWEs, where the available control apparatus demonstrates its limits and incapacity, with potential catastrophic outcomes; (2) empower society, by properly disseminating knowledge so that a culture of fire management is socially understood and accepted. Prevent EWEs or cope with them is not a task of fire agencies but requires the involvement of all society mainly the one that lives in fire-risk areas.

References

[1] I.I. d'Agriculture (I.I.A), Enquête internationale sur les incendies de forêt, Rome, 1933.
[2] A.P. Fischer, T.A. Spies, T.A. Steelman, C. Moseley, B.R. Johnson, J.D. Bailey, A.A. Ager, P. Bourgeron, S. Charnley, B.M. Collins, Wildfire risk as a socioecological pathology, Front. Ecol. Environ. 14 (2016) 276–284.
[3] B. Gabbert, Peshtigo Fire: 137 Years Ago Today, Wildfire Today - Wildfire News Opin, 2008. https://wildfiretoday.com/2008/10/08/peshtigo-fire-137-years-ago-today/.
[4] J. Lewis, October 8, 1871: Peshtigo, Wisconsin, Is Consumed by Fire - Forest History Society, For. Hist. Soc. (2011). https://foresthistory.org/october-8-1871-peshtigo-wisconsin-is-consumed-by-fire/.
[5] J. Cohen, The wildland-urban interface fire problem: a consequence of the fire exclusion paradigm, For. Hist. Today. Fall (2008) 20–26, 20-26.
[6] W.D. Muir, Forest Fire Control in the United States of America, Sidney, 1941, https://foresthistory.org/wp-content/uploads/2017/01/Forest_Fire_Control_Muir_1941_md.pdf.
[7] FHS, U.S. Forest service fire suppression, For. Hist. Soc. (2017). https://foresthistory.org/research-explore/us-forest-service-history/policy-and-law/fire-u-s-forest-service/u-s-forest-service-fire-suppression/.
[8] C. Fowler, E. Konopik, The history of fire in the southern United States, Hum. Ecol. Rev. 14 (2007) 165–176, https://doi.org/10.2307/24707703.
[9] M.N. McGeary, Gifford Pinchot: Forester-Politician, Princeton University Press, 2015.
[10] S.J. Pyne, Vestal Fire: An Environmental History, Told through Fire, of Europe and Europe's Encounter with the World, University of Washington Press, 1997.
[11] S. Pyne, Foreword - passing the torch, in: M. Malvin (Ed.), Natl. Prescr. Fire Use Surv. Rep., 2018, pp. 3–5. https://www.stateforesters.org/wp-content/uploads/2018/12/2018-Prescribed-Fire-Use-Survey-Report-1.pdf.
[12] S.J. Pyne, Fire in America: A Cultural History of Wildland and Rural Fire, University of Washington Press, 2017.
[13] G. Pinchot, A Primer of Forestry, US Dep, 1903.
[14] W. Bergoffen, 100 Years of Federal Forestry - Agriculture Information Bulletin No. 402, Washington, 1976, https://foresthistory.org/wp-content/uploads/2017/01/1-60.pdf.

[15] J. Roberts, The Best of Intentions - sometimes the people in charge of keeping us safe know just enough to put us in even greater danger, Sci. Hist. Inst. (2015). https://www. sciencehistory.org/distillations/magazine/the-best-of-intentions.

[16] G.H. Aplet, Evolution of wilderness fire policy, Int. J. Wilderness. 12 (2006) 9—13. http:// ijw.org/wp-content/uploads/2006/12/Apr-2006-IW-vol-12-no-1small.pdf#page=10.

[17] A. Berry, Forest Policy up in Smoke: Fire Suppression in the United States, 2007. http:// www.perc.org/wp-content/uploads/old/Forest Policy Up in Smoke.pdf.

[18] D.M. Smith, Sustainability and Wildland Fire: The Origins of Forest Service Wildland Fire Research, US Department of Agriculture, Forest Service, Forest Health Protection, 2017.

[19] C. Du Bois, Systematic Fire Protection in the California Forests, US Government Printing Office, Washington, 1914.

[20] D.M. Smith, The missoula fire sciences laboratory: a 50-year dedication to understanding wildlands and fire, Gen. Tech. Rep. RMRS-GTR-270. Fort Collins, CO US Dep. Agric. For. Serv. Rocky Mt. Res. Station. 62 (2012) 270.

[21] D. Vanhoozer, 9/13/17 — the "10 a.m. Policy": the U.S. Forest service and wildfire suppression, Stevens Hist. Res. Assoc. (2017). https://www.shraboise.com/2017/09/91317-10-m-policy-u-s-forest-service-wildfire-suppression/.

[22] S.B. Show, B. Clarke, Forest Fire Control, Food and Agriculture Organization of the United Nations, Rome, Italy, 1953, 99pp. plus Gloss.

[23] V.C.A.A., Bushfires in Our History, 1851—2009, 2011. http://www.bushfireeducation.vic. edu.au/verve/_resources/Bushfires_in_our_History.pdf.

[24] C. Hanrahan, What a Century of Bushfire Data Teaches Us about How to Save Lives This Summer, ABC - NEWS, 2018. https://www.abc.net.au/news/2018-12-27/bushfire-history-australia-data-saving-lives/10606144.

[25] R.H. Luke, A.G. McArthur, Australia, Forestry and Timber Bureau., Commonwealth Scientific and Industrial Research Organization (Australia), Division of Forest Research., Bushfires in Australia, Australian Government Pub. Service, 1978.

[26] FHS, 1905 Use book. The use of the national forest reserves, For. Hist. Soc. (2019) n.d. https://foresthistory.org/research-explore/us-forest-service-history/u-s-forest-service-publications/general-publications/1905-use-book/.

[27] A.D. Folweiler, Theory and Practice of Forest Fire Protection in the United States, 1937, p. 163. https://babel.hathitrust.org/cgi/pt.?id=mdp.39015006889557;view=1up;seq=171.

[28] A.D. Folweiler, Forest Fire Prevention and Control in the United States, 1938, p. 163. http://agris.fao.org/agris-search/search.do?recordID=US201300348866.

[29] C.E. Van Wagner, Six decades of forest fire science in Canada, For. Chron. 66 (1990) 133—137, https://doi.org/10.5558/tfc66133-2.

[30] R.S. McAlpine, B.J. Stocks, C.E. Van Wagner, B.D. Lawson, M.E. Alexander, T.J. Lynham, Forest fire behavior research in Canada, in: Proc. Int. Conf. for. Fire Res., 1990, pp. 19—22.

[31] D.X. Viegas, G. Bovio, A. Ferreira, A. Nosenzo, B. Sol, Comparative study of various methods of fire danger evaluation in southern Europe, Int. J. Wildland Fire 9 (1999) 235, https://doi.org/10.1071/WF00015.

[32] C. Sirca, M. Salis, B. Arca, P. Duce, D. Spano, Assessing the Performance of Fire Danger Indexes in a Mediterranean Area, 2018, p. 563, https://doi.org/10.3832/IFOR2679-011. http://iforest.sisef.org/11.

[33] E.S. Artsybashev, Forest fires and their control, in: For. Fires Their Contro, Oxonian Press Pvt. Ltd., New Delhi, Calcutta, 1983, p. 160. https://www.cabdirect.org/cabdirect/abstract/19840696287.

[34] J.G. Goldammer, A. Sukhinin, I. Csiszar, The current fire situation in the Russian Federation: implications for enhancing international and regional cooperation in the UN framework and the global programs on fire monitoring and assessment, Int. For. Fire News. 32 (2005) 13—42.

[35] J. San-Miguel-Ayanz, The European forest fire information system(EFFIS) as a support to forest fire prevention, in: IV Int. Conf. Strateg. for. Fire Prev. South. Eur., Bordeaux, France Jan. 7th—9th, 2013.

[36] European Commission, Forest Fires Sparking Firesmart Policies in the EU, Publicatio, European Commission, Luxembourg, 2018, https://doi.org/10.2777/248004.

[37] C. MONTIEL-MOLINA, Fire use practices and regulation in Europe: towards a fire framework directive, in: 5th Int. Wildl. Fire Conf. Sun City, South Africa 9e13 May, 2011. http://www.wildfire2011.org/material/papers/Cristina_Montiel-Molina_2.pdf.

[38] E. and M. National Academies of Sciences, A Century of Wildland Fire Research, National Academies Press, Washington, D.C., 2017, https://doi.org/10.17226/24792.

[39] S. McCaffrey, E. Toman, M. Stidham, B. Shindler, Social Science Findings in the United States, Wildfire Hazards, Risks and Disasters, 2015, pp. 15—34, https://doi.org/10.1016/B978-0-12-410434-1.00002-6.

[40] S.L. Stephens, L.W. Ruth, Federal forest-fire policy in the United States, Ecol. Appl. 15 (2005) 532—542, https://doi.org/10.1890/04-0545.

[41] S.J. Pyne, Problems, paradoxes, paradigms: triangulating fire research, Int. J. Wildland Fire 16 (2007) 271, https://doi.org/10.1071/WF06041.

[42] M.A. Melvin, National Prescribed Fire Use Survey Report 2018 National Prescribed Fire Use Survey Report I, 2018. www.prescribedfire.net.

[43] W.F.E. C, The National Strategy: The Final Phase in the Development of the National Cohesive Wildland Fire Management Strategy, 2014. Washington, DC.

[44] S. Landi, Organizzazione e tecnica della lotta contro gli incendi boschivi, Arti Grafi, Cittaducale, 1977.

[45] D. Bowman, Tasmanian Fires: Impacts and Management Lessons (2019), The Fire Centre, Fire Cent. Res. Hub - Univ. Tasmania, Aust., 2019, https://firecentre.org.au/2019-tasmanian-fires-impacts-and-management-lessons/.

[46] J. San-Miguel-Ayanz, T. Durrant, R. Boca, G. Libertà, A. Branco, D. de Rigo, D. Ferrari, P. Maianti, T.A. Vivancos, H. Costa, F. Lana, Advance EFFIS Report on Forest Fires in Europe, Middle East and North Africa 2017, 2018, https://doi.org/10.2760/476964.

[47] CCIAA, Atti del Convegno nazionale "L'incremento del patrimonio forestale e la sua difesa dal fuoco", (Bergamo, n.d).

[48] A. Hofmann, La ricerca scientifica,tecnologica ed applicata nella lotta contro il fuoco in Italia, in: C.C.I.A.A. Bergamo (Ed.), Atti Del Convegno Naz. "L'incremento Del Patrim. For. e La Sua Dif, Dal Fuoco, Bergamo, 1967, pp. 167—187.

[49] E. Stechishen, E. Little, M. Hobbs, W. Murray, Productivity of Skimmer Air Tankers, 1982. http://www.cfs.nrcan.gc.ca/bookstore_pdfs/12092.pdf.

[50] I.T. Loane, J.S. Gould, Aerial suppression of bushfires: cost-benefit study for Victoria, Aer. Suppr. Bushfires Cost-Benefit Study Victoria (1986). https://www.cabdirect.org/cabdirect/abstract/19860611619.

[51] Australian Academy of sciences, What tools can we use against a raging bushfire? How we fight bushfires, Aust. Acad. Sci. (2019). https://www.science.org.au/curious/earth-environment/how-we-fight-bushfires.

[52] D. Ascoli, G. Bovio, Prescribed Burning in Italy: Issues, Advances and Challenges, 2013, p. 79, https://doi.org/10.3832/IFOR0803-006. http://iforest.sisef.org/6.

[53] V. Leone, A. Signorile, V. Gouma, N. Pangas, S. Chronopoulous, Obstacles in prescribed fire use in Mediterranean countries: early remarks and results of the Firetorch project, in: Proc. DELFI Int. Symp. for. Fires Needs Innov. Athens (Greece), 1999, pp. 132—136.

[54] C. Montiel, D.T. Kraus, Best Practices of Fire Use: Prescribed Burning and Suppression: Fire Programmes in Selected Case-Study Regions in Europe, European Forest Institute, 2010.

[55] J. Whittaker, M. Taylor, Community preparedness and responses to the 2017 NSW bushfires: research for the New South Wales rural fire service, Melbourne, Aust, Bushfire Nat. Hazards CRC (2018).

[56] G.M. Byram, Combustion of forest fuels, in: K.P. Davis (Ed.), Forest Fire: Control and Use, 1959, pp. 61—89.

[57] D. Perrakis, M. Cuz, M. Alexander, S. Taylor, J. Beverly, No title, in: Introd. To Can. Conifer Pyrom., Presented at the 6th International Fuels and Fire Behaviour Conference, Marseille, France, 2019. https://www.researchgate.net/publication/332950937_Intro_to_ Canadian_Conifer_Pyrometrics_Presented_at_the_6th_International_Fuels_and_Fire_ Behaviour_Conference_April-May_2019_Marseille_France.

[58] F. Tedim, V. Leone, M. Amraoui, C. Bouillon, M. Coughlan, G. Delogu, P. Fernandes, C. Ferreira, S. McCaffrey, T. McGee, J. Parente, D. Paton, M. Pereira, L. Ribeiro, D. Viegas, G. Xanthopoulos, F. Tedim, V. Leone, M. Amraoui, C. Bouillon, M.R. Coughlan, G.M. Delogu, P.M. Fernandes, C. Ferreira, S. McCaffrey, T.K. McGee, J. Parente, D. Paton, M.G. Pereira, L.M. Ribeiro, D.X. Viegas, G. Xanthopoulos, Defining extreme wildfire events: difficulties, challenges, and impacts, Fire 1 (2018) 9, https:// doi.org/10.3390/fire1010009.

[59] A. Ekin, It eats everything' — the new breed of wildfire that's impossible to predict | Horizon: the EU Research & Innovation magazine | European Commission, in: Horiz. - EU Res. &Innovation Mag., 2019. https://horizon-magazine.eu/article/it-eats-everything-new- breed-wildfire-s-impossible-predict.html.

[60] C. Santin, S. Doerr, Wildfires in Mediterranean Europe will increase by 40% at 1.5°C warming, say scientists, Conversat — Acad. Rigour, Journalistic Flair (2018). https:// theconversation.com/wildfires-in-mediterranean-europe-will-increase-by-40-at-1-5-c- warming-say-scientists-104270.

[61] N.D. Weinstein, Unrealistic optimism about future life events, J. Personal. Soc. Psychol. 39 (1980) 806.

[62] A. Filkov, T. Duff, T. Penman, reportdetermining Threshold Conditions for Extreme Fire Behaviour: Interim Report Describing Outcomes From Phase 1 Of The ProjecT from Phase 1 of the Project, (n.d.). http://naturalhazardscrc.com.au/sites/default/files/managed/ downloads/determining_threshold_conditions_for_extreme_fire_behaviour_annual_ report_2017-2018_final_19.pdf (accessed May 27, 2019).

[63] CTI, Avaliação dos Incêndios ocorridos entre 14 e 16 de outubro de 2017 em Portugal Continental, Relatório Final, Lisboa, 2018. https://www.parlamento.pt./Documents/2018/ Marco/RelatorioCTI190318N.pdf.

[64] P.J. Murphy, The art and science of fire management, in: M.E. Alexander, G. Birgrove (Eds.), Proc. First West Fire Counc. Annu. Meet. Work. Kananaskis Village, Alberta, 1988, pp. 23—26.

[65] A.D. Folweiler, A.A. Brown, 1900, Fire in the Forests of the United States, 1946. http:// agris.fao.org/agris-search/search.do?recordID=US201300368428.

What can we do differently about the extreme wildfire problem: An overview

Fantina Tedim [1,2], Sarah McCaffrey [3], Vittorio Leone [4],
Giuseppe Mariano Delogu [5], Marc Castelnou [6,7], Tara K. McGee [8], José Aranha [9]
[1]Faculty of Arts and Humanities, University of Porto, Porto, Portugal; [2]Charles Darwin University, Darwin, NWT, Australia; [3]Rocky Mountain Research Station, USDA Forest Service, Fort Collins, CO, United States; [4]Faculty of Agriculture, University of Basilicata (retired), Potenza, Italy; [5]Former Chief Corpo Forestale e di Vigilanza Ambientale (CFVA), Autonomous Region of Sardegna, Italy; [6]Bombers Generalitat.DGPEiS. DI., Barcelona, Spain; [7]University of Lleida, Lleida, Spain; [8]Department of Earth and Atmospheric Sciences, University of Alberta, Edmonton, AB, Canada; [9]Centre for the Research and Technology of Agro-Environmental and Biological Sciences (CITAB), University of Trás-os-Montes and Alto Douro, Vila Real, Portugal

13.1 Introduction

Fire is a natural process that has shaped the history of Earth long before human presence; imagining a "*world without fires is like a sphere without roundness*" ([1], p.599). Evidence that massive and intense fires naturally occurred throughout the Holocene [1–3] demonstrates that extreme wildfires events (EWEs) are not a recent reality. Although many ecosystems are resilient to such intense fires (e.g., in Yellowstone Park, within the first decade after the fire occurred in 1988, plant and animal communities were well on their way to their pre-fire composition; [3a]), recent tragic fires in Portugal, Greece, and the United States highlight the vulnerability of individuals and communities to such fires.

Even with increased wildfire suppression resources and improved training and co-ordination, by definition, EWEs cannot be stopped from spreading unless and until weather conditions or fuel load change [4–7]. In extreme conditions, it is only when fireline intensity, rate of spread, and spotting activity and distance decrease that firefighters can effectively control the fire [8]. As Ingalsbee [5] states, "*The hard work of brave firefighters gets all the credit in the news media for corralling large wildfires, but this is after the fires have largely stopped spreading on their own, in a process analogous to the Lilliputians "capturing' Gulliver after he had laid down and fallen asleep*" ([5]: p.558).

Although there is some awareness of the limits of fire control capacity, even in the most prepared regions in the world, this is not always fully acknowledged by either the fire management community [7] or by society, media, and decision-makers.

Extreme Wildfire Events and Disasters. https://doi.org/10.1016/B978-0-12-815721-3.00013-8

For example, in European countries, an emergency vision still prevails, and the standard response is a quick and hard attack to suppress the flames. Public awareness programs are usually focused on preventing fire outbreaks by fuel management initiatives, supported by coercive and punitive measures, with little attention to improve citizens' awareness on how to prepare for a wildfire and how to enhance their safety [9]. In comparison, outreach efforts in the United States, Canada, Australia, and New Zealand, along with a focus on preventing fires, concentrate on reducing the hazard through fuels management, and providing information on how individuals can mitigate their fire risk with changes to their property and evacuation planning.

One might expect that tragic wildfires would be transformative events, out of a desire not to repeat the event, but this is rarely the case. Lessons learned from postfire inquiries tend to concentrate on operational issues and response actions leading authorities to focus on matching the growing wildfire threat with greater suppression forces, particularly aerial resources [7]. For instance, after the 2017 fire season where 112 civilians were killed in nine fires in Central Portugal, one of the main measures adopted by the Portuguese government was to increase aerial resources [10]. Similarly, in response to significant 2017 and 2018 wildfire losses in Portugal and Greece, the European Union (EU) created, for the 2019 fire season, an aerial firefighting fleet (seven firefighting airplanes and six helicopters) as part of the new rescEU system to respond to natural disasters, which is expected to be further reinforced in future years [11].

These increases occur despite growing questions as to when and where aerial resources are effective, both in terms of intrinsic operational limits of their control capacity and whether there are sufficient crews on the ground to complement aerial efforts [12]. Aerial limitations are particularly relevant in relation to an EWE where conditions that can restrict the use of aerial equipment due to the inability to operate are more likely, including strong convective phenomena, rapid air gusts, and large amounts of smoke, or (for scoopers planes) when wave height in sea, lakes, or basins exceeds the threshold of 1 m (Canadair 215 and 415).

Although there are growing questions about the effectiveness of increased firefighting resources as a mean of improving outcomes, the response does fit into the historical approach to fire as a *"War on Fire"* (*WoF*) [13] or *"war against fire"* [14]. However, there is a growing realization that this paradigm, with its focus on fire control and exclusion, only created a *"wildfire paradox"* [15], increasing the fire hazard in many places through fuel buildup. It also creates a *"firefighter trap,"* ironically a concept used most in the business world to describe the tendency to fix an immediate problem instead of addressing the underlying causes that created it [16]. *"Systems often get trapped in cycles of self-reinforcing feedback whereby the problem symptoms continue to grow yet the solution remains the same"* ([15], p.6).

Increasing suppression costs are likely in part a reflection of the conviction that more resources will increase the ability to control all categories of fires, including EWEs. Vested interests and media tendencies to highlight the negative events, telling only the story of fire as a catastrophic agent, reinforce existing policy [13]. However, simply using a reactive approach to suppress all wildfires is unlikely to succeed in an era of heightened wildfire activity [17] as extreme weather conditions and drought, coupled with high fuel loads, challenge wildland fire response programs and land management policies [7].

A pressing question then is how to shift the wildfire management paradigm away from the tendency to address the immediate symptom to a focus that addresses the underlying causes that contribute to high-impact fires. Such a shift in wildfire management paradigm is needed for societies to better *"thrive with fire"*, including EWEs, with fewer losses and fatalities.

13.2   Looking for a new paradigm of wildfire management: Existing ideas and proposals

WoF, the modern focus on suppressing all wildfires, was initially based on a view from European forestry schools that saw all fires as a threat to society and ecosystems; the approach to fire management was one of full suppression and control of wildfire on the landscape. Overtime, the *WoF* approach has gradually introduced some slight changes in its original singular focus on excluding fire from the landscape (e.g., by allowing for use of prescribed burning and the potential for not immediately suppressing every fire but instead managing them to reduce fire risk or foster ecological benefits), but the primary focus remains on the immediate problem of fire control, without providing effective actions to solve the *"wildfire paradox"* and the *"firefighting trap"*. However, the escalating impacts of wildfires in many places provides evidence of the weaknesses of the *WoF* focus on addressing the symptoms and highlights the need to develop an alternative approach that may better address underlying factors and lead to a more sustainable coexistence with wildfires.

A growing body of literature advocates for a shift in wildfire management and proposes new perspectives [13,18−22]. The suggestions on how to shift the current fire paradigm are diverse: Some focus on the integration of concepts such as the dual role of fire and the socioecological nature of wildfires; some are centered on specific types of intervention such as strategic fuel management; some focus on the establishment of frameworks to promote and orientate a change; and others propose a vision of a new paradigm. A nonexhaustive summary of these perspectives is provided in the following sections.

13.2.1   The dual role of fire

WoF paradigm is based on the notion of fire exclusion, but since the 1960s research has provided mounting evidence that along with its potential negative impacts, fire also has a crucial ecological role in many ecosystems [23−27]. Depending on the spatial and temporal context of occurrence, fire can be a hazard or a benefit as an ecological process necessary for many ecosystems' health and biodiversity [28−31].

Recognizing the crucial ecological role of fire, Ingalsbee [5] proposes the ecological fire management paradigm based on the need to adopt an integrative (prevention and response activities) and holistic thinking.

The acceptance of the dual role of fire is an important step on the way to solve the wildfire *"wicked problem"* and cope with EWEs.

13.2.2   Strategic fuel management

Another set of recommendations focus on fuels management. The concept of fire-smart management of forest landscapes, proposed by Hirsch et al. [32], focused on fuel management to create more fire-resistant (less flammable) and/or fire-resilient (able to better recover) environments at the scales of landscape and forest stands. The effects and effectiveness as well as the economics of treating fuels are well understood [33—36]. The main purpose of fuel reduction strategies (isolation, structural modification, and type conversion [34]) is not to decrease fire occurrence but to modify fire behavior, by abating intensity, and severity. Fuel reduction can both facilitate suppression actions (if the fire intensity does not exceed the control capacity) and limit fire size, but its effects are influenced by the spatial features of managed patches and their temporal maintenance [33,37]. A successful integration of fire and land management programs and objectives can result in treatments that restore ecosystems as well as treating fuels.

As fuel management tends to be limited in the amount, location, and type of permitted treatments, another suggested measure is integrating unplanned burned areas as a fuel treatment and an ecological process, as a good option in remote areas where no assets are at risk [34]. In the wildland—urban interface (WUI), fuel management has the primary objective of protecting structures and people [37]. Some authors recognize as a priority reducing fuel loadings and safely reintroducing wildland fire to expedite forest restoration [38].

Although strategic fuel management at an appropriate temporal and spatial scale has insights into improving fire outcomes, its focus on modifying fire behavior remains part of the *WoF* notion that fire can always be controlled.

13.2.3   The socioecological nature of wildfires

In 1984, Martell [39] proposed *"fire impact management policies"* whereby decisions concerning wildfire suppression and use of prescribed fires should be based on the use of sound social, economic, and ecological principles [39]. A growing body of literature has demonstrated the interrelated nature of the human and biophysical processes in shaping fire [18,40—46]. A growing number of scholars argue that wildfires cannot be understood or managed by just an ecological approach or a social approach separately considered, as they result from a complex network of interrelationships between natural and anthropic (social, political, legal, technical, economic, and psychological) systems [20,47—52].

These scholars argue that only a holistic and integrated approach will enable a clear understanding of the complex interdependencies between fire, landscape, climate, and communities' values, attitudes, and behaviors. Considering wildfire in a context of socioecological or coupled human and natural systems promotes *"qualitative social change to fully resolve all the human factors driving the wildfire paradox"* ([5], p. 559). However, while making the critical point of the need for developing an approach to better consider how social factors contribute to poor wildfire outcomes, these arguments nevertheless tend to provide a simplistic view of social dynamics, often making fairly general recommendations based on little empirical social evidence.

13.2.4   Integrated fire management and community-based fire management

Some have argued that Integrated Fire Management (IFM) is the best approach to improve wildfire outcomes and promote a paradigm shift [53,54]. IFM, a concept initially used in describing fire management approaches in less developed regions [55], recognizes that fire has been traditionally used and continues to be used as a land management tool in rural areas where, in many cases, the vast majority of wildfires have human causes and motivations [56]. IFM was thus initially developed to describe the integration of fire suppression with fire use actions (i.e., prescribed burning, and traditional fire use) in developing overall fire management strategies [56].

In the EU, IFM is currently defined as "*a combination of prevention and suppression strategies stemming from social, economic, cultural and ecological evaluations. Beyond the sole consideration of fire prevention and fire suppression, integrated fire management links the four steps of emergency crisis management, i.e., mitigation, preparedness, response and recovery. IFM is a concept for planning and operational systems aiming at minimizing the damage from and maximizing the benefits of fire*" ([52], p.19.).

While this current approach is more holistic, considering biological, environmental, cultural, social, economic, and political interactions [29,57—59], it nevertheless maintains a primary focus on the *WoF* emphasis on emergency management as the main actor in improving outcomes. A shift of governance model is not advocated, nor does it provide a specific answer for EWEs. In effect, IFM as it has been conceptualized is just another attempt to improve the outcomes of *WoF*.

Another approach used in some parts of the world, that does focus on more than fire response and management, is Community-Based Fire Management (CBFiM) which has been defined as "*a type of land and forest management in which a locally resident community (with or without the collaboration of other stakeholders) has substantial involvement in deciding the objectives and practices involved in preventing, controlling or utilizing fires*" ([60], p. 4). This approach moves beyond formal organizations to include local communities in the fire management process, inclusive of using fire for a range of beneficial reasons (e.g., to manage fires for controlling weeds, reducing the impact of pests and diseases, generating income from nontimber forest products, pasture renovation, fuel reduction) and in preparedness and suppression of wildfires [61]. The activities and knowledge that communities generally practice and apply are primarily those associated with planning and supervision of activities, joint action for prescribed fire and fire monitoring and response, applying sanctions, and providing support to individuals to enhance their fire management tasks [61—63]. This approach is very interesting as it covers two aspects that we valorize: the double role of fire and a bottom-up approach. It has been applied in several geographical contexts, usually with effective results. While it has potential to reduce the occurrence of EWEs, it nevertheless fails to provide a systematic approach to addressing differential conditions in different locations.

13.2.5   Comprehensive strategies and frameworks

Another set of approaches to address escalating wildfire impacts involve the development of frameworks that identify key areas that need to be addressed. In the United

States the National Cohesive Wildland Fire Management Strategy (https://www. forestsandrangelands.gov/strategy/) sets broad management goals for federal land management agencies to work collaboratively among all stakeholders and across all landscapes to improve wildfire outcomes by addressing three goals: Resilient Landscapes, Fire Adapted Communities, Safe and Effective Wildfire Response. The National Strategy accepts fire as a natural process necessary for the maintenance of many ecosystems and proposes *"to safely and effectively extinguish fire when needed; use fire where allowable; manage natural resources; and as a nation, to live with wildland fire"* [64].

The US National Cohesive Wildland Fire Management Strategy inspired Alcasena et al. [65] to propose a similar wildfire management comprehensive strategy for Southern European countries, which includes four main objectives: (1) Create fire-resilient landscapes (through the option of fuel management); (2) restore the cultural fire regimes (through human ignition prevention and natural fire reintroduction); (3) support an efficient fire response (through fire suppression); and (4) generate fire-adapted human communities (promoting community action) [66].

O'Neill and Handmer [67], assessing evidence from the 2009 Black Saturday fires in Australia, call for a transformation in wildfire risk management to avoid similar future disasters. They identify four areas of intervention: (1) Diminish the hazard through prevention of initial ignition and through fuel reduction; (2) reduce the exposure of infrastructures and buildings by avoiding very high hazard areas for vulnerable uses and improving building codes to make houses safer and resistant in extreme conditions; (3) reduce the vulnerability of people by acknowledging and addressing individual fragilities through formal fire safety training, including mental preparedness training, for those at risk; (4) increase the adaptive capacity of institutions both through mechanisms such as insurance, and in changing the culture within institutions.

All the proposed frameworks address important areas of intervention, but while their broadness does allow for tailoring to local conditions, it fails to address the challenges that EWEs pose and provides little incentive to change from the default practice, or doing what one knows how to do best: suppression. Only O'Neill and Handmer [67] mention the need to change the culture within the institutions, pointing to the need to change the governance model.

13.2.6 Build wildfire resilience

Finally, recent years have seen increased focus on the notion of resilience in a number of science, policy, and management arenas, including wildfire. Resilience is one of the components considered in the risk-to-resilience continuum framework proposed by Smith et al. [31] focused on risk, adaptation, mitigation, and resilience for reducing community vulnerability and improving the resilience of landscapes.

Schoennagel et al. [17] argue for the need to develop adaptive resilience to address wildfires and reduce future vulnerability. Although framed in the language of resilience, the pivotal aspects of their approach are familiar tweaks to the *WoF* paradigm: Fuels reduction and managing more wild and prescribed fires. Although they do make a general recommendation for planning residential development to withstand inevitable wildfire, they provide little detail or empirical evidence to support the recommendation.

Olson et al. [13] proposed a fire resilience paradigm based on learning to live with fire and on the comanagement of risk among different actors to create a sustainable approach to wildland fire management. In this paradigm, fire-resilient communities require resistant structures (building codes, use of more fire-resistant building materials, buffer zones), restricting the places where embers can enter and ignite structures, establishing safe zones to protect people in the community, and identifying evacuation routes. In addition, lands involving communities should be treated to make them fire-resilient as well. Once communities and adjacent lands have been made fire-resilient, the challenge is to maintain this pattern over time [13]. While providing more detail on important actions outside of those that could be effective on fire behavior, including the notion of co-management, the proposal nevertheless maintains an outsider or top-down approach to the problem with no clear idea of governance change.

In addition, although they are framed around the notion of resilience, the different proposals land on distinct goals and actions and provide no clear sense of what they mean by resilience; recent scholars have highlighted how the focus on resilience can be problematic as it is an elastic concept and thus an empty signifier that can be used with different meanings to justify very diverse goals [68−70].

13.2.7   The road ahead

Most of the suggestions for how to improve wildfire outcomes tend to propose only small modifications of the existent theory or practice or provide general goals with few specifics or a road map on how to achieve that goal. The more specific suggestions about how to improve future wildfire outcomes tend to lay at the ends of a continuum, ranging from general concepts (e.g., the dual role of fire, IFM) to specific activities (e.g., strategic fuels management) that still are fairly narrowly focused on modifying potential fire behavior to facilitate suppression practices, an approach well within the *WoF* paradigm.

Others propose more general frameworks (e.g., cohesive strategy, socioecological approaches) that begin to allow for a broader thinking, but the balance of the focus remains on the biophysical aspects of fire behavior modification through fuel management, while the social side is absent or tends toward very general recommendations with little empirical grounding of what is needed. With specific suggestions on the ecological side and only generalizations on the social, it is not surprising that the default focus will remain on actions that the current *WoF* paradigm knows how to address well.

A paradigm shift requires a thorough rethinking of practices that is built on an understanding of fire that moves beyond the fire management community and individual scientific disciplines, to assess alternative definitions of how the wildfire problem is created, perceived, and addressed. The growing recognition that fire exclusion can exacerbate rather than improve the long-term risk and is increasingly ineffective at decreasing negative outcomes, particularly for EWEs, has led to much discussion about the need to reconsider the current paradigm. The existing proposals and ideas continue to focus on the symptoms of a fire occurrence, trying to reduce the likelihood and intensity of wildfires in general, rather than addressing the conditions that impact outcomes whether or not the wildfire occurs. By accepting that, by definition, an EWE overwhelms fire response, and focusing on what is needed to improve outcomes for an

EWE, we can shift the paradigm from a *WoF* to one of learning how to live, indeed thrive, with the natural process of fire. Focusing on EWEs is a useful starting point to develop a new paradigm as, although they represent a very low percentage of fires, they are the evidence of the *"wildfire iceberg"* [71] that the current paradigm cannot solve, and represent the highest threat to society.

In addition, fire science has been performed in *"isolated silos,"* including institutions, organizational structures, and research foci; these *"silos"* promote research, management, and policy actions that tend to only focus on targeted aspects of the wildfire problem [31], whereas a new wildfire paradigm requires a more holistic view (i.e., a 360 degrees approach). Ultimately, effectiveness in wildfire management will require putting into practice interdisciplinary contributions [72].

Wildfires are recognized as a *"wicked problem"* [18,31,73], and EWEs are at the top of the scale. To understand this problem, it is necessary to break it into parts and understand each of them from an interdisciplinary and comprehensive way. We argue that the best way is to identify key actions needed at the three distinct temporal parts of the wildfire cycle: before fire outbreak, during the fire, and after the fire extinction. This is the baseline of our vision of how to *"thrive with fire"* to create a new paradigm in wildfire management: Shared Wildfire Governance (*SwG*).

13.3 The Shared Wildfire Governance paradigm and framework

The *SwG* paradigm is based on two main principles: (1) Eliminating fire in many ecosystems is not possible or desirable. While it is crucial to reduce wildfire risk, it is equally crucial to allow for the benefits of fire that stimulates vegetation regeneration, contributes to a diversity of vegetation types, provides critical species habitat, and sustains nutrient cycling; (2) Throughout the wildfire life cycle, outcomes reflect interactions between both natural and social elements, making necessary the use of a socioecological approach in improving future wildfire outcomes.

The general objectives of *SwG* paradigm are to (1) empower community engagement in fire adaptation processes; (2) enhance ecological benefits of fire; (3) enhance civilian and firefighter safety; (4) address human causes and motives of wildfires; (5) reduce exposure and vulnerability of humans, structures, and landscapes; (6) reduce potential losses and damage; (7) enhance effective fire response; and (8) enhance effective recovery.

13.3.1 The arguments of Shared Wildfire Governance paradigm and framework

The *SwG* framework is focused on identifying the range of key actions necessary to *"thrive with fire"*. It incorporates the existing *WoF* focus on response (addressing the symptom) into a more holistic framework that better addresses the underlying causes of poor wildfire outcomes. The starting point is how we perceive the nature of wildfires, including EWEs. Identifying a range of actors and actions that need to be considered and the objective of those actions provides a framework that has cross-cultural value. While fire and the potential threats it represents are the same across the world, specific dynamics and vulnerabilities of the socioecological systems affected by wildfires will vary.

The *SwG* framework is supported by the following arguments:

(1) **Distinction of stages in wildfire process to orient action.** The terms *Before fire outbreak*, *During the fire*, and *After the fire extinction* are easily understandable as they have an intrinsic meaning in themselves; this translates the situations of presence/nonpresence of the wildfire, requiring different levels of attention, action, and promptness. We purposely avoid the use of the terms prevention and mitigation as global umbrellas as they assume different meaning in distinctive cultural, disciplinary, and operational contexts. The *SwG* framework also captures the full range of areas where actions are necessary to improve wildfire outcome, both supporting the benefits of fire and reducing future losses.

(2) **Societal engagement to cope with EWEs.** The usual conviction, supported by the *WoF* paradigm, that firefighters can protect people and assets in every condition is still alive in many countries but has proven to be ineffective in many situations and can contribute to increased risk, not just in terms of fuels buildup but also in terms of public safety. Firefighters cannot always assist all people under threat, frequently people are on their own, sometimes with little knowledge of the wildfire hazard characteristics or potential protective actions they could take. Therefore, a key goal is helping populations at risk develop a sustained capacity to *"thrive with fire,"* by increasing the involvement of all parts in reducing hazardous conditions and developing effective mechanisms to prepare for, cope with, and recover from fire impact [74].

(3) **Collaborative work between agencies and citizens using a shared governance approach.** It requires empowering all members of society potentially affected by wildfire to have a voice in the decision-making process; the main principles are partnership, equity, accountability, and ownership [75]. Effective fire management is essentially a social process that requires knowledge, choices, agreements, collaboration, and prioritization. Research specific to wildfires has shown that community partnerships between agencies and citizens are critical for identifying and implementing place-based solutions that effectively reduce wildfire vulnerability [76].

(4) **Cross-cultural and scalar applicability.** The *SwG* framework is based on the general processes that influence how wildfires interact with human systems that at a high level are independent of the cultural and socioeconomic context where they occur. How the processes will play out in different settings will vary; some dynamics may be more relevant than others in a given location. Specific priority and focus of actions will need to be adapted to the context and the hazard characteristics as well as to the resources, skills, and objectives of the area. The framework is also designed to be used at multiple scales, to develop specific plans in a community or inform institutional processes at a regional/national level.

(5) **Connectivity and synergies between actions.** Abating the wildfire *"wicked problem"* cannot be obtained by a single measure; each action addresses specific objectives and also influences the efficiency of other measures. The identification of objectives for an action can help identify the best option for specific circumstances and potential synergies between actions.

(6) **Monitoring efficacy.** Assessing outcomes of actions is critical to identifying potential trade-offs of different actions and enable adaptive response when actions are not as effective as expected.

(7) **Abating *"silos"*.** Integrating citizens, fire agencies, decision-makers, and scientists is critical to eliminating narrow and potentially counterproductive actions, which are more likely to occur when taken by *"isolated silos."* Developing a fully interdisciplinary understanding of the phenomenon and its interactions with the social context is critical for discovering better ways to adapt to the wildfire scenario with fewer costs and higher benefits.

(8) **Need for good data to inform decisions.** The critical thinking that is needed to inform a new approach requires a higher level of knowledge and of accurate information. EWEs are complex phenomena that require more and better information in terms of spatial and temporal accuracy to reduce their threat. Wildfire is determined by many drivers of natural and human origin for which there is little accurate available data at the moment; thus there is a poor foundation for the development of alternative approaches [77]. Despite all the technological evolution and the volumes of data available, it does not mean that the most accurate data in terms of the adequate scale and timing are available. Particularly important in determining how to most effectively adapt is understanding the trends and patterns of EWEs across the world and the maximum values of the physical properties (e.g., fireline intensity, rate of spread) a wildfire can reach.

13.3.2 The description of Shared Wildfire Governance framework

The framework that supports *SwG* paradigm (Fig. 13.1 and Table 13.1) is built on four levels, organized around the three successive temporal stages (first level): *Before fire outbreak, During the fire, After the fire extinction.* In each stage, fields of action (second level) are identified along with specific actions (third level). For each action, specific objectives are clearly identified (fourth level). Considering objectives as an integral part of the framework reinforces the need that any action must be understood in its purpose to be effective. Providing detail on the range of actions and objectives that need to be considered can help shift both thinking and actions from current default activities of the *WoF* paradigm, by helping people understand the purpose and need for specific actions and provide additional motivation to transform intentions into actions [78].

In the stage *Before fire outbreak,* seven fields of action are identified: *Ignition avoidance* (A1); *Wildfire hazard* (A2); *Leverage the benefits of fire* (A3); *Assets vulnerability* (A4); *Human vulnerability* (A5); *Emergency planning* (A6); and *Outreach and Coordination* (A7).

Most of the known causes of wildfires have an anthropic origin. Thus, the actions in *Ignition avoidance* (A1) cover four main domains: (1) Development of awareness, (2) surveillance, (3) selective accessibility restrictions, and (4) causes and motives. The purposes are to develop programs and policies to reduce unwanted ignitions (resulting from negligence or accidents or criminal behaviors), decrease risky and unwise behaviors, and enhance the capacity to identify causes and motives.

In *Wildfire hazard* (A2) field of action, two main domains are considered: (1) Identify key factors contributing to local wildfire hazard and (2) forest and fuel management. Considering that the term hazard is currently used with different meanings in wildfire literature, in *SwG* framework, hazard is considered as a *"process, phenomena or human activity that may cause loss of life, injury or other health impacts, property damage, social and economic disruption or environmental degradation damage"* [79].

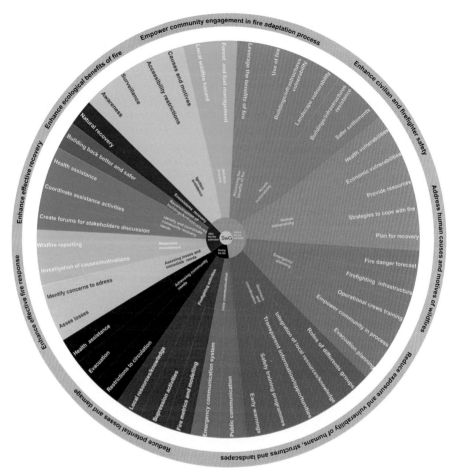

Figure 13.1 Overview of *SwG* framework where the four levels are represented. The shape of the figure puts in evidence a continuum in the life cycle of the wildfire phenomenon split in three different stages (*Before, During* and *After fire*). The majority of possible actions can be taken in *Before fire* outbreak stage to reduce the probability of unwanted ignitions and to reduce losses and damage. The number of possible actions and of their interplay well depicts the complexity of the phenomenon and, conversely, confirms the incapacity of the *During the fire* stage to act on such complexity. In the figure only the general objectives are represented. For detailed information on the specific objectives see Table 13.1.

As fire intensity level is independent of likelihood for operational use, it is critical to describe wildfire hazard in terms of potential intensity rather than likelihood. Better understanding the hazard helps assess how the physical properties of fire can threaten people and assets. Improving the ability to model potential fire hazard is also important to better orient forest and fuel management activities undertaken to reduce the fuel load in order to modify fire behavior and optimize future firefighting interventions (e.g., the Z.A.L. *"Zones d'Appui à la Lutte"* in Corsica, France) [80].

Table 13.1 The *SwG* framework.

Fields of action	Actions	Specific objectives
A. BEFORE THE FIRE OUTBREAK		
A1. Ignition avoidance	**1) Development of awareness:** i) Establishment of education programs/sensitization campaigns about dangerous behaviors conducive to negligent or accidental fires ii) awareness based on past fires experience	Reduce unwanted fire outbreaks
	2) Surveillance: i) Dissuasive purposes to avoid fire outbreaks caused by risky behaviors	Contain risky and unwise behaviors
	3) Selective accessibility restrictions: i) Access control to forest and wildland areas during high-risk periods	Reduce fire outbreaks
	4) Causes and motives: i) Identify and monitor the sources of accidental fire outbreaks (arching in electric lines; lack of maintenance of the transmission line right-of-way; rail-road; vehicles; works; weapon; explosives); ii) attribute responsibility to the actors (agencies, public administration, agricultural syndicates) to promote good practices to avoid unwanted ignitions; iii) investigate the causes and motives of intentional fire outbreaks; iv) provide appropriate law enforcement resources to monitor and investigate suspected arson	Decrease exposure to unexpected wildfires Enhance the capacity to identify ignition sources and develop policies and programs to address causes and motivations Reduce the incidence of criminal behaviors Reduce the occurrence of unwanted fire outbreaks due to accident and negligence
A2. Wildfire hazard	**1) Identify key factors contributing to local wildfire hazard:** i) Understand wildfire process and the physical properties that create threat to human systems; ii) assess current wildfire hazard; iii) model potential fire hazard	Understand how to modify fire behavior Identify strategies to reduce hazard, decrease losses, and enhance safety
	2) Forest and fuel management: i) Create land use mosaic; ii) modification of forest structure (thinning, pruning, reducing seedlings density, elimination of fuel ladders, perimetral isolation) at territorial and stand scales; iii) grazing; iv) wise use of traditional burning patterns; v) develop agriculture of production and of conservation; vi) land use conversion	Reduce fuel load and promote biodiversity Modify fire behavior Optimize firefighting intervention

Table 13.1 The *SwG* framework.—cont'd

Fields of action	Actions	Specific objectives
A3. Leverage the benefits of fire	**1) Assess the benefits of fire using environmental and social parameters:** i) Understand fire as an ecological and land management tool; ii) develop plans for fire use given environmental and social conditions	Promote environmental services and ecologically healthy landscapes
	2) Use of fire: i) Prescribed burning; ii) work with those that have know-how and need to use fire as a land management (e.g., pasture renovation) and conservation tool; iii) enhance residual fire use skills and traditional fire knowledge (TFK)	Enhance the use of fire as an ecological and land management tool in the best ecological and safest conditions Recover TFK and/or promote fire use skills
A4. Assets vulnerability	**1) Evaluate buildings and infrastructures vulnerability:** i) Understand buildings' loss of tenability conditions; ii) identify key local structural practices to improve (e.g., roof type, electrical lines); iii) adopt differential measures to cope with different categories of fires	Reduce losses Enhance people safety
	2) Evaluate and reduce adjacent landscape vulnerability: i) Mapping accurate information and georeferenced activities to clarify vulnerabilities; ii) create a system of firebreaks, opened areas, thinned forest areas, and fuel managed areas close to settlements	Reduce assets vulnerability to losses and damage
	3) Increase buildings and infrastructures resistance: i) Implement building codes; ii) increase fire resistance of structures, particularly to ember attack (e.g., use resistant materials; install sprinkling systems); iii) renovations and retrofits for existing buildings; iv) improve architectural and engineering design to decrease potential for adverse impacts; v) create defensible space and keep clean the home ignition zone taking into account different paths of simulated fire; vi) smart gardening (at mesoscale level)	Reduce assets vulnerability to losses and damage Enhance people safety

Continued

Table 13.1 The *SwG* framework.—cont'd

Fields of action	Actions	Specific objectives
	4) Creation of safer settlements: i) Improve safety conditions in settlements to reduce the threat of fire and facilitate appropriate protective action (e.g., evacuation, shelter in place); ii) land use plans appropriate to local conditions that help decrease exposure; iii) ensure planning at three different scales: At landscape/territory scale; at mesoscale scale (urbanization), and at microscale level (single house)	Enhance people safety Decrease exposure to unexpected wildfires
A5. Human vulnerability	**1) Identify health vulnerabilities:** i) Identify physical and mental disabilities	Enhance physical and mental capacities
	2) Identify individual and business economic vulnerabilities: i) Develop plans to reduce vulnerability, improve response, and promote rapid recovery after a fire; ii) business continuity planning	Decrease potential economic impacts Enhance recovery
	3) Provide resources: i) Information and financial resources to facilitate adoption of appropriate protective measures	Enhance awareness and skills and provide resources to reduce losses and damage
	4) Identify and plan for appropriate strategies to cope with actual fire: i) Evacuation; ii) passive shelter; iii) active defense; iv) prepare response plan (including storage of family documents and photos, goods with high personal value in a safer location or/and easy to move); v) create safety zones for firefighters	Enhance safety Enhance an efficient response
	5) Plan for long-term community and individual recovery: i) Work across multiple scales and stakeholder groups to identify roles and responsibilities, likely needs, and long-term desired outcomes; ii) insurance	Increase resilience of human communities Reduce economic losses Improve recovery efficiency

Table 13.1 The *SwG* framework.—cont'd

Fields of action	Actions	Specific objectives
A6. Emergency planning	**1) Fire danger forecast**: i) Ignition forecast, ii) weather conditions	Inform fire organization readiness levels Provide information to activate public warnings Ensure rapid and tactic deployment of firefighting resources
	2) Firefighting infrastructure: i) Identification of roles/responsibilities of different organizations; ii) identify minimum needed equipment/personnel; iii) identify and map critical infrastructure (e.g., fixed water points, road network); iv) strategic predeployment of resources; v) adopt a cost—benefit approach for prioritizing decisions	Enhance efficiency of firefighting resources
	3) Operational crews training: i) Coordination of human and material resources; ii) simulation exercises	Enhance crew's performance
	4) Empower individuals and communities/stakeholders in planning process: i) Work with stakeholder groups to identify local concerns and community prioritization of values to protect; ii) engage in strategic scenario planning with local and stakeholders (simulate and plan for EWEs to predefine decisions to be taken in each context); iii) identify roles/responsibilities of different organizations; iv) identify optimal strategies/tactics for different conditions	Engage community in fire adaptation process Enhance firefighters responding capacity to intervene in a fire and avoid it becomes an EWE Minimize potential impact on communities
	5) Evacuation planning: i) Identification of roles/responsibilities of different organizations/individuals; ii) identification of protective actions and potential considerations (evacuation and sheltering options, evacuation routes); iii) development of preferred and backup response plans	Increase response capacity Enhance safety of citizens and responders

Continued

Table 13.1 The *SwG* framework.—cont'd

Fields of action	Actions	Specific objectives
A7. Outreach and coordination	**1) Identify roles and responsibilities of different individuals, groups/ organizations for all activities and at all time points:** i) Ensure coordination, reduce overlap, identify gaps; ii) reassess and reconsider existing structures to identify potential synergies and alternative approaches to work	Build relationships/ trust Creation of synergies among actions Enhance the effectiveness of the governance model by adapting or changing it Increase shared fire governance
	2) Integration of local resources/ knowledge: i) Enhance the collaboration among stakeholders; ii) develop shared understanding of key needs/desired outcomes	Engage community in fire adaptation process
	3) Provide transparent information and interactive opportunities to discuss: i) Local fire ecology; ii) fire behavior; iii) self-protection measures, home preparation, and evacuation; iv) fire response	Empower community to understand critical local fire needs and considerations Identify, assess, and implement appropriate protective actions
	4) Formal safety training programs for civilians and firefighters: i) Mental preparedness; ii) development of skills	Empower community members to effectively undertake protective actions Enhance personal safety of civilians and firefighters
	5) Early warnings: i) Red Flag days, alert to high hazard, stay tuned to communication channels for warning messages	Optimize the appropriate protective action (evacuation, shelter)
B. DURING THE FIRE		
B1. Crisis communication	**1) Public communication/Evacuation warning:** i) Inform about the category of expected wildfires; ii) detailed information on weather changes, and fire spreading and behavior; iii) recommended protective actions; iv) information about traffic conditions	Enhance safety Disseminate real and objective information on fire events in progress Influence people attitudes and behaviors

Table 13.1 The *SwG* framework.—cont'd

Fields of action	Actions	Specific objectives
	2) Operational emergency communication system: i) Creation and maintenance of a communication network	Ensure and enhance efficiency of communication network among operational activities Establish alternative communication networks to maintain the communication flow
B2. Firefighting activities	**1) Fire weather and behavior metrics and modeling:** i) Fire analysis to predict expected fire behavior and potential evolution of fire; ii) provide operational staff with current information to inform decision	Integrate fire analysts in the firefighting activities Inform the operational room and all operative crews about the physical characteristics of fire in different locations Increase efficiency of suppression operations
	2) Suppression activities: i) Ensure appropriate human and material resources; ii) determine appropriate strategies and tactics (direct attack, indirect attack, suppression fire); iii) identify key logistical needs (e.g., water points, access, lookout towers); iv) ensure proper mop-up to reduce rekindle	Contain or modify wildfire behavior
	3) Integration of local resources/ knowledge: i) Wherever possible, integrate in crews rural people having a good traditional knowledge of fire	Ensure local resources and knowledge of factors that could influence the fire are taken into account Ensure local views of key values at risk are incorporated in response.
B3. Addressing community needs	**1) Restrictions to circulation:** i) Minimize traffic congestion; ii) control access to evacuated areas	Enhance safety and allow a better circulation of operational vehicles

Continued

Table 13.1 The *SwG* framework.—cont'd

Fields of action	Actions	Specific objectives
	2) Evacuation: i) Timely, clear notifications; ii) assistance for those with evacuation impediments (transportation, health, etc.); iii) support those who are protecting homes	Enhance evacuation safety and protection of assets
	3) Health assistance: i) Managing respiratory issues, and trauma; ii) assistance of injured people	Assistance of injured crews and people Act on acute physical and mental health outcomes
C. AFTER THE FIRE EXTINCTION		
C1. Assessing losses and immediate needs	**1) Establishment and implementation of protocols to assess losses:** i) Establish criteria for the compensation of damage in a safer and better organized space; ii) orientate the money to better preventions; iii) assess the distortions of the compensation damages in the future fires	Make transparent the process of recovery with private and public financial aid
	2) Identify immediate concerns to address: i) Landscape damage and potential impacts (e.g., floods, soil erosion, landslides); ii) housing and communication needs; iii) potential health issues; iv) damage to critical infrastructure	Minimize immediate impacts to community Reestablish normal life conditions
C2. Response assessment	**1) Investigation of causes/motivations**: i) Improve the methodology to identify ignition points (e.g., method of physical evidences) and assess causes and motives	Adapt wildfire management policies and prioritize actions
	2) Wildfire reporting: i) For lessons learning activities related with mitigation and suppression activities	Improve wildfire management
C3. Identify and coordinate community recovery needs	**1) Create forums for stakeholder discussion:** i) Identify key needs/desired recovery outcomes; ii) identify potential equality/social justice issues that may need to be addressed	Engage community in fire adaptation/ transformation process
	2) Coordinate assistance activities: i) Manage donations; ii) animal care; iii) postfire housing; iv) volunteers; v) funding and grant support; vi) case management	Ensure effective and legal use of resources

Table 13.1 The *SwG* framework.—cont'd

Fields of action	Actions	Specific objectives
	3) Health assistance: i) Medical assistance and trauma care	Improve mental and physical health outcomes
C4. Reconstruction of buildings and infrastructure	**1) Building back better and safer**: i) Rebuild to standards likely to improve future fire outcomes; ii) facilitate the reconstruction with orientated and simplified administrative procedures; iii) facilitate access to credit; iv) create dedicated line of credit	Reestablish normal life conditions Decrease potential for negative outcomes in future fire events
C5. Ecosystem recovery	**1) Permit natural recovery based on resilience:** i) Evaluate the extent and intensity of salvage logging to be coherent with ecological exigencies of species; ii) allow natural recovery and recognize the temporal scales involved with ecosystem evolution; iii) active reseeding and replanting should be conducted only under limited conditions and only when necessary; iv) preserve capabilities of species to naturally regenerate; v) avoid structural postfire restoration	Enhance recovery and biodiversity Decrease potential for negative outcomes in future fire events

Leverage the benefits of fire (A3) has two main domains of action: (1) Assessing benefits of fire using environmental and social parameters and (2) use of fire. Together these two fields of actions have the goal of better understanding and working with the ecological nature of fire by understanding how prescribed and traditional use of fire can both modify the hazard and support desired ecological processes such as creating landscape mosaics, modification of forest structure (thinning, pruning, reducing seedlings density, elimination of fuel ladders, perimetral isolation) at territorial and stand scales, pastoral practices, and wise use of traditional burning practices.

In *Assets vulnerability* (A4), the main four actions are: (1) Evaluate buildings and infrastructures vulnerability in different scenarios of fire intensity; (2) evaluate and reduce adjacent landscape vulnerability; (3) increase buildings and infrastructure resistance; and (4) creation of safer settlements. The main goal of these activities is to reduce the vulnerability of the built environment to decrease potential damage and losses and enhance safety. We highlight the need for appropriate land use plans to reduce exposure, rather than the often recommended notion of restricting residential development in high-risk areas. Although the latter may be superficially appealing it is highly problematic and often unrealistic to put into practice due to a number of considerations, including: (1) The area available for needed housing is reduced; (2) in many locations areas at risk can be fairly extensive; (3) preventing people from living in an area likely only displaces them to areas prone to other equally serious hazards; (4) the measure can have major equity considerations, as the wealthy will always manage to build in high-risk areas that are desirable (amenity risk!!), while in other places the high-risk areas are the only ones those with less income can afford to live. It is important that assessing assets vulnerability occurs at three different scales: Landscape, mesoscale (local government or community), and microscale (the single house or building) (Caballero, personal communication).

In the field of *Human vulnerability* (A5), areas to address include: (1) Identifying health vulnerabilities; (2) Identifying individual and business economic vulnerabilities; (3) providing resources to facilitate adoption of appropriate protective measures; (4) identifying and planning for appropriate strategies to cope with actual fire; and (5) planning for long-term community and individual recovery. The purpose of the actions is to identify and reduce key human vulnerabilities by enhancing capacities and skills to take appropriate protective action to reduce potential losses; enhance safety from a number of angles beyond evacuation, including for firefighters; improve recovery; and reduce the long-term vulnerability of humans, structures, and landscapes.

Emergency planning (A6) covers activities developed by both emergency responders and citizens to prepare for actions during a wildfire. Key areas of focus are: (1) Fire danger forecast; (2) firefighting infrastructure; (3) operational crews training; (4) empowering individuals and community/stakeholders in the planning process; and (5) evacuation planning. All stakeholders working together before an event to identify key concerns and reach agreement on and plan for appropriate response activities (for both expected and worst case scenarios), including likely response to the fire itself, is a key part of a shift toward a *SwG* approach. Evacuation planning comprises identification of the range of appropriate protective actions and considerations (traffic routes, evacuation and sheltering options, identifying the preferred course of action and decision trigger points for taking action) and includes development of alternative plans if the preferred response is not possible. The goal of this field of action is to enhance the capacity for safe and efficient organizational and community response.

The last field of action in the stage *Before the fire outbreak* is *Outreach and coordination* (A7). In many ways this field is the most crucial element of the *SwG*, as its activities are critical to a shared governance model. The main actions are: (1) Identify roles and responsibilities of different groups/organizations to facilitate citizens and organizations working together to develop solutions and enhance synergies; (2) integration of local resources/knowledge to understand key needs and desired outcomes in relation to wildfires; (3) provide transparent information and interactive opportunities to discuss crucial subjects (e.g., fire behavior, likely fire response, self-protection measures); (4) formal safety training programs for civilians and firefighters; and (5) early warnings. The purpose of this field of action is to build shared governance, empower communities, and optimize the appropriate protective actions to enhance the efficiency.

In *During the fire* stage, three fields of action are identified: *Crisis communication* (B1); *Firefighting activities* (B2); and *Addressing community needs* (B3). In *Crisis communication* (B1) we consider public communication/evacuation warning activities such as providing information to the public about fire behavior and spread, as well as expected weather changes, particularly changes of wind direction and speed. In many recent wildfires most of the fatalities occurred after a sudden change of wind. This field would also involve proactive information about traffic conditions and evacuation needs, following well-identified best practices (Chapter 8). Another aspect of crisis communication is ensuring the operational emergency communication system is robust enough to ensure coordination between actors in the field and operational command. Effective communication is critical in increasing safety of crews and civilians and effective fire response activities. Although effective in the majority of situations, in more complex fires, particularly an EWE, the Incident Command System can be challenged to adapt to the complexity of the number of actors and needed activities. In that case, an emerging problem is the excess of crossed communications that creates the so called *"infotoxica-tion"* [81]. This adds to the many and most evident fragilities of the WoF paradigm.

Firefighting activities (B2) integrates three main actions. First, fire weather and behavior metrics and modeling to predict expected fire behavior and potential evolution of fire to provide operational staff with current information to inform decisions about the most appropriate fire response strategies and tactics, including fire use. Second, suppression activities which include ensuring human and material resources, determining strategies and tactics, identifying logistical needs, and reducing rekindles by proper mop-up. Third, integration of local resources and knowledge to improve the capacity to cope with fire. The goals of this field of action are the integration of fire analysts in firefighting activities, informing about the physical characteristics of fire, modification of fire behavior, and ensuring that local resources and knowledge are considered.

The third field of action is *Addressing community needs* (B3) with three actions: (1) Restrictions to circulation, (2) evacuation, and (3) health assistance. This area, critical as response to the fire, is supporting safety of those affected by the fire and includes providing timely and clear notifications of fire location and evacuation concerns, traffic management, assistance for those with health issues, and support for a range of evacuation related needs, including shelters for evacuees and assistance for those with evacuation impediments or who cannot evacuate.

In *After the fire extinction* stage the areas of intervention are *Assess losses and immediate needs* (C1); *Response assessment* (C2); *Identify and coordinate community recovery needs* (C3); *Reconstruction of buildings and infrastructures* (C4); and *Ecosystem recovery* (C5). It is important to note that better outcomes from the various activities during this stage are more likely to occur when they have been taken into account in discussions and planning during the *Before the fire outbreak* period.

Assessing losses and immediate needs (C1) ideally is done with protocols developed before the fire. Areas to assess include immediate environmental concerns (e.g., increased soil erosion, flood and landslide potential), damage to critical infrastructure, housing and communication needs, and potential health issues. To facilitate the evaluation of costs and the recovery process, the establishment and implementation of protocols supported by geospatial information is particularly important. The purpose is to minimize immediate impacts to community, reestablish normal life conditions, and make transparent the process of recovery with private and public financial aid.

Response assessment (C2) covers the investigation of causes and motives (1) to determine the origin of fire outbreaks, and wildfire reporting (2) to learn lessons, inform fire management policies, and prioritize actions to improve future wildfire management. This field of action focuses on analyzing suppression response activities, strategies and tactics used, coordination, and communication, as well as fatalities, losses, and other significant impacts.

Identify and coordinate community recovery needs (C3) is an action often overlooked in the current paradigm but is critical in learning how to better *"thrive with fire."* Here a key action is to (1) create forums for stakeholder discussion to identify key needs and desired recovery outcomes, and identify potential equality or social justice issues that may need to be considered in the recovery process. Two other action areas in this field refer to (2) coordinating assistance activities (e.g., manage donations, funding and grant support, case management) and (3) providing health assistance. The purpose of these efforts is to engage the community in the fire adaptation process, ensure effective use of resources, and address mental and physical health outcomes.

Reconstruction of assets and infrastructures (C4) is another action area where planning during the *Before the fire outbreak* period can play a critical role in better long-term outcomes. A goal here is to *"build back better"* in ways that reduce potential for future losses. Other important actions in enhancing reconstruction and recovery are developing efficient administrative procedures and facilitate the access to a line of credit. Their purpose is to foster a recovery that decreases potential for negative outcomes in future events, while allowing for the reestablishment of normal life conditions as rapidly as possible.

Finally, *Ecosystem recovery* (C5) reflects actions to address environmental effects of the fire, in a way that takes into account desired future conditions and allows for the temporal scales involved with ecosystem evolution. This may mean addressing immediate environmental risks (such as flood and water quality impacts), as well as actions and planning that will enhance natural recovery and maintenance of biodiversity, while decreasing potential for negative outcomes in future events.

13.3.3 The strengths of Shared Wildfire Governance paradigm

The *WoF* is based on protocols to respond to wildfires that have proven effective on normal fires but collapse when the fire behavior characteristics become more extreme. This may have been less a concern when human populations were small, but as this approach no longer meets the current societal and environmental conditions, there is a need for creative thinking to find ways to look at the wildfire problem from a fresh perspective to develop alternative and perhaps unorthodox approaches and solutions.

The *SwG* represents a potential turning point in wildfire management (Table 13.2). The main differences between *WoF* and *SwG* paradigms is in the way they see the wildfire problem (including EWEs), how the social context is considered, the relation between citizens and organizations related with fires, the relation between practice and science, and the holistic approach to addressing the wildfire challenge. As it can be

Table 13.2 Comparative analysis of *WoF* and *SwG* paradigms of wildfire management.

WoF	SwG
Fire always a threat. Fire use criminalized. Exclude any fire role	Acceptance of the dual role of fire
Self-perpetuating. Reactive approach centered on addressing immediate symptoms but ignoring the roots of the problem.	Adaptive, improving outcomes over time. Addresses underlying causes. Proactive and long-term approach to empower citizens and fire services to plan and work together to adapt and transform human—wildfire interactions.
Primary focus on one point in temporal cycle – how to make response more effective	Focuses on actions and interactions across temporal cycle – how to have better outcomes at all points in time
Values that drive decisions are primarily on protecting people and property, with a secondary emphasis on protecting environmental values	Values that drive decisions are improving the beneficial outcomes of living with fire, while decreasing negative outcomes for a range of human values (life safety, property, environmental, cultural, etc.)
Approach to all fires with the same standard response, irrelevant of fire behavior and potential impacts. Unable to cope with fires that exceed control capacity	Recognizes that the same response is not appropriate for different categories of fires, particularly an EWE. Decreases likelihood of and prepares to respond adaptively to EWEs minimizing losses, injured people, and fatalities
Focus on isolated actions. Organizational activities and individual initiatives with limited interactions	Focus on shared governance. Building trust between citizens and organizations
Top-down approach: Isolated from citizens	Interactive approach: Takes advantage of both bottom-up and top-down strengths
Normative approach with rigid protocols. Difficulty in adapting to extreme conditions and integrating lessons learned. Can be efficacious in normal fires, but collapses in case of more complex emergencies.	A more flexible approach that can adjust to different event complexities. Adaptive approach: Continuous process of learning/research-adaptation/ transformation action
Paternalistic approach. Organizations value their competence and knowledge over that of citizens. Fosters dependence on specific organizations	Shared governance approach. Values diverse interests and competencies of all parties affected by fire. Fosters self-reliance and mutual assistance processes
Tries to reduce uncertainty through standardized system of response	Reduces uncertainty by empowering stakeholders to work together to share knowledge, plan scenarios, and reach agreement of how to respond if a fire occurs

Continued

Table 13.2 Comparative analysis of *WoF* and *SwG* paradigms of wildfire management.—cont'd

WoF	SwG
Imposition of actions with limited consideration of citizen and community knowledge, skills, and capacities to cope with wildfires	Identifies skills, competences, and resources from individuals and communities and integrates them at all stages of addressing wildfire risk
Focus on activities and equipment that have limited use for other purposes than fire response	Integrative approach allows for synergies between actions to reduce the negative impacts of the fire and other societal needs
Focus on short-term and very narrow risk management approach	Risk-benefit governance approach that acknowledges and works to address the full diversity of potential risks (e.g., temporal, spatial, social/ecological)
Limited trust between citizens and organizations	Building trust between citizens and organizations
Social issues, namely the influence of social context, poorly understood and considered	Social issues as a crucial component. The social context determines diverse types of actions
Integration of scientific evidence on a silo's basis that overlooks complexity of system	Science based action that takes into account a broad range of system components and interactions
Weak transparency and evaluation process based on simple metrics and not on accuracy and effectiveness of interventions	Transparent evaluation based on the outcomes of the collaborative work using a system of monitoring effectiveness with very clear parameters
Criteria to evaluate cost and efficiency of the intervention self-reinforce the individualistic approach of the organizations	Criteria to evaluate cost and efficiency based on the capacity of institutions to work together
Metrics of progress: Number of fires, burned area, fire size	Metrics of progress: Full range of costs and potential benefits

seen in Fig. 13.2, in the initial *WoF* model the vast majority of activities focused on the *During the fire* stage. In the past decades minor adjustments have added more actions to *Before* and to a lesser extent to *After fire extinction*, but the majority of focus remains on response. In comparison, in the *SwG* model *During fire* actions are a relatively minor portion of an effective effort to decrease negative societal impacts of wildfires.

The *SwG* paradigm proposes a more robust and long-term perspective to address the increasing societal wildfire impacts, by addressing not just the symptoms but also the range of societal elements that contribute to negative outcomes when humans interact

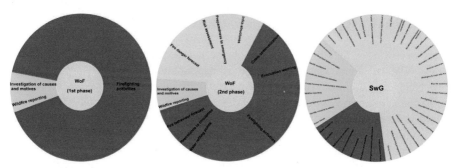

Figure 13.2 Comparison between *WoF* and *SwG* paradigms' actions. In red, activities developed *During the fire* stage are represented; in green, the actions taken *Before the fire outbreak*; and in blue, the actions in *After the fire extinction* stage. The relatively higher number and importance of activities taken in the *Before the fire outbreak* stage well reflects the effort of *SwG* to cope with EWEs, which exceed any capacity of control, the main focus of the *WoF* approach.

with the natural process of fire. Moving beyond the current focus on preventing fires while improving our ability to fight fires through improved equipment, better understanding of fire behavior, and fuels treatments, *SwG* focuses on building a shared governance approach to determining how to best *"thrive with fire"* in a given location. This paradigm is better able to address the fact that EWEs overwhelm current control response capacity as it values all parties working together to plan and identify: Key desired outcomes, the range of preparedness actions needed before a fire, and how different fires, including an EWE, will be responded to should they occur. We believe that overtime such an approach means that humans cannot just *"learn to live with fire"* but can *"thrive with fire"*.

The current approach also *"siloes"* the thinking about wildfire in multiple ways (fire response organizations vs. community, *during fire* as separate from *before* and *after* actions, and between scientific disciplines). *SwG* supports learning, action, and evaluation in a continuous and interactive dialog by those potentially affected by wildfires and highlights the necessity of all entities, including fire agencies, to work together across all three temporal stages (i.e., before, during and after the fire). Although this approach may initially feel daunting, it has more potential to reduce the uncertainty of our current situation by taking into account the social context where fires occur and taking into account the diverse values and needs of different actors (individuals, communities, private and governmental organizations etc.). The *SwG* framework emphasizes proactive strategies that combine multiple and complementary measures, each of them with an identified objective, an approach more likely to lead to better outcomes than the reactive approach of the *WoF*. Focusing on identifying the full range of ways to address the range of needs and issues posed by both damaging and beneficial fires, within the context of the natural environments and socioeconomic systems in which they occur, the *SwG* framework also holds more potential to find synergies with nonfire concerns, increasing the likelihood of developing effective and implementable solutions.

13.4   Next steps

Every decision we take has an impact on the landscape that we will inhabit in the future. What burns and what does not burn is shaping tomorrow's landscapes and fires. If we change our approach from defensive to proactive and creative, we can include the landscape of tomorrow as decision driver of today. That has implications in our actions before, during, and after a fire. Many locations around the world have already, at some level, started moving away from the primary focus on fighting fires to actions that fit more into the *SwG* framework. However, the lack of new language or detailed framework means that often positive intentions default to language and actions that are more reflective of the *WoF* paradigm than a shared governance approach to wildfire management. A brief look at the graphic in Fig. 13.2 shows how the emphasis of work shifts from primary response activities *During* a fire to activities primarily *Before* and to a lesser extent *After* the actual fire; it is a clear difference in approach. We hope that providing more details around the range of factors that need to be considered for learning how not just to respond to, but to live and thrive with fire, will increase the ability to focus more attention, energy, and resources on the activities that are more likely to lead to better fire outcomes, for both humans and the environment, over the long term.

References

[1] J.G. Pausas, J.E. Keeley, A burning story: the role of fire in the history of life, Bioscience 59 (2009) 593−601, https://doi.org/10.1525/bio.2009.59.7.10.

[2] W.H. Romme, D.G. Despain, Historical perspective on the Yellowstone fires of 1988, Bioscience 39 (1989) 695−699, https://doi.org/10.2307/1311000.

[3] S.H. Millspaugh, C. Whitlock, P.J. Bartlein, Variations in fire frequency and climate over the past 17 000 yr in central Yellowstone National Park, Geology 28 (2000) 211, https://doi.org/10.1130/0091-7613(2000)28<211:VIFFAC>2.0.CO;2.

[3a] L.L. Wallace (Ed.), After the fires: the ecology of change in Yellowstone National Park, Yale University Press, New Haven, Connecticut, US.

[4] F. Tedim, V. Leone, M. Amraoui, C. Bouillon, M. Coughlan, G. Delogu, P. Fernandes, C. Ferreira, S. McCaffrey, T. McGee, J. Parente, D. Paton, M. Pereira, L. Ribeiro, D. Viegas, G. Xanthopoulos, F. Tedim, V. Leone, M. Amraoui, C. Bouillon, M.R. Coughlan, G.M. Delogu, P.M. Fernandes, C. Ferreira, S. McCaffrey, T.K. McGee, J. Parente, D. Paton, M.G. Pereira, L.M. Ribeiro, D.X. Viegas, G. Xanthopoulos, Defining extreme wildfire events: difficulties, challenges, and impacts, Fire 1 (2018) 9, https://doi.org/10.3390/fire1010009.

[5] T. Ingalsbee, Whither the paradigm shift? Large wildland fires and the wildfire paradox offer opportunities for a new paradigm of ecological fire management, Int. J. Wildland Fire 26 (2017) 557, https://doi.org/10.1071/WF17062.

[6] J. Williams, L. Hamilton, R. Mann, M. Rounsaville, H. Leonard, O. Daniels, D. Bunnell, The mega-fire phenomenon: toward a more effective management model, A Concept Pap (2005).

[7] J. Williams, Exploring the onset of high-impact mega-fires through a forest land management prism, For. Ecol. Manage. 294 (2013) 4−10, https://doi.org/10.1016/J.FORECO.2012.06.030.

[8] M. Finney, I.C. Grenfell, C.W. McHugh, Modeling containment of large wildfires using generalized linear mixed-model analysis, For. Sci. 55 (2009) 249−255, https://doi.org/10.1093/forestscience/55.3.249.

[9] G.M. Delogu, Dalla parte del fuoco, ovvero, Il paradosso di Bambi, Il maestrale, 2013. https://books.google.pt./books?id=KfX2ngEACAAJ.

[10] ANEPC, Diretiva Operacional Nacional no 2-DECIR, Dispositivo Especial de Combate a Incêndios Rurais 2019, Lisboa, 2019, http://www.prociv.pt./bk/PROTECAOCIVIL/LEGISLACAONORMATIVOS/OUTROSNORMATIVOSDIRETIVAS/Documents/ANEPC_DON_2_DECIR_2019_www.pdf.

[11] European Commission, rescEU: EU Establishes Initial Firefighting Fleet for Next Forest Fire Season, Eur. Comm. Press Release, 2019. http://europa.eu/rapid/press-release_IP-19-2553_en.htm.

[12] S. Bassi, M. Kettunen, E. Kampa, S. Cavalieri, Forest fires: causes and contributing factors in Europe, Eur. Parliam. Brussels (2007).

[13] R.L. Olson, D.N. Bengston, L.A. DeVaney, T.A.C. Thompson, Wildland Fire Management Futures: Insights from a Foresight Panel, 2015, https://doi.org/10.2737/NRS-GTR-152.

[14] S.J. Pyne, The Misplaced War against Fire, Proj. Synd. - World's Opin. Page, 2013. https://www.project-syndicate.org/commentary/the-arizona-fire-in-context-by-stephen-j–pyne?barrier=accesspaylog.

[15] S.F. Arno, J.K. Brown, Overcoming the paradox in managing wildland fire, West, Wildlands 17 (1991) 40−46. https://www.frames.gov/catalog/34743.

[16] R.D. Collins, R. de Neufville, J. Claro, T. Oliveira, A.P. Pacheco, Forest fire management to avoid unintended consequences: a case study of Portugal using system dynamics, J. Environ. Manag. 130 (2013) 1−9, https://doi.org/10.1016/J.JENVMAN.2013.08.033.

[17] T. Schoennagel, J.K. Balch, H. Brenkert-Smith, P.E. Dennison, B.J. Harvey, M.A. Krawchuk, N. Mietkiewicz, P. Morgan, M.A. Moritz, R. Rasker, M.G. Turner, C. Whitlock, Adapt to more wildfire in western North American forests as climate changes, Proc. Natl. Acad. Sci. U. S. A 114 (2017) 4582−4590, https://doi.org/10.1073/pnas.1617464114.

[18] D. Paton, P.T. Buergelt, F. Tedim, S.M. McCaffrey, Wildfires: international perspectives on their sociale ecological implications, in: D. Paton, P.T. Buergelt, S.M. McCaffrey, F. Tedim (Eds.), Wildfire Hazards, Risks, and Disasters, Elsevier, Oxford, UK, 2015, pp. 1−14.

[19] S.H. Doerr, C. Santín, Global trends in wildfire and its impacts: perceptions versus realities in a changing world, Philos. Trans. R. Soc. Biol. Sci. 371 (2016) 20150345, https://doi.org/10.1098/rstb.2015.0345.

[20] M.A. Moritz, E. Batllori, R.A. Bradstock, A.M. Gill, J. Handmer, P.F. Hessburg, J. Leonard, S. McCaffrey, D.C. Odion, T. Schoennagel, A.D. Syphard, Learning to coexist with wildfire, Nature 515 (2014) 58−66, https://doi.org/10.1038/nature13946.

[21] D.E. Calkin, J.D. Cohen, M.A. Finney, M.P. Thompson, How risk management can prevent future wildfire disasters in the wildland-urban interface, Proc. Natl. Acad. Sci. USA 111 (2014) 746−751, https://doi.org/10.1073/pnas.1315088111.

[22] F. Tedim, V. Leone, Enhancing resilience to wildfire disasters: from the "war against fire" to "coexist with fire Disaster resilience: an integrated approach (pp. 362-383), in: D. Paton, D. Johnston (Eds.), Resil. An Integr. Approach, Charles C Thomas Publisher, Springfield, IL, US, 2017, pp. 362−383.

[23] H.H. Biswell, Research in wildland fire ecology in California, in: 2nd Tall Timbers Fire Ecol. Conf., Tallahassee, FL, 1963, pp. 63−97. https://talltimbers.org/wp-content/uploads/2018/09/Biswell1963_op63.pdf.

[24] R.E. Keane, E. Karau, Evaluating the ecological benefits of wildfire by integrating fire and ecosystem simulation models, Ecol. Model. 221 (2010) 1162–1172, https://doi.org/10.1016/J.ECOLMODEL.2010.01.008.

[25] S. Gómez-González, C. Torres-Díaz, G. Valencia, P. Torres-Morales, L.A. Cavieres, J.G. Pausas, Anthropogenic fires increase alien and native annual species in the Chilean coastal matorral, Divers. Distrib. 17 (2011) 58–67, https://doi.org/10.1111/j.1472-4642.2010.00728.x.

[26] R.W. Mutch, Wildland fires and ecosystems–A hypothesis, Ecology 51 (1970) 1046–1051, https://doi.org/10.2307/1933631.

[27] M. Milne, H. Clayton, S. Dovers, G.J. Cary, Evaluating benefits and costs of wildland fires: critical review and future applications, Environ. Hazards 13 (2014) 114–132, https://doi.org/10.1080/17477891.2014.888987.

[28] North Atlantic Fire Science Exchange, The Nature Conservancy: Maintaining Fire's Natural Role — North Atlantic Fire Science Exchange, 2015. http://www.firesciencenorthatlantic.org/maps-tools-1/2015/4/7/the-nature-conservancy-maintaining-fires-natural-role.

[29] R. Myers, Living with Fire-Sustaining Ecosystems & Livelihoods through Integrated Fire Management Global Fire Initiative, Tallahassee, 2006. http://nature.org/fire.

[30] A. Shlisky, J. Waugh, P. Gonzalez, M. Gonzalez, M. Manta, H. Santoso, E. Alvarado, A.A. Nuruddin, D.A. Rodríguez-Trejo, R. Swaty, Fire, ecosystems and people: threats and strategies for global biodiversity conservation, Arlingt. Nat. Conserv. (2007).

[31] A.M.S. Smith, C.A. Kolden, T.B. Paveglio, M.A. Cochrane, D.M. Bowman, M.A. Moritz, A.D. Kliskey, L. Alessa, A.T. Hudak, C.M. Hoffman, J.A. Lutz, L.P. Queen, S.J. Goetz, P.E. Higuera, L. Boschetti, M. Flannigan, K.M. Yedinak, A.C. Watts, E.K. Strand, J.W. van Wagtendonk, J.W. Anderson, B.J. Stocks, J.T. Abatzoglou, The science of firescapes: achieving fire-resilient communities, Bioscience 66 (2016) 130–146, https://doi.org/10.1093/biosci/biv182.

[32] K. Hirsch, V. Kafka, C. Tymstra, R. McAlpine, B. Hawkes, H. Stegehuis, S. Quintilio, S. Gauthier, K. Peck, Fire-smart forest management: a pragmatic approach to sustainable forest management in fire-dominated ecosystems, For. Chron. 77 (2001) 357–363, https://doi.org/10.5558/tfc77357-2.

[33] M.A. Finney, Design of regular landscape fuel treatment patterns for modifying fire growth and behavior, For. Sci. 47 (2001) 219–228, https://doi.org/10.1093/forestscience/47.2.219.

[34] P.M. Fernandes, Fire-smart management of forest landscapes in the Mediterranean basin under global change, Landsc. Urban Plan. 110 (2013) 175–182, https://doi.org/10.1016/j.landurbplan.2012.10.014.

[35] M.H. Taylor, K. Rollins, M. Kobayashi, R.J. Tausch, The economics of fuel management: wildfire, invasive plants, and the dynamics of sagebrush rangelands in the western United States, J. Environ. Manag. 126 (2013) 157–173, https://doi.org/10.1016/J.JENVMAN.2013.03.044.

[36] C.L. Tubbesing, D.L. Fry, G.B. Roller, B.M. Collins, V.A. Fedorova, S.L. Stephens, J.J. Battles, Strategically placed landscape fuel treatments decrease fire severity and promote recovery in the northern Sierra Nevada, For. Ecol. Manage. 436 (2019) 45–55, https://doi.org/10.1016/J.FORECO.2019.01.010.

[37] E.D. Reinhardt, R.E. Keane, D.E. Calkin, J.D. Cohen, Objectives and considerations for wildland fuel treatment in forested ecosystems of the interior western United States, For. Ecol. Manag. 256 (2008) 1997–2006, https://doi.org/10.1016/J.FORECO.2008.09.016.

[38] K. Barnett, S. Parks, C. Miller, H. Naughton, K. Barnett, S.A. Parks, C. Miller, H.T. Naughton, Beyond fuel treatment effectiveness: characterizing interactions between fire and treatments in the US, Forests 7 (2016) 237, https://doi.org/10.3390/f7100237.

[39] D.L. Martell, Fire impact management in the boreal forest region of Canada, in: Conf. Proc. Resour. Dyn. Boreal Zo. Assoc. Can. Univ. North. Stud. Thunder Bay, Ontario August, 1984.

[40] S.J. Pyne, Vestal Fire: An Environmental History, Told through Fire, of Europe and Europe's Encounter with the World, University of Washington Press, Seattle, WA, U. S, 2000.

[41] S.J. Pyne, Problems, paradoxes, paradigms: triangulating fire research, Int. J. Wildland Fire 16 (2007) 271, https://doi.org/10.1071/WF06041.

[42] A. Granström, Fire management for biodiversity in the European boreal forest, Scand. J. For. Res. 16 (2001) 62−69, https://doi.org/10.1080/028275801300090627.

[43] W.J. Bond, J.E. Keeley, Fire as a global 'herbivore': the ecology and evolution of flammable ecosystems, Trends Ecol. Evol. 20 (2005) 387−394, https://doi.org/10.1016/J.TREE.2005.04.025.

[44] V. Clément, Les feux de forêt en Méditerranée : un faux procès contre Nature, Espace Géogr. 34 (2005) 289, https://doi.org/10.3917/eg.344.0289.

[45] C.G. FLINT, A.E. LULOFF, Natural resource-based communities, risk, and disaster: an intersection of theories, Soc. Nat. Resour. 18 (2005) 399−412, https://doi.org/10.1080/08941920590924747.

[46] M.R. Coughlan, A.M. Petty, Linking humans and fire: a proposal for a transdisciplinary fire ecology, Int. J. Wildland Fire 21 (2012) 477, https://doi.org/10.1071/WF11048.

[47] J. Liu, T. Dietz, S.R. Carpenter, M. Alberti, C. Folke, E. Moran, A.N. Pell, P. Deadman, T. Kratz, J. Lubchenco, Complexity of coupled human and natural systems, Science (80-) 317 (2007) 1513−1516.

[48] T.A. Spies, E.M. White, J.D. Kline, A.P. Fischer, A. Ager, J. Bailey, J. Bolte, J. Koch, E. Platt, C.S. Olsen, D. Jacobs, B. Shindler, M.M. Steen-Adams, R. Hammer, Examining fire-prone forest landscapes as coupled human and natural systems, Ecol. Soc. 19 (2014), https://doi.org/10.5751/ES-06584-190309 art9.

[49] P. Arnould, C. Calugaru, Incendies de forêts en Méditerranée: le trop dit, le mal dit, le non-dit, Forêt Méditerranéenne, XXIX 3 (2008) 281−296.

[50] F. Tedim, V. Leone, G. Xanthopoulos, A wildfire risk management concept based on a social-ecological approach in the European Union: fire Smart Territory, Int. J. Disaster Risk Reduct. 18 (2016) 138−153, https://doi.org/10.1016/J.IJDRR.2016.06.005.

[51] T. Steelman, U.S. wildfire governance as social-ecological problem, Ecol. Soc. 21 (2016), https://doi.org/10.5751/ES-08681-210403 art3.

[52] A.P. Fischer, T.A. Spies, T.A. Steelman, C. Moseley, B.R. Johnson, J.D. Bailey, A.A. Ager, P. Bourgeron, S. Charnley, B.M. Collins, Wildfire risk as a socioecological pathology, Front. Ecol. Environ. 14 (2016) 276−284.

[53] I.U. of F.R. Organizations, Global Fire Challenges in a Warming World Summary Note of a Global Expert Workshop on Fire and Climate Change, Occasional, IUFR, Vienna, 2018. https://www.profor.info/sites/profor.info/files/IUFRO_op32_2018.pdf.

[54] European Commission, Forest Fires Sparking Firesmart Policies in the EU, Publicatio, European Commission, Luxembourg, 2018, https://doi.org/10.2777/248004.

[55] J.G. Goldammer, P.G.H. Frost, M. Jurvélius, E.M. Kamminga, T. Kruger, S.I. Moody, M. Pogeyed, Community participation in integrated forest fire management: experiences from Africa, Asia and Europe, in: Communities Flames; Proc. An Int. Conf. Community

Involv. Fire Manag. Balikpapan, Indones. (25–28 July 2001), Bangkok FAO Reg. Off. Asia Pacific. RAP Publ., Citeseer, 2002, pp. 33–52.

[56] J.S. Silva, F. Rego, P. Fernandes, E. Rigolot, Towards Integrated Fire Management - Outcomes of the European Project Fire Paradox, Joensuu, Finland, 2010, http://www. repository.utl.pt./bitstream/10400.5/15236/1/REP-FIRE Paradox-efi_rr23.pdf.

[57] F. Rego, E. Rigolot, P. Fernandes, C.M. Joaquim, S. Silva, Towards integrated fire management, EFI Policy Brief 4 (2010). http://gfmc.online/wp-content/uploads/Fire-Paradox-Policy-Brief-Integrated-Fire-Management-ENG.pdf.

[58] J. Williams, D. Albright, A.A. Hoffmann, A. Eritsov, P.F. Moore, J.C. Mendes De Morais, M. Leonard, J. San Miguel-Ayanz, G. Xanthopoulos, P. Van Lierop, Findings and implications from a coarse-scale global assessment of recent selected mega-fires, in: 5th, International Wildl. Fire Conf., FAO, Sun City, South Africa, 2011.

[59] S. Aguilar, C. Montiel, The challenge of applying governance and sustainable development to wildland fire management in Southern Europe, J. For. Res. 22 (2011) 627–639, https://doi.org/10.1007/s11676-011-0168-6.

[60] D. Ganz, R.J. Fisher, P.F. Moore, Further Defining Community-Based Fire Management: Critical Elements and Rapid Appraisal Tools, in: n.d. http://www.neevia.com (accessed July 30, 2019).

[61] Food and Agriculture Organization of the United Nations, Community-based fire management : a review, in: Food and Agriculture Organization of the United Nations, 2011.

[62] R. Vélez, Community based fire management in Spain, For. Prot. Work. Pap. (2005).

[63] D. Mukhopadhyay, Community based wildfire management in India, in: Proc. IV Int. Wildl. Fire Conf., 2007, pp. 13–17.

[64] W.F.E. C, The National Strategy: The Final Phase in the Development of the National Cohesive Wildland Fire Management Strategy, 2014. Washington, DC.

[65] F.J. Alcasena, A.A. Ager, J.D. Bailey, N. Pineda, C. Vega-García, Towards a comprehensive wildfire management strategy for Mediterranean areas: framework development and implementation in Catalonia, Spain, J. Environ. Manag. 231 (2019) 303–320, https:// doi.org/10.1016/J.JENVMAN.2018.10.027.

[66] M. Salis, M. Laconi, A.A. Ager, F.J. Alcasena, B. Arca, O. Lozano, A. Fernandes de Oliveira, D. Spano, Evaluating alternative fuel treatment strategies to reduce wildfire losses in a Mediterranean area, For. Ecol. Manage. 368 (2016) 207–221, https://doi.org/10.1016/ j.foreco.2016.03.009.

[67] S.J. O'Neill, J. Handmer, Responding to bushfire risk: the need for transformative adaptation, Environ. Res. Lett. 7 (2012) 014018, https://doi.org/10.1088/1748-9326/7/1/ 014018.

[68] K. Tierney, Resilience and the neoliberal project, Am. Behav. Sci. 59 (2015) 1327–1342, https://doi.org/10.1177/0002764215591187.

[69] J. Weichselgartner, I. Kelman, Geographies of resilience: challenges and opportunities of a descriptive concept, Prog. Hum. Geogr. 39 (2015) 249–267, https://doi.org/10.1177/ 0309132513518834.

[70] D.E. Alexander, Resilience and disaster risk reduction: an etymological journey, Nat. Hazards Earth Syst. Sci. 13 (2013) 2707–2716, https://doi.org/10.5194/nhess-13-2707-2013.

[71] M.P. Thompson, D.G. MacGregor, C.J. Dunn, D.E. Calkin, J. Phipps, Rethinking the wildland fire management system, J. For. 116 (2018) 382–390, https://doi.org/10.1093/ jofore/fvy020,

[72] M.P. Thompson, D.G. MacGregor, D. Calkin, Risk management: core principles and practices, and their relevance to wildland fire, Gen. Tech. Rep. RMRS-GTR-350, 2016.

Fort Collins, CO U.S. Dep. Agric. For. Serv. Rocky Mt. Res. Station. 29 P. 350, https://www.fs.usda.gov/treesearch/pubs/50913.

[73] M.S. Carroll, K.A. Blatner, P.J. Cohn, T. Morgan, Managing fire danger in the forests of the US inland northwest: a classic "wicked problem,, in public land policy, J. For. 105 (2007) 239−244, https://doi.org/10.1093/jof/105.5.239.

[74] D. Paton, N. Okada, S. Sagala, Understanding preparedness for natural hazards: cross cultural comparison, IDRiM J 3 (2013) 18−35.

[75] D. Swihart, T.P. O'Grady, Shared Governance : A Practical Approach to Reshaping Professional Nursing Practice, first ed., HCPro, 2006.

[76] S. McCaffrey, Community wildfire preparedness: a global state-of-the-knowledge summary of social science research, Curr. For. Reports 1 (2015) 81−90, https://doi.org/10.1007/s40725-015-0015-7.

[77] UNDRR, Global Assessment Report on Disaster Risk Reduction 2019, UNDRR, 2019, ISBN 978-92-1-004180-5.

[78] D. Paton, Disaster risk reduction: psychological perspectives on preparedness, Aust. J. Psychol. (2018), https://doi.org/10.1111/ajpy.12237.

[79] UNDRR, Terminology on Disaster Risk Reduction, United Nations off. Disaster Risk Reduct, 2017. https://www.unisdr.org/we/inform/terminology.

[80] R. Française, Le plan de protection de forêts et des espaces naturels contre les incendies − Cahier II Documents techniques et graphiques 2013/2022, 2012. http://www.corse-du-sud.gouv.fr/IMG/pdf/FIN_PPFENI_CII_A4_YC_2014010001.pdf.

[81] M. Castellnou, M. Miralles, J. Pallàs, Special Session Three: rethinking Awareness, between firefighter safety and safety strategy, in: Proc. 13TH Int. Wildl. FIRE Saf. SUMMIT 4TH Hum. Dimens. Wildl. FIRE, International Association of Wildland Fire, Missoula, Montana, USA, Boise, Idaho, USA, 2015, pp. 85−86.

Index

'*Note:* Page numbers followed by "f" indicate figures and "t" indicate tables.'

A

Aerial firefighting, 119, 121–122
Agency for Integrated Rural Fires (AGIF), 126
Agricultural land abandonment, 155–156
Aircraft-crew fatalities, 95
Angstrom Index, 61
Asphyxiation, 95
Australia, 3–4, 102–103, 202–203
  biomass burning, 55–56
  farmland abandonment, 204
  late evacuations, 95
  Life Loss database, 91
  South Australia, 45
  Southeastern Australia, 45, 190
  Victoria, 45
    bushfires, 45–46
Awareness, 99, 143, 190–192, 227–228

B

Bastrop County Complex Fire, 178
BehavePlus, 33, 34t, 43, 43t
BESAFE
  after the fire, 106
  building safety, 103
  citizens safety behavior, 103
  coexistence, 103
  components, 104, 104f
  before the fire, 104–105
  during the fire, 105–106
Biodiversity, 188, 254
Biomass burning, 55–56
Black Friday fire, 14
Black Saturday fires, 3–4, 94–95, 97–98
  Australia, 179
  planning and preparedness, 99
  risk awareness, 99
  Victoria, 179
Black Sunday fire, 3–4

Black Thursday fire, 3–4, 219
Build Back Better (BBB), 176, 181–182
Building codes, 159, 205
Buildings loss, 97–98
Buildup Index (BUI), 61–62
Bureau of Land Management (BLM), 223
Burnover, 76, 96–97

C

California Department of Forestry, 123
Camp Fire, 4–5, 93
Canada, 5, 58
  fire management, 188
  Fort McMurray Horse River wildfire, 48–49, 166
  unmanageable fire, 63
  wildfire control, 218–219
  wildfire fatalities, 91–92
Canadian Fire Weather Index (FWI), 61–62, 220–221
Caramulo
  fire evolution, 35, 36f
  Olival Novo Accident, 35–36
  São Marcos accident, 36
  Silvares, 35
Carr Fire, 4–5
Change agents, 169
Chile, 5
Chimney Tops 2 Fire, 4–5
Climate change, 3, 76, 189
  projections, 62–63
  robust projections, 63
Climate types, 56
Code Forestier, 76
Coexist with fire, 145–146, 155, 209–211
Common Agricultural Policy (CAP), 202
Commonwealth Scientific and Industrial Research Organization (CSIRO), 221

Communal forest-fire committees (CCFF), 76–77
Communication, 46–47, 143
Community-based fire management (CBFiM), 237
Community Wildfire Protection Plans (CWPP), 205
Como-Janelli Fire, 16
Controlled burning, 76
Crisis communication, 169–170

D
Daily severity rating (DSR), 61–62
Dance Township Fire, 5
Dead Man Zone, 101
Delphi method, 142
Diablo winds, 4–5
Diffusion of innovations
  change agents, 169
  new practices
    compatibility, 168
    complexity, 168
    observability, 168
    relative advantage, 168
    trialability, 168
  preventive innovations, 168
Disaster recovery
  Build Back Better (BBB), 176
  disillusionment phase, 175–176
  early restoration (ER), 176
  empower local communities (ELC), 177
  FEMA/SAMHSA phases, 175–176, 176f
  honeymoon phase, 175–176
  Linking Relief, Rehabilitation, and Development (LRRD), 176
  reconstruction phase, 175–176
Drought Code (DC), 61–62
Drought Period (DP), 59–60
Duff Moisture Code (DMC), 61–62

E
Early restoration (ER), 176, 179–181
Eastern Attica Fire, 6, 42, 142, 205–206
Ecosystems, 56–57, 189
El-Niño (EN), 59
EM-DAT, 92, 93f
Empower local communities (ELC), 177
ENSO, 59
Entrapment, 96–97

Environment Code, 76
Escalos Fundeiros, 37
Escape routes, 100
Europe
  Eastern Europe, 135
  farmland abandonment, 204
  forest destruction, 117
  Mediterranean countries, 119
  North Europe, 6
  Southern Europe, 188, 238
  wildfire fatalities, 91
  wildfire safety policies, 102
European Forest Fire Information System (EFFIS), 221–222
Extreme wildfire cases, 32t
  Australia, 45–46
  Canada, 48–49
  Greece, 41–42
  Italy, 42–44
  Portugal, 31–41
  United States of America (USA), 46–47
Extreme wildfires events (EWEs), 187, 203, 233
  definition, 7–11, 10f–11f
  definition rationale
    disaster, 15–16
    duration, 13–14
    physical properties, 11–13
    place-dependent, 15
    wildfire size, 14–15
  France, 74–77
  Portugal, 77–81
  safety enhancement. *See* Safety enhancement
  standardized definition, 6–7, 8t–9t
  United States of America (USA), 81–85
  worldwide problem, 3–6

F
Farmland abandonment, 204
Federal forest-fire exclusion policies, 187
Fine Fuel Moisture Code (FFMC), 61–62
Fire-Adapted Community program, 205
Fire-causing technologies, 155–156
Fire danger rating, 61
  operational and research purposes, 62
Fire exclusion policies, 136–137
Firefighting, 4, 99

approaches
  development history, 117–120
  differences, 120
  effectiveness and efficiency, 127–128, 128f
  emphasis and capacity, for indirect attack, 120–121
  extreme wildfires, 128–130
  Greece, 123–124
  ground *vs.* aerial firefighting, 121–122
  Italy, 124–125
  land management *vs.* urban/civil protection, 122–123
  Portugal, 125–126
  professional *vs.* volunteer, 121
  United Kingdom, 126–127
Dead Man Zone, 101
escape routes, 100
fatalities, 99–100
margin of safety, 101
policy, 187–188
rate of travel, 100–101
Firefighting trap, 144–145, 190, 201, 235
Fire impact management policies, 236
Fire incidence seasonality, 56
Fire intensity, 13, 98, 203
Fireline intensities (FLIs), 12
Fire of Regadas, 37
Fire paradox, 120, 190–192
Fire shelter, 99
Fire Smart Communities approach, 204
Fire Smart Territory (FST)
  advantages, 206–207
  components, 208–209
  definition, 207–208
  SWOT analysis, 209–211
Firewise Community program, 205
Flame length (FL), 12
Flat Top Complex Wildfires, 5
Forest Code, 76–77
Forest fire industry, 225
Fort McMurray Horse River Fire, 5, 16, 48–49, 166, 179–180
Fosberg's Fire Weather Index, 61
Fourmile Canyon Fire, 180
France, 17, 73–74, 219–220
  area burned, 193
  fire and fire regulations, 75–77
  land management, 74–75
  landscape, 74
  property, 74–75
  spatial planning, 75

G
Gatlinburg wildfires, 46–47
Global Fire Monitoring Center (GFMC), 92–93
Grass Valley Fire, 97–98
Great Chicago Fire, 217–218
Great Smoky Mountains National Park (GSMNP), 46
Greece, 5–6, 91, 202–203
  Artemida, 42
  firefighting approaches, 123–124
  fire suppression policy, 138
  lessons, 42
  Palaiohori, 41
  Smerna, 42
  Zaharo, 42
Guejito Fire, 97–98

H
Hazard-specific vacuum, 166
Heat waves (HW), 60
Houses loss, 98
House-to-house fire spread, 97–98

I
Incident Command System (ICS), 118–119
Indirect attack, 120–121
Initial Spread Index (ISI), 61–62
Institute for the Conservation of Nature and Forests (ICNF), 79–80, 126
Integral of instantaneous energy release (IIER), 13
Integrated Fire Management (IFM), 188, 237
Interregional Fire Suppression Crews, 118–119
Isolated silos, 242
Italy, 91
  firefighting approaches, 124–125
  Laconi Fire, 44
  Peschici Fire, 43, 43t
  resistance to change, 224–227

J
Japan, farmland abandonment, 204
Joint Fire Science Program (JFSP), 223

K
Kilmore East Fire, 6
Köppen-Geiger climate classification, 56

L
Laconi Fire, 44
Land-use regulations, 161–162
Large-scale climatic patterns, 59
Last-minute evacuation, 97
LCES protocol, 99–100
Linking Relief, Rehabilitation, and
        Development (LRRD), 176
Living with fire, 206f, 209–211
Local government paradox, 202
Local ordinances, 159

M
Matheson fire, 5
Mediterranean Europe, 121–122, 155–156
Misguided fire policy
    causes of ignition, 189
    coefficient of variation (CV), 190, 190f
    ecosystem services, 189
    fire control operations, 189
    firefighting trap, 190
    France, 193
    mean slope coefficients, 190
    observations and spline smoothing of area
        burned, 190–192, 192f
    perpetuation of, 193–195
    Portugal, 193
    wildfire paradox, 190–192
    yearly burned area, 190, 191t
Mitigation and preparedness, 156
    diffusion of innovations, 168–169
    efficacy, 165
    emergency response, 160
    evacuation decisions, 166–168
    experience, 164–165
    individual/household (micro) scale,
        162–163
    intervening/mid-level factors (mesoscale),
        161–162
    natural hazards
        definition, 157
        individual decision-making, 157–158
        societal stages of response, 158–160
    nonwildfire considerations, 166
    risk and crisis communication

        interactive processes, 169–170
        local context, 170
        trust, 170
    risk interpretation, 163–164
    scale, 160–161
    societal (macroscale), 161
    wildfire-specific considerations, 165–166
Mount Carmel Fire, 16

N
National Authority for Emergency and Civil
        Protection (ANEPC), 126
National Forest Fire Protection System
        (NFFPS), 80
National Parks Service (NPS), 223
National Plan for the Defense of Forest
        Against Fires (NPDFAF), 80–81
National Republican Guard (GNR), 126
National Wildfire Coordinating Group
        (NWCG), 118–119
Natural hazards, 166
    definition, 157
    individual decision-making, 157–158
    societal stages of response
        building codes, 159
        local ordinances, 159
        loss absorption, 158
        nonstructural mitigation, 159
        structural mitigation, 158
Nesterov Index, 61
Nonstructural mitigation, 159
Normal fires, 17
North America, 55–56, 102–103, 204
North Europe, 6
Nuns Fire, 4–5

O
October 15th fires, Central Portugal, 44f
    fire evolution, 38, 39f
    Lousã, 39
    Oliveira do Hospital, 39–40
    Seia, 38–39
    Sertã, 40
    Vouzela, 40–41
Olival Novo Accident, 35–36

P
Paiute Forestry, 84
Pastoral fire, 76

Pedrógão Grande Fire, 5, 37, 38f, 93–95
Peschici Fire, 43, 43t
Picões fire, 32
  BehavePlus, 33, 34t
  Cilhade, 32
  estimated propagation, 33, 34f
  precipitation, 33
  Quinta das Quebradas, 33
  relative humidity, 33
  topography and wind, 33
Plan d'occupation des sols (POS), 75
Plan local d'urbanisme (PLU), 75
Portugal, 5, 73–74, 91, 155–156, 202–203
  burned area, 193, 193f
  Caramulo, 35–36
  fire and fire regulations, 79–81
  firefighting approaches, 125–126
  firestorms of 2017, 36–41
  landscape, 77–78
  number of fires, 193, 193f
  Picões fire, 32–35
  prevention and fuel management, 81
  property, 77–78
  spatial planning
    disarticulation, 79
    excessive fragmentation, 78–79
    updated land/property cadaster, 79
  suppression capacity, 193, 193f
Portuguese National Authority for Civil
    Protection (ANPC), 126
Precipitation, 6, 33, 63
Preparedness. See Mitigation and
    preparedness
Protection and Assistance Intervention
    Group (GIPS), 126
Protection Service of Nature and the
    Environment (SEPNA), 126
Pyrocumulonimbus (PyroCb), 3–4, 45
Pyrometrics, 101, 227–228

R
Rainforests, 56–57
Rate of spread (ROS), 12, 145
Red Tuesday fire, 3–4
Regional Municipality of Wood Buffalo
    (RMWB), 48
Reinforcement Brigade for Forest Fires
    (BRIF), 119–120
Resident and community recovery, 175, 182

disaster recovery, 175–177
wildfire recovery, 177–182
Residential development patterns (RDPs),
    10–11
Richardson Fire, 16
Risk communication, 169–170
Risk interpretation, 163–164
Rodeo-Chediski fire, 179
Rural-urban interface (RUI), 141
Russia, 61, 92–93

S
Safe development paradox, 201
Safe separation distance (SSD), 100
Safety enhancement
  Australia, 102–103
  BESAFE
    after the fire, 106
    building safety, 103
    citizens safety behavior, 103
    coexistence, 103
    components, 104, 104f
    before the fire, 104–105
    during the fire, 105–106
  causes of death, 95
  circumstances
    attitudes and behaviors, 98–99
    buildings loss, 97–98
    burnover, 96–97
    entrapment, 96–97
    last-minute evacuation, 97
    people vulnerability, 98–99
  Europe, 102
  mechanisms, 96
  North America, 102–103
  operational staff, 99–101
  wildfire disasters, 91–95, 93f, 94t
Safety zones, 100
São Marcos accident, 36
Savanna, 55–56
Sea-surface temperature (SST), 58–59
Set-aside paradox, 202
Shared Wildfire Governance (SwG)
  abating "silos", 242
  collaborative work, 241
  connectivity and synergies, 241
  cross-cultural and scalar applicability,
    241
  description, 243f

Shared Wildfire Governance (SwG)
    (*Continued*)
    assessing losses and immediate needs,
        253–254
    assets vulnerability, 251–252
    distinction of stages, 241
    ecosystem recovery, 254
    emergency planning, 252
    firefighting activities, 253
    during the fire stage, 253
    framework, 244t–251t
    human vulnerability, 252
    identify and coordinate community
        recovery needs, 254
    ignition avoidance, 242
    leverage the benefits of fire, 251
    reconstruction of assets and
        infrastructures, 254
    response assessment, 254
    good data to inform decisions, 242
    monitoring efficacy, 241
    societal engagement to cope, 241
    strengths, 254–257, 255t–256t, 257f
    wildfire hazard, 242–243
Slave Lake wildfire, 177
Social media, 47
SO Index (SOI), 59
Southern Oscillation (SO), 59
Spain, 63, 91, 119–120, 225
Spatial planning
    France, 75
    Portugal
        disarticulation, 79
        excessive fragmentation, 78–79
        updated land/property cadaster, 79
    United States of America (USA), 83
Special Rural Fire Fighting Device
        (DECIR), 126
Spotting activity, 12–13, 97–98
SST. *See* Sea-surface temperature (SST)
"Stay or Go" policy, 99
Strategic problem-solving approach, 142
Strengths, Weaknesses Opportunities,
        Threats (SWOT) analysis, 138,
        139t–140t, 209–211, 210t
Structural mitigation, 158
Substance Abuse and Mental Health
    Services Administration
        (SAMHSA), 175–176

Suffocation, 95
Suppression policy, 217–219

T
Teleconnection, 59
Tennessee Emergency Management Agency
        (TEMA), 46
Terrestrial biomes, 56
Tillamook Burn, 4
Transforming fire management policies, 195

U
United Kingdom, firefighting approaches,
        126–127
United Nations Office for Disaster Risk
        Reduction (UNISDR), 175
United States of America (USA), 4, 73–74,
        202–203
    civilian wildfire fatalities, 91
    farmland abandonment, 204
    federal forest-fire exclusion policies, 187
    fire and fire regulations, 83–85
    fire suppression policy, 138
    Gatlinburg wildfires, 46–47
    Interregional Fire Suppression Crews,
        118–119
    landscape, 81–82
    property, 81–82
    spatial planning, 83
Utilized agricultural area (UAA), 77

V
Vegetation flammability, 57
Vulnerability, 98–99

W
War against fire, 135, 139t–140t, 234
War on Fire (WoF), 234–235
18 Watch Out, 99–100
Weather and climate
    blockings, 58–59
    Canadian Fire Weather Index (FWI),
        61–62
    defined, 55
    drought period (DP), 59–60
    existence, 56–57
    extinction, 57–58
    fire and climate patterns, 55–56
    fire danger rating, 61–62